KU-450-191
WITHDRAWN
FROM STOCK

# A FLEXIBLE SYSTEM OF ENZYMATIC ANALYSIS

# A FLEXIBLE SYSTEM OF ENZYMATIC ANALYSIS

OLIVER H. LOWRY

*Department of Pharmacology*
*Washington University School of Medicine*
*St. Louis, Missouri*

JANET V. PASSONNEAU

*National Institute of Neurological*
*Diseases and Stroke*
*National Institutes of Health*
*Bethesda, Maryland*

1972

ACADEMIC PRESS New York and London

ACADEMIC PRESS, INC.
111 Fifth Avenue, New York, New York 10003

*United Kingdom Edition published by*
ACADEMIC PRESS, INC. (LONDON) LTD.
24/28 Oval Road, London NW1

LIBRARY OF CONGRESS CATALOG CARD NUMBER: 78-187229

PRINTED IN THE UNITED STATES OF AMERICA

# CONTENTS

## Chapter 3 **Glassware**



## Chapter 4 **Typical Fluorometric Procedures for Metabolic Assays**

## Chapter 5 Measurement of Enzyme Activities with Pyridine Nucleotides



## Chapter 6 Improvement, Modification, Adaptation, Trouble Shooting, and Development of New Methods



## Chapter 7 Preparation of Tissues for Analysis



## Chapter 8 Enzymatic Cycling



## Chapter 9 A Collection of Metabolite Assays

## Chapter 11 **Dissection and Histological Control**



## Chapter 12 **The Quartz Fiber Fishpole Balance**



## Chapter 13 **Histochemical Analysis**



# Part III **APPENDIX**

## **Appendix**

# PREFACE

This is a multipurpose manual of laboratory methods. It offers a systematic scheme for the analysis of biological materials from the level of the whole organ down to the single cell and beyond. It is intended as a guide to the development of new methods, to the refinement of old ones, and to the adaptation in general of methods to almost any scale of sensitivity.

The incentives for the methodology in this book came from the rigorous demands of quantitative histochemistry and cytochemistry. These demands are specificity, simplicity, flexibility, and, of course, sensitivity—all likewise desirable attributes of methods for other purposes.

The specificity is provided by the use of enzyme methods. Simplicity is achieved by leading all reactions to a final pyridine nucleotide step. Flexibility and part of the sensitivity result from the use of fluorometry for measuring the oxidation or reduction of the pyridine nucleotide. Greater sensitivity is provided by chemical amplification by means of "enzymatic cycling."

The book is in a sense narrow. It does not have the comprehensive character of either Bergmeyer's "Methods of Enzymatic Analysis" or of Glick's "Quantitative Histochemistry." Instead, with a few exceptions, only methods and tools actually used by the authors and their long-suffering colleagues and trainees are presented. It is hoped, however, that the principles presented can be exploited for new analyses when needed and therefore compensate for the lack of breadth.

The book is intended for novices as well as experts who are familiar with other analytical styles. It contains, therefore, rudimentary information, perhaps oversimplified, on such matters as enzyme kinetics and the use of buffers. It is divided into three parts: a general section, one on quantitative histochemistry, and an appendix containing information that may be useful to have at the bench.

The general section is comprised of nine chapters, the first three of which deal with the properties of the pyridine nucleotides, kinetics, and glassware. Chapter 4 describes in considerable detail the methods for three metabolites. In each case 5 or 6 protocols, covering a 100,000-fold range of sensitivity, are given, together with the principles involved in making the sensitivity changes. Chapter 5 is a similar presentation for three enzymes. Chapter 6, "Improvement, Modification, Adaptation, Trouble Shooting, and Development of New Methods," is meant to be the "heart" of the book. We hope most users will read this chapter. The remaining chapters in this section deal with the preparation of tissues for analysis, the enzymatic cycling methods, and a compendium of thirty-six metabolite assays.

The quantitative histochemistry section is comprised of four chapters which include information on the preparation of frozen-dried material and dissection of samples for analysis, the fishpole balance for weighing samples, and, finally, the generalities of analysis with emphasis on the "oil well technique."

It will be seen that no presentation is made of methods dependent on radioactive tracers. There is no implication from this that radioactive methods are not at least as important as direct chemical methods. On the contrary, we would argue that one complements the other. One common misconception worth dispelling is that radioactive methods are inherently more sensitive than chemical ones. For example, to count carrier-free $^{14}C$ in 2 hours with an error of not over 3% requires about $5 \times 10^{-14}$ moles. Direct chemical assays with a single enzymatic cycling step can provide 50 times this sensitivity and with double cycling 50,000 times this sensitivity.

Unless noted otherwise, abbreviations for chemical compounds will follow the recommendations of the *Journal of Biological Chemistry*. The symbols $M$, m$M$, $\mu M$, n$M$, $10^{-8}$ $M$, etc., will be used exclusively to indicate *concentration*, i.e. molar, millimolar, micromolar, nanomolar, $10^{-8}$ molar, etc. *Amounts* of material will be indicated by the terms moles, mmoles, $\mu$moles, $10^{-12}$ moles, etc. For example, an enzyme velocity given as 20 $\mu M$/min would mean 20 $\mu$moles per liter per minute (20 $\mu$molar/ minute). Wavelengths will be given as nm (nanometers or m$\mu$). Enzyme activities will be given in international units (U), i.e. $\mu$moles/minute.

We wish to acknowledge the ideas and criticism generously given by many colleagues. Special thanks must go to Drs. Helen B. Burch, David. B. McDougal, and Franz M. Matschinsky and to Demoy W. Schulz, Joyce G. Carter, and Joseph G. Brown, and to a long series of patient postdoctoral fellows who have struggled with us to develop and use this methodololgy.

Oliver H. Lowry
Janet V. Passonneau

PART I

# GENERAL PRINCIPLES

## CHAPTER 1

# PYRIDINE NUCLEOTIDES

All of the methods of enzymatic analysis to be considered depend on pyridine nucleotides. In 1935 Negelein and Haas in Warburg's laboratory described a method for the determination of glucose-6-P dehydrogenase activity based on the increase in absorption in the near-ultraviolet as TPNH was produced. This is apparently the first publication of a method of this type. Greengard (1956) was the first to describe fluorometric pyridine nucleotide methods for measuring metabolites. Since 1935 an enormous number of enzymes and metabolites have been measured with the aid of DPN and TPN. In fact, with the help of auxiliary enzymes almost every substance of biological interest could be measured with a pyridine nucleotide system.

The usefulness of DPN and TPN for analytical purposes depends on some of their unusual properties. (1) They serve as the natural oxidizing and reducing agents in a wide variety of specific enzyme systems. With the appropriate enzyme as catalyst, they can selectively oxidize or reduce a particular substrate in the presence of innumerable other compounds. If one or more auxiliary enzymes are used, it is a rare substance which cannot be specifically oxidized or reduced in this way. (2) The reduced forms of the nucleotides, DPNH and TPNH, are fluorescent whereas the oxidized forms are not. Because the fluorescence can be measured accurately at concentrations down to $10^{-7}$ *M*, oxidation or reduction of a pyridine nucleotide can be measured with great sensitivity. (3) The reduced forms can be completely destroyed in acid without affecting the oxidized forms. Conversely, the oxidized forms can be destroyed in alkali without affecting the reduced forms. This means that at the end of a reaction the excess pyridine nucleotide of the reagent can be destroyed, leaving the fraction that has been oxidized or reduced intact for subsequent measurement by procedures which provide even greater sensitivity than the native fluorescence. (4) At this point two possibilities are available. Both the oxidized and reduced forms can be converted to highly fluorescent forms in strong alkali. This permits accurate measurements at concentrations down to $10^{-8}$ *M*. Alternatively, much greater sensitivity can be provided by "enzymatic cycling" in which the pyridine nucleotide acts as the catalytic intermediate for a two-enzyme system (Chapter 8).

To make the fullest analytical use of the pyridine nucleotides it is desirable to be thoroughly familiar with those properties which have analytical relevance. In this chapter, these properties will be described, together with a discussion of fluorometry and its limitations. In addition, the preparation, standardization and storage of pyridine nucleotides will be detailed.

## Spectral Absorption and Fluorescence of Reduced Pyridine Nucleotides

DPNH and TPNH have identical absorption bands in the near-ultraviolet which peak at 340 nm. $DPN^+$ and $TPN^+$ do not absorb at this wavelength. Therefore changes in oxidation or reduction can be measured in the spectrophotometer.

Some of the light which is absorbed is reemitted as fluorescence. Although only a small fraction of the absorbed light is emitted, this can be measured to give much greater sensitivity than can be easily attained from measurement of the absorption itself. In the spectrophotometer (i.e., with absorption measurements) it is the nonabsorbed light that is registered on the phototube, and very small decreases in this transmitted light are hard to measure. The ordinary spectrophotometer requires for precision a decrease in the transmitted light of 5% or more, and there is little to be gained by increasing the intensity of the incident light or sensitivity of the phototube. In the fluorometer, the photocell sees only the emitted fluorescence and this is directly proportional to the intensity of the incident light. Thus, the sensitivity can be increased almost without limit by increasing the intensity of the light source, the amplification, or the phototube sensitivity (by using a photomultiplier tube).

Although spectrophotometry does not readily offer the sensitivity of fluorometry, it is exceedingly useful for standardization purposes and for measurements with concentrations of DPNH and TPNH above the range suitable for fluorometry. The molar extinction coefficient ($\varepsilon$) for DPNH or TPNH at 340 nm is 6270. This means that a 0.1 m*M* (100 $\mu M$) solution would have an optical density (O.D.) of 0.627. This corresponds to 23.6% transmission ($I$) through a 1 cm light path. [With a 1 cm light path, log $(I_0/I)$ = O.D., or for a 100 $\mu M$ solution, log (100/23.6) = 0.627.]

With ordinary equipment accurate measurements can be made with DPNH concentrations of 10–200 $\mu M$ (O.D., 0.06–1.2). With volumes of 0.5 ml this corresponds to 5–100 $\times 10^{-9}$ mole. With special equipment (microcells or special differential amplifiers) much smaller amounts of DPNH can be measured. But fluorometry offers a simpler and easier way to achieve high sensitivity.

In the fluorometer the analytically useful range of concentration with DPNH (TPNH) is from 0.1 to 10 $\mu M$, equivalent to 0.1–10 $\times 10^{-9}$ mole in a 1 ml volume. Thus, the upper limit for fluorometry is about the lower limit for accurate spectrophotometry (with conventional equipment). Fluorometric sensitivity can be further increased, and extended to oxidized as well as reduced pyridine nucleotides, by treatment with strong alkali (see below).

## Limitations of Fluorometry

For the great advantages of fluorometry a certain price has to be paid. It has a few peculiarities and limitations which should be clearly recognized.

### Quenching

Any substance which absorbs the exciting light, including the pyridine nucleotide itself, will reduce the intensity of the exciting light as it passes through the solution. This will thereby diminish the emitted light, i.e., will "quench" the fluorescence. In consequence, the emitted light is proportional to the concentration of the substance measured only so long as the absorption of the exciting light is negligible. It is for this reason that the upper concentration limit for DPNH measurement is set at about 10 $\mu M$ (corresponding to an O.D. of 0.063 or a 5% absorption of light at the middle of a 1 ml fluorometer tube).

A second source of error is absorption of part of the *emitted* light by colored substances which might be present. Fortunately, this is seldom a problem. Because the fluorescence is in the visible region, the presence of absorbing materials can be detected by eye.

### Fluorescence Enhancement

A number of cations can enhance the fluorescence of pyridine nucleotides in alkaline solution. For example, the presence of 1 m$M$ $Mg^{2+}$ at pH 11.5 will enhance the native fluorescence of DPNH by 50% and that of TPNH by 300%. The enhancement with metals is not analytically useful, since it is affected by the order of mixing and the fluorescence is not stable with time. Enhancement by metals should therefore be avoided by either adding an excess of EDTA or keeping the pH at 10 or below for reading.

### Effect of Enzymes on Native Fluorescence

The fluorescence of DPNH or TPNH may be affected (usually enhanced) by the enzymes they serve as coenzyme. In most situations this is not a problem. It can become significant when high levels of enzyme are used or if the enzyme has a very strong effect. The difficulty is that the enhancement is not the same at all coenzyme levels, since only that fraction of the coenzyme which is in combination with the enzyme is affected. Consequently proportionality is lost at some point, usually at a low level, because high molar concentrations of enzyme are seldom used. The safe procedure is to test the system with several levels of DPNH or TPNH in the working range and if necessary calculate analytical results from a standard curve.

### Effect of pH

There appears to be no effect of pH per se on the fluorescence of DPNH (TPNH) from pH 6 to pH 13. However pH's below 7 are ordinarily avoided because of instability of the reduced forms. At pH's above 10.5, $DPN^+$ ($TPN^+$) begins to be converted into a fluorescent product. Therefore, high pH is avoided when the oxidized nucleotides are present.

### Temperature Effects

Another potential source of error is the large negative temperature coefficient which is characteristic of fluorescence in general. In the case of pyridine nucleotides this amounts to 1.6% per degree. If standards and samples are all read at the same temperature, there is of course no problem, but differences in temperature must be avoided.

### Fluorescence Blanks

An ultimate limitation to fluorescence sensitivity is set by blank readings. Most reagents are either slightly fluorescent or contain fluorescent impurities. With $H_2O$ alone it is difficult to reduce the instrumental reading below a value equivalent to the fluorescence of 0.05 $\mu M$ DPNH. Part of this is due to failure of the optical system to exclude all scattered light from the phototube; part may be due to slight fluorescence of the light filters; and part is possibly due to the Raman spectrum of the $H_2O$ itself.

An increase in scattered light, whether from a scratched tube or from turbidity in the solution will increase the blank even in the best fluorometer. In a poorly designed instrument, or with improper filters the effect will be much greater.

It is unnecessary to tolerate a blank with distilled $H_2O$ greater than the equivalent of 0.1 $\mu M$ DPNH. If the $H_2O$ blank is higher than this the fault can be with the fluorometer, the filters, the $H_2O$, the vessel used to store the $H_2O$, or the fluorometer tubes (scratched, dirty, or dusty).

LIMITS OF USEFUL SENSITIVITY

In many instances the useful sensitivity of a fluorescence measurement is set by factors outside of instrumentation. As sensitivity is increased, contamination becomes an increasing problem. Certain substances such as inorganic phosphate or ammonia or lactate are so common that it becomes impracticable to measure them at concentrations below $10^{-6}$ *M*. In general, at the highest sensitivity there is certain to be danger of contamination with traces of fluorescent material or specks of dust. The greatest useful direct sensitivity is achievable if the assay is based on the difference between two readings made before and after the addition of the specific reagent. It is helpful if the time lapse is not too great and essential that the instrument be highly stable.

## Fluorescence Reference Standards

All fluorescence readings have to be made by comparison with some reference solution. A stable working standard such as quinine sulfate is ordinarily used, and all readings made against it. Such standards ought not to serve as a basis for calculation, since they may change with time and may not have the same temperature coefficient as DPNH. In each analytical procedure a true standard is carried through the entire process.

Some problems have been encountered in preparing stable quinine standards of the concentrations desired. The standards are routinely prepared in test tubes like those used for fluorometric analyses. If the tubes are sealed with corks, the solutions evaporate, and require constant rechecking. If the tubes are sealed by heating in the flame, the quinine solutions frequently do not fluoresce as much as expected. The following procedure is recommended to avoid these problems.

The Pyrex tubes (10 × 75 mm) are heated and pulled in an oxygen–gas flame so that there is a long, narrow neck. At this stage the tubes are filled with 50% $HNO_3$ and heated for 5 or 10 min in a boiling water bath. The tubes are then rinsed thoroughly three or four times with distilled water, filled with distilled water, and again heated. The water is shaken out and quinine sulfate (0.002 to 0.2 $\mu$g/ml) prepared in 0.01 *N* $H_2SO_4$ is added to the tubes in a volume of 1 ml. The narrow neck of the tube can then be easily sealed in the flame without risk to the quinine sulfate solution. Standards prepared in such a manner yield the expected fluorescence and are relatively stable with time.

The ratio between the fluorescence readings with quinine and DPNH varies somewhat with the wavelength of the exciting light and the transmission characteristics of the secondary light filter. With a tungsten light source and the secondary filter recommended, quinine is about 30 times more fluorescent than DPNH or TPNH on a molar basis. This means that a 1 $\mu M$ DPNH solution is equivalent to approximately 0.013 μg/ml of quinine sulfate. (Quinine fluorescence is strongly quenced by halide ion, therefore standards must be prepared in $H_2SO_4$ rather than HCl.)

## Excitation and Emission Wavelengths and Light Filters

The fluorescence spectra of DPNH and TPNH appear to be identical in position and magnitude in pure solution. Their fluorescence is not light-sensitive. Because the absorption maximum is 340 nm, this is ordinarily the optimal wavelength for excitation. There may be situations, however, in which the ratio of blank to sample fluorescence can be reduced by using a somewhat longer wavelength. The usual fluorometer actually provides excitation with the 365 nm Hg line. This is no disadvantage as far as sensitivity is concerned, since the useful sensitivity limit is determined by the blank fluorescence and not by the absolute output of fluorescent light. With a Hg light source a very satisfactory primary filter is Corning No. 5860, which has peak transmission (about 25%) at 360 nm. With the weaker ultraviolet light provided by a tungsten light source Corning filter No. 5840 with peak transmission (about 70%) at 355 nm may be preferable, especially for weak solutions.

The fluorescence emission maximum is at 460 nm. Here again the best wavelength for measurement may not always be at the peak. The actual choice can be made empirically by finding the optimal ratio between sample and blank fluorescences. For general purposes the best filter combination we have found consists of a combination of Corning filters No. 4303 and No. 3387. Because No. 3387 is slightly fluorescent, it is placed nearest the photocell, where it is shielded from most of the scattered light by the other filters. Otherwise somewhat higher blank readings will be obtained.

## Conversion of Pyridine Nucleotides in Alkali into Fluorescent Products

As discovered by Kaplan *et al.* (1951), the destruction of $DPN^+$ and $TPN^+$ in strong alkali results in the formation of highly fluorescent products. There is no difference between the products from the two nucleotides in regard to absorption spectrum, fluorescence spectrum, or fluorescence intensity. DPNH and TPNH, which are very stable in alkali, can also be converted into these products if they are first oxidized to $DPN^+$ and $TPN^+$.

The absorption spectrum has its peak at 360 nm, i.e., it is shifted only 20 nm from the peak for the native *reduced* nucleotides. The fluorescence spectrum is narrower than in the case of DPNH and TPNH but the peak is in the same position (460 nm; Lowry *et al.*, 1957). The fluorescence under recommended conditions is about 10 times greater than the native fluorescence of the reduced nucleotides. The same fluorometer filters used for the native fluorescence can ordinarily be used to measure the fluorescence induced in alkali.

The induced fluorescence product yield increases almost linearily with NaOH concentration from 0.05 *N* to 6 *N*, even though the nucleotides are destroyed completely over the whole range. This appears to be the consequence of two competing reactions. One results in the destruction of $DPN^+$ ($TPN^+$) without producing fluorescence. This reaction is *not* accelerated by increasing the NaOH concentration above 0.05 *N*. The other reaction converts $DPN^+$ ($TPN^+$) to a fluorescent product. This second reaction is accelerated in proportion to the NaOH concentration. The two competing reactions must have almost equal temperature coefficients, since the same fluorescence yield is obtained, to within 1%, at 38°, 45°, and 60°. In 6 *N* NaOH the maximum fluorescence is developed within 1 hour at 25°, 30 min at 38°, or 10 min at 60°. If the tubes are heated to hasten development of fluorescence, they need to be cooled exactly to room temperature before reading, because the fluorescence has a negative temperature coefficient of 1.3% per degree. Once the fluorescence has been developed, the intensity is the same between pH 10.5 and pH 15 (10 *N* NaOH).

The fluorescent product is light-sensitive, particularly in strong alkali. In 6 *N* NaOH the product is sufficiently light-sensitive to make readings a little difficult when using a strong light source. This difficulty can be minimized either by keeping the exciting light to a minimum or by diluting before reading. Dilution is effective because light sensitivity is remarkably influenced by the strength of the NaOH. Thus, sensitivity to light is 35 times less in 1 *N* NaOH than in 6 *N* NaOH. Presumably the light-sensitive molecules represent only a small fraction of the total and are increased as the square of the NaOH concentration. It is common practice to develop fluorescence in 0.1 or 0.2 ml of 6 *N* NaOH in the fluorometer tube and then dilute with 1 ml of $H_2O$ to read.

It saves a step, however, to develop fluorescence directly in 1 ml of 6 *N* NaOH in the fluorometer tube, and to reduce the exciting light to the point where fading is not a problem. If fading should occur because of over long exposure to the exciting light, the contents of the tube can be remixed and read again. The second reading will be close to the original reading, since only a small proportion of the total fluorescent product is exposed to the light beam. The development of fluorescence directly in 1 ml of NaOH offers an advantage in regard to the blank (see below).

Even ordinary room illumination can cause slow fading in 6 *N* NaOH. Therefore, it is desirable to shield the samples from direct illumination during fluorescence development, and to keep them in a dark cupboard if they are not to be read fairly promptly. In the dark there is no detectable decrease in fluorescence for at least 15 hr.

One other precaution should be mentioned. Strong NaOH is viscous and has a high specific gravity. If 6 *N* NaOH is run slowly into a tube containing the sample there will be incomplete mixing at the interface. This means that some of the nucleotide will be exposed to weaker alkali than intended, and will start to be destroyed with a low fluorescence yield. Therefore, mixing should be as prompt as possible, and where there is a choice, the sample should be added to the NaOH. This permits immediate mixing, whereas, the reverse involves a delay while the large viscous volume is running into the tube.

Because of the viscosity of the NaOH it is necessary to mix especially well. A second mixing after heating may be indicated.

#### Analytically Useful Range

Reproducible measurements can be made down to 0.01 $\mu M$ concentrations or less and up to nearly 10 $\mu M$. The upper limit is set by the point of departure from linearity due to substantial absorption of the exciting light by the pyridine nucleotide product. The lower limit is set by the fluorescent blank of the NaOH solution. This can easily be kept down to the equivalent of 0.01 $\mu M$ $DPN^+$. Fresh NaOH solutions often have a much higher blank than this, especially if they have come into contact with organic matter (or rubber stoppers!). This blank fluorescence is light-sensitive and can be eliminated by leaving the bottle in direct sunlight for a few hours.

When working at low levels, the absolute value of the blank is less important than its reproducibility. The effect of the blank can often be minimized if a reading can be made on individual tubes of reagent before the sample addition. This is possible if fluorescence is developed directly in 1 ml of 6 *N* NaOH (see above).

#### Interfering Substances

Because of the high sensitivity, possible interfering substances can usually be diluted out. However, certain reactive substances must be kept at rather low concentrations during fluorescence development. These include pyruvate and $\alpha$-ketoglutarate, which decrease the fluorescence yield at concentrations above 0.1 m*M*, and glucose, which inhibits at concentrations much above 1 m*M*. The presence of 6 m*M* (0.02%) $H_2O_2$ in the strong alkali will reduce interference from all of these carbonyl compounds. In any event, it is desirable to measure standards under conditions identical to those for the unknown samples.

### Increased Fluorescence from DPNH and TPNH with Strong Alkali

When reduced pyridine nucleotides are treated with $H_2O_2$ in 6 *N* NaOH they are oxidized and yield the same highly fluorescent products that are obtained without $H_2O_2$ from the oxidized forms. This 10-fold increase in fluorescence may be of advantage not only when the concentration is too low to measure accurately by the native fluorescence, but also when a blank is present that would be disturbing without the enhancement.

Oxidation of DPNH (TPNH) by $H_2O_2$ is inappreciable at pH's between 8 and 12 but proceeds quite rapidly in 6 *N* NaOH. There even appears to be some oxidation in 6 *N* NaOH without $H_2O_2$. The time to produce full fluorescence is somewhat longer than in the case of $DPN^+$ and $TPN^+$, because of the extra time needed for the oxidation. With 3 m*M* $H_2O_2$, full fluorescence is obtained in 2 hr at 25°, 1 hr at 38°, and 15 min at 60°. With higher levels of peroxide, partial destruction occurs; e.g., there is 4% loss with 10 m*M* $H_2O_2$, and 40% with 100 m*M*. As in the case of $DPN^+$ and $TPN^+$, tubes which have been heated must be cooled exactly to room temperature before readings are made.

The $H_2O_2$ is added to the 6 *N* NaOH from a 3% (0.9 *M*) solution, which is prepared within a week from 30% $H_2O_2$. If the stronger $H_2O_2$ is added directly to the 6 *N* NaOH, a precipitate will form which may be difficult to dissolve. The 6 *N* NaOH–$H_2O_2$ solution is not very stable, and should be prepared within an hour of use if kept at room temperature, and within a few hours if kept in ice water.

Anything that reacts with $H_2O_2$ in alkali could interfere if the level is too high. Increasing the $H_2O_2$ concentration can often compensate.

## Stability and Selective Destruction of Reduced and Oxidized Forms of Pyridine Nucleotides

It is of extreme analytical importance that oxidized and reduced pyridine nucleotides are differentially sensitive to destruction by acid and alkali. DPNH and TPNH are rapidly destroyed in acid under conditions that leave $DPN^+$ and $TPN^+$ completely intact. Conversely $DPN^+$ and $TPN^+$ are readily destroyed in alkali without discernible loss of DPNH or TPNH. This makes it possible at the end of a reaction to eliminate the excess pyridine nucleotide, while preserving the fraction that has been oxidized or reduced for subsequent measurement.

The following describes the influence of pH, temperature, and certain other factors on the rates of destruction. This permits selection of conditions appropriate to a particular analytical situation.

## Destruction of DPNH and TPNH in Acid and Their Stability in Alkali

The rate of destruction of reduced pyridine nucleotides is a linear function of hydrogen ion concentration over a wide pH range (Fig. 1-1). The time calculated for 99% destruction of DPNH at 23° is 1.2 min at pH 2

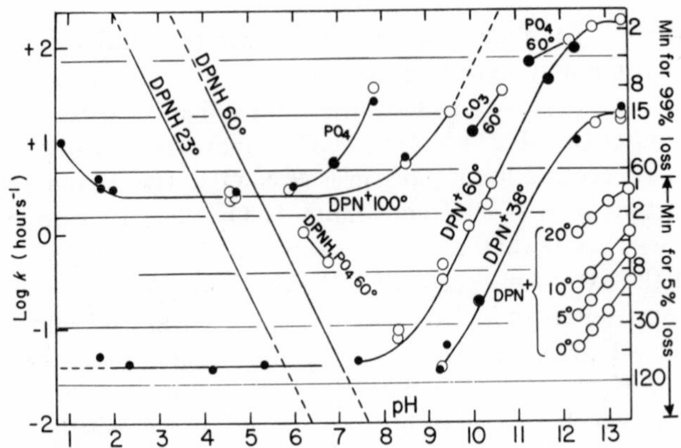


*Fig. 1-1.* Rates of destruction of $DPN^+$ and DPNH as a function of pH value and temperature. The pH values were all measured at 25°. For the $DPN^+$ observations, solutions with pH values below 3 contained sulfuric acid or sulfuric acid plus sodium sulfate. Unless otherwise indicated buffers for the rest of the samples were acetate, 0.04–0.2 *M*, at pH 4.0–5.2; Tris, 0.1 *M*, at pH 7.3–8.3; 2-$NH_2$-2-methyl-1,3-propanediol, at pH 8.3–9.3 and 2-$NH_2$-2-methylpropanol at pH 9.3–10.5. NaOH was used above pH 11.5. Adapted from Lowry *et al.* (1961) with additional data from Burch *et al.* (1967).

(0.01 *N* NCl) and 2 hr at pH 4. The rate is 3 or 4 times faster at 38° and 20 times faster at 60°. TPNH has not been studied so thoroughly but it is somewhat more easily destroyed than DPNH. At 30°, for example, it is destroyed 80% faster than DPNH.

At pH 2, $DPN^+$ is roughly 100,000 times more stable than DPNH. Thus, an enormous excess of DPNH (TPNH) can be eliminated with no detectable loss of $DPN^+$ ($TPN^+$). The destruction of course, follows first-order kinetics. Thus, if 99% is destroyed in 1.2 min, 99.99% should be destroyed in 2.4 min, etc.

Except at low concentration in small volumes (see below) DPNH becomes exceedingly stable as the pH increases. It can, in fact, be heated at 100° for 60 min in 0.1 *N* NaOH without significant loss. The chief danger of loss above pH 8 is from oxidation. This is accelerated by impurities in the solution and is an important factor in storage of DPNH solutions (see below).

At pH 7, in most solutions, the rate of destruction at 25° is about 0.2% per hour and this would increase to 2% per hour at pH 6. Phosphate accelerates the destruction of DPNH. For example, a concentration of 0.1 *M* phosphate increases the destruction rate 3.5-fold at pH 6.2 and 7-fold at pH 6.8, whether the temperature is 25°, 38°, or 60°. EDTA has little effect on the phosphate acceleration. The loss is only partially due to oxidation of DPNH to $DPN^+$. This phenomenon may be of practical importance in the choice of buffer for analytical reactions in this pH range.

## Stability of DPNH and TPNH During Storage

Table 1-1 presents a protocol of a storage test conducted with 0.4 m*M* and 40 m*M* DPNH at various alkaline pH's at two temperatures.

It is seen that strong solutions of DPNH appear to keep reasonably well at −20° at all pH values tested, but undergo large losses at 4°, the loss increasing with increasing pH. Conversely, weak DPNH solutions are fairly stable at 4° from pH 9 to 11, but were destroyed at −20° at pH 10.5 and above. At −85° (not shown in the table) no loss was observed at any of the four pH's.

**TABLE 1-1**

Percentage of DPNH Remaining After Storage for 80 Days[a]

| | % Remaining | | | |
|---|---|---|---|---|
| | DPNH, 0.4 m*M* | | DPNH, 40 m*M* | |
| pH | 4° | −20° | 4° | −20° |
| 9.1 | 92 | 90 | 75 | 86 |
| 10.5 | 89 | 74 | 66 | 93 |
| 11.2 | 91 | 57 | 35 | 91 |
| 12.7 | 74 | 8 | 29 | 88 |

[a] The media were, respectively, 100 m*M* 2-amino-2-methyl-1, 3-propanediol–HCl buffer; 75 m*M* $Na_2CO_3$: 25 m*M* $NaHCO_3$; 40 m*M* $Na_3PO_4$:40 m*M* $K_2HPO_4$; and 50 m*M* NaOH.

The results can be explained if one assumes that impurities in the DPNH preparation, or conceivably the DPNH itself, catalyze the oxidation, especially at more alkaline pH values. This would account for greater loss in stronger samples at 4°. At −20° impurities would be very concentrated in the residual liquid phase and would accelerate oxidation of weak DPNH samples, but strong samples would be protected by precipitation of much of the DPNH. At −85°, there should be no liquid phase at all, and consequently no chance for catalyzed oxidation. If this is the correct explanation, stability could probably be achieved at any temperature below −40°.

TPNH has not been tested as extensively as DPNH but appears to possess similar stability characteristics. Thus, a 0.5 m*M* solution of TPNH in 0.02 *N* NaOH (pH 12.3) showed no loss in a week at 4° or −85°, but a 13% loss at −20°.

### DPNH Oxidation in Small Volumes

For some unknown reason when DPNH is present in small volumes at high dilution it becomes sensitive to oxidation even at neutral pH. The sensitivity increases approximately as the inverse square root of the volume. Thus it is proportionately 10 times faster in 10 $\mu$l than in 1 ml. The percentage oxidation also appears to increase roughly as the inverse cube root of the concentration. Thus it was found to be four times faster at 1 $\mu M$ than at 100 $\mu M$ concentration. As an example, in one experiment a 5 $\mu$l volume of 2 $\mu M$ DPNH solution was 10% oxidized to $DPN^+$ in an hour at room temperature.

The oxidation does not appear to be the result of metal catalysis, at least it is not prevented by either EDTA or dithiothreitol. Fortunately, it is almost completely prevented by ascorbic acid at 1 or 2 m$M$ concentration.

## Destruction of $DPN^+$ and $TPN^+$ in Alkali and Stability in Acid

The effects of pH on the oxidized pyridine nucleotides are almost the mirror image of those on the reduced nucleotides (Fig. 1-1). The stability is high and independent of pH from pH 2 to pH 6.5. Above pH 7 the rate increases, until it is 5000-fold faster at pH 12.5–13. The results are those expected if $DPN^+$ is converted from a relatively stable form to a much more unstable form by removal of one proton from an acid group with p$K_a$ of about 10. At 60° the rate reaches a plateau at about pH 12.5. At lower temperatures the maximum rate occurs above 13.

$TPN^+$ has not been studied as extensively as $DPN^+$, but it appears to be slightly more stable (15%) than $DPN^+$ at alkaline pH's (Burch *et al.*, 1967). This may be contrasted with the somewhat lower stability of TPNH as compared to DPNH noted above.

### Effect of Temperature

The destruction of $DPN^+$ is strongly affected by temperature. In the temperature range of 38°–100°, the rate of $DPN^+$ destruction increases 2.8-fold for each 10° rise in temperature (11% per degree). At least in the pH 12–13 range, the temperature coefficient is even larger below 20° (Fig. 1-1). The rate increases more than 4-fold between 0° and 10° at pH 12.5. This is of analytical importance since it makes it possible to stop an enzyme reaction with alkali at 0° without suffering serious losses of $DPN^+$ or $TPN^+$. This is of particular significance for the preparation of tissue extracts without loss of either the reduced or oxidized pyridine nucleotides (Burch *et al.*, 1967).

### Effects of Salts

The rate of destruction of $DPN^+$ is increased by a number of salts. At pH 10 and 60°, for example, the rate is increased 40% by 0.2 $M$ NaCl and

10% by 0.2 *M* sodium maleate. Carbonate has a greater effect. At 60° the rate was found to be increased 380%, 500%, and 630% by 0.1 *M*, 0.2 *M*, and 0.5 *M* carbonate buffer. This acceleration is not affected by 1 m*M* EDTA. Phosphate accelerates DPN$^+$ destruction in the neighborhood of pH 7 (Fig. 1-1).

### Destruction of DPN$^+$ and TPN$^+$ as an Analytical Step

When DPNH or TPNH is to be measured at the end of an analysis by the fluorescence induced in strong alkali, or by cycling, it is necessary to destroy the excess DPN$^+$ or TPN$^+$. If the alkaline fluorescence method is to be used, the destruction needs to be done with minimal fluorescence production. A pH between 11.7 and 12.3 is optimal. In this range destruction can be accomplished with fluorescence of the order of 1% of that produced in 6 *N* NaOH. The proper pH can be obtained with NaOH (an excess of 0.005 to 0.02 *N*). However, it is usually easier to use a buffer, which permits a wide latitude in adjusting the pH of well-buffered solutions. For this purpose a mixture of $K_2HPO_4$ and $Na_3PO_4$ is recommended. During heating a final ratio of $HPO_4^{2-}$ : $PO_4^{3-}$ between 2 : 1 and 1 : 3 is satisfactory. $Na_3PO_4$ may precipitate out of 6 *N* NaOH if the concentration is too high. The concentration of $P_i$ in the strong alkali should therefore be kept below 0.05 *M*. In adjusting the pH it is not infrequently overlooked that $NH_4^+$ is an acid, with a $pK_a$ of 9.4, and that $(NH_4)_2SO_4$, perhaps added with an enzyme solution, must be taken into account in calculating the amount of NaOH or buffer to be added.

If the DPNH (TPNH) is to be measured by cycling, the amount of fluorescence produced during destruction is ordinarily of little importance and for safety's sake stronger alkali (e.g., 0.1 *N* or 0.2 *N* NaOH) can be used in the destruction.

Approximate half-times for destruction of DPN$^+$ at 60° are 2 and 0.5 min at pH 11.7 and pH 12.3, respectively. From these figures, the time necessary to destroy 99.9% would be 20 min at pH 11.7 and 5 min at pH 12.3.

### Hydrazine Interference

In some analytical reactions involving DPN$^+$ it is desirable to trap the product with hydrazine. The hydrazine does not interfere with measurement of DPNH by native fluorescence, but can cause serious difficulty for indirect measurement. A less objectionable trapping agent might be found, but so far attempts to do so have not been successful. It is, however, possible to counteract the effects of hydrazine under the two circumstances in which it causes trouble.

If hydrazine is present when weak alkali is used to destroy excess DPN$^+$, it can cause partial destruction of the DPNH present. For example, in 0.2 *M* carbonate buffer at pH 9.7 the presence of 5 m*M* hydrazine results in the loss of 40% of the DPNH during heating for 10 min at 60°. Curiously, the loss is less with higher levels of hydrazine (20 m*M*). The loss is also less in NaOH. The minimum loss is in 0.01 *N* NaOH (10% with 5 to 20 m*M* hydrazine during 10 min heating at 60°).

The loss can be prevented by the addition of 2-amino-2-methylpropanol (AM-propanol). This is a buffer with $pK_a$ of 9.9, but even as the free base it protects DPNH from destruction during heating in weak NaOH. Thus, addition of 100 m*M* AM-propanol permits heating of DPNH (60° for 10 min) without loss in 0.15 *N* NaOH containing 5–20 m*M* hydrazine. Even 10 m*M* AM-propanol protects DPNH (0.1 to 2 μ*M*) when heated in 0.02 *N* NaOH containing 5 m*M* hydrazine.

A second difficulty with hydrazine is that it can cause complete destruction of the fluorescent product formed in strong alkali from $DPN^+$ or DPNH. For example, 1 m*M* and 0.1 m*M* hydrazine in 6 *N* NaOH were found to completely destroy the fluorescent product at 23° with half-times of 7 min and 40 min, respectively. In this case AM-propanol is not effective. Instead, $Cu^{2+}$ added to the strong NaOH will destroy the hydrazine and can thus protect the fluorescent product. In order for $Cu^{2+}$ to be effective it is important to use enough to destroy the hydrazine rapidly, preferably before much of the fluorescent product has been formed. Thus, when measuring DPNH, $H_2O_2$ is omitted from the strong NaOH and is not added until the $Cu^{2+}$ has been allowed to react for at least 10 min at room temperature. With levels of hydrazine up to 0.25 m*M*, 0.4 m*M* $Cu^{2+}$ gives complete protection. Since $Cu^{2+}$ accelerates the decomposition of $H_2O_2$, the latter is increased from the usual 0.03% to 0.05%. With 1 m*M* hydrazine the $Cu^{2+}$ is increased to 2 m*M*, and the $H_2O_2$, again added separately after the $Cu^{2+}$, is increased to 0.1%.

## Preparation, Standardization, and Storage of Pyridine Nucleotide Solutions

$DPN^+$, $TPN^+$, and TPNH powders are stored desiccated at −20°; DPNH can be stored at 0–25° if it is kept in the dark.

Primary standardization of stock solutions is made in the spectrophotometer. A procedure for this purpose is given below in the case of each of the nucleotides.

$DPN^+$ and $TPN^+$ solutions can be stored frozen for months without significant change in concentration. DPNH and TPNH solutions ordinarily need to be restandardized more frequently unless stored at very low temperature (see below).

### General Standardization Protocol

It is convenient to pipette an exactly measured volume of the appropriate reagent into each of three cuvettes. One of the three cuvettes serves as a blank the other two provide duplicate assays. In cells of 4 or 5 mm inner width a convenient volume is 400 or 500 $\mu$l. This permits easy mixing without introducing a rod or inverting the cell. The nucleotide is added in a small volume to give a concentration of 50–100 $\mu$M (e.g., 5 $\mu$l of 10 m$M$ nucleotide added to 500 $\mu$l of reagent). The cuvettes are all read at 340 nm, the enzyme is added, and the solutions are well mixed. Readings are made at suitable intervals until there is no further change in optical density. Under the conditions recommended each reaction should be complete within 10 min. The calculations are based on a molar extinction coefficient of 6270 at 340 n$M$ for DPNH and TPNH. The concentration (m$M$) of the standard solutions is equal to

$$\Delta\text{O.D.} \times \frac{1000}{6270} \times \frac{\text{total volume}}{\text{volume of nucleotide solution}}$$

where ΔO.D. is the average change in optical density corrected for any change in the blank. The exact volumes of the pipettes are used in the calculation, although approximate volumes are given in the directions.

### $DPN^+$ and $TPN^+$ Solutions

These keep well dissolved in $H_2O$. To prepare stock solutions of about 100 m$M$ concentration, dissolve 70 mg of $DPN^+$ or 80 mg of $TPN^+$ per milliliter. It is convenient for frozen storage to prepare 1 or 2 ml in an 8 ml screw cap tube. Portions of the stock solution are diluted to approximately 10 m$M$ and 1 m$M$ with water. For purposes of further calculation these dilutions should be made quantitatively, e.g., 200 $\mu$l and 20 $\mu$l volumes measured with calibrated constriction pipettes are each diluted with 2 ml of $H_2O$, measured with accurate volumetric pipettes. All three dilutions can be stored at $-20°$ and are indefinitely stable. However, since there may be some evaporation on long storage, the solutions should be restandardized at intervals when used as a basis of calculation in fluorometric tests.

#### *Standardization of $DPN^+$*

*Reagent.* Tris-HCl buffer, pH 8.7 (80 m$M$ Tris base, 20 m$M$ Tris-HCl); ethanol, 1 $M$ (about 5% by volume); EDTA, 1 m$M$.

Dilute yeast alcohol dehydrogenase to a concentration of 3 mg/ml in 20 m$M$ Tris-HCl buffer (pH 8.1). If the assay is conducted with 400 or 500 $\mu$l of reagent, a volume of 2 $\mu$l of this enzyme solution should complete the reaction in 10 min or less.

#### *Standardization of $TPN^+$*

*Reagent.* Tris-HCl buffer, pH 8.1 (50 m$M$ Tris base, 50 m$M$ Tris-HCl); glucose-6-P, 1 m$M$.

Yeast glucose-6-P dehydrogenase is diluted to a concentration of 8 U/ml in 20 m$M$ Tris-HCl, pH 8.1. Of this 2 $\mu$l are used in the assay when the reagent volume is 500 $\mu$l.

DPNH AND TPNH SOLUTIONS

Although, as seen earlier, DPNH and TPNH are extremely stable toward direct destruction in alkali, they are susceptible to oxidation even at low temperature. Therefore storage of reduced nucleotide solutions without serious loss presents much greater problems than with $DPN^+$ and $TPN^+$. The storage properties described above suggest that unless a freezer at −40° or below is available, DPNH and TPNH solutions should be prepared at concentrations no greater than 5 m*M*, at a pH of 9–11, and be stored at 4°. If a low-temperature freezer is available stronger solutions can be prepared and probably stored indefinitely without loss.

If it is desirable to keep $DPN^+$ or $TPN^+$ to a minimum in reduced nucleotide solutions it is recommended that these be prepared in a carbonate buffer at pH 10.6 (80 m*M* $Na_2CO_3$ : 20 m*M* $NaHCO_3$). In this case, $DPN^+$or $TPN^+$ initially present, or that which may accumulate during storage, can be destroyed by heating 3 min at 100°. Heating is not otherwise recommended for reactions to be followed fluorometrically, since there is some increase in blank fluorescence in the process.

To prepare an approximately 5 m*M* DPNH or TPNH solution use 8 or 9 mg of powder per milliliter of the carbonate buffer.

*Standardization of DPNH*

*Reagent.* Phosphate buffer, pH 7.0 (60 m*M* $Na_2HPO_4$, 40 m*M* $NaH_2PO_4$); pyruvate, 1 m*M*.

Lactate dehydrogenase (heart or skeletal muscle) is diluted to a concentration of 0.5 mg/ml in 20 m*M* phosphate buffer (pH 7.0).

For 500 μl of reagent, 1 μl of this enzyme solution should suffice.

*Standardization of TPNH*

*Reagent.* Tris-HCl buffer, pH 8.1 (50 m*M* Tris base, 50 m*M* Tris-HCl); ADP, 100 μ*M*; α-ketoglutarate, 5 m*M*; ammonium acetate, 5 m*M*.

Glutamate dehydrogenase (beef liver) is diluted to a concentration of 5 mg/ml with 20 m*M* Tris-HCl (ph 8.1). Of this 5 μl should be sufficient for an assay with 500 μl of reagent. DPNH can also be standardized in this reagent, rather than in the pyruvate-lactate dehydrogenase system, since glutamate dehydrogenase reacts with either coenzyme.

**Fluorometers**

The two most desirable features of a fluorometer are stability and low background reading. If the readings generated are stable and reproducible, small increments in reading can be measured with precision. If the background reading is low, high sensitivity can be used when needed. Any fluorometer provided with a photomultiplier tube is likely to have more than enough sensitivity, but this will not be usable to full capacity if the instrumental blank is too high.

There are two basic types of fluorometers, those which isolate the light wavelengths for excitation and fluorescence with filters, and those which do so spectroscopically. The latter are much more expensive, and, because they require a very intense light source (usually a xenon arc lamp), are likely to be unstable. They are also ordinarily designed to use square cuvettes rather than test tubes, and are therefore far less convenient for analyzing many samples that require multiple readings. For all these reasons, for the present purpose filter type instruments are much to be preferred. These are either direct reading instruments, such as that produced by the Farrand Optical Company, or else they operate on a null point principle, in which the fluorescence light is balanced against a small fraction of the exciting light itself. This type is available from the Turner Company. The null point principle has the advantage that variations in the light source are cancelled out. Almost all of our experience, however, has been with the direct reading Farrand instrument which has the advantage that sensitivity is continuously variable over the whole range, instead of in steps only (null point instruments could, of course, be made with continuously variable sensitivity).

Most filter fluorometers are provided with a mercury arc light source. This can be a cause of trouble due to instability. The light output is quite temperature-sensitive, and in some lamps the arc has a tendency to migrate. If difficulty is encountered on this score, we recommend that stability be increased by substituting for the arc lamp an incandescent lamp (General Electric No. 2331). This is designed for 6 V, but by running it on 8 V it emits two or three times as much ultraviolet light. (The overvoltage shortens the lifetime to 15 or 20 hr.) The lamp can be operated from a storage battery supplied with a battery charger so adjusted as to approximately balance the drain from the lamp. Even at 8 V the ultraviolet output is much less than with an arc lamp, but in spite of this a photomultiplier tube can provide all the sensitivity that can be used.

The fluorometer may be furnished with a turntable to permit insertion of several tubes into the instrument at once. Unless this device is exceedingly well designed, it will increase the variability of readings. This is because a very slight difference in tube position will have a big effect on the emitted light which reaches the sensitive area of the phototube. It is recommended that the turntable be fixed rigidly in position and the tubes inserted and read one at a time.

The background reading is largely determined by how carefully the exciting light is screened from the phototube. The ratio between the exciting and emitted light for a 0.1 $\mu$M DPNH solution is more than 100,000 to 1. A light leak of 1 part in a million would be significant. If the reading of a tube filled with high-grade distilled water is greater than the equivalent of 0.1 $\mu M$ DPNH, a light leak should be suspected. It may be possible to seal shut the offending joints.

All of the analytical directions in this book have been written on the assumption that the fluorometer is designed to use tubes with 1 ml sample volumes. If larger or smaller volumes are required the directions will need to be adjusted accordingly.

CHAPTER 2

# A KINETICS PRIMER FOR THE TISSUE ANALYST

A rudimentary working knowledge of enzyme kinetics is of great value in the use of enzymes as analytical tools. An analysis may fail if too little enzyme is used, due to incomplete reaction, but it may also fail if too much enzyme is used. This might not be true if absolutely pure, single action, enzymes were available. In this case a tremendous excess of enzyme could be employed and no harm done. But even crystalline enzymes are frequently contaminated with interfering activities, or the enzyme itself may have a disturbing side action. Fortunately, if just the "right amount" is employed, enzyme preparations that are only partially purified can often be used successfully. The "right amount" of enzyme (the least amount that will do the job) can be rather exactly defined. In order to determine how much enzyme should be used, kinetic information is needed. The subject of enzyme kinetics often fills the neophyte with fear. It is the hope here to present simple kinetics in such a way that even the relatively inexperienced tissue analyst can proceed with confidence.

## NONENZYME KINETICS

Before discussing enzyme kinetics it may be useful to present some fundamental concepts applicable to all chemical reactions.

### "ORDER" OF THE REACTION

Chemical reactions are defined as "first order", "second order", or "higher order" if there are one, two, or more reactants. The decay of a radioactive element is a first-order reaction; the combination of an alcohol with an acid to form an ester is a second-order reaction. The hydrolysis of an ester is also second order, but because one of the reactants ($H_2O$) is present in enormous excess, the reaction has the mathematical form of a first-order reaction. It is called "pseudo first order" or usually simply "first order." There is also a (pseudo) "zero-order" reaction, which may sound impossible. Only reactions following zero-, first-, or second-order kinetics are of interest here. Of these we shall mainly emphasize those of first order.

## First-Order Reactions

Consider the first-order reaction

$$A \xrightarrow{k} B$$

Its velocity ($v$) can be described by the equation:

$$v = k[A] \tag{2-1}$$

As the reaction proceeds the velocity will decrease exactly as [A] decreases (Fig. 2-1). When A is half gone the velocity will be half of the initial velocity, etc. [A] is said to diminish by a "die-away" curve or by "logarithmic decrement"; $k$ is a "first-order rate constant". Note its meaning. It is equal to the fraction of A which would be converted to B in unit time *if the velocity did not fall off*. For example, if $k = 0.3$ per minute, the velocity *at any instant* is

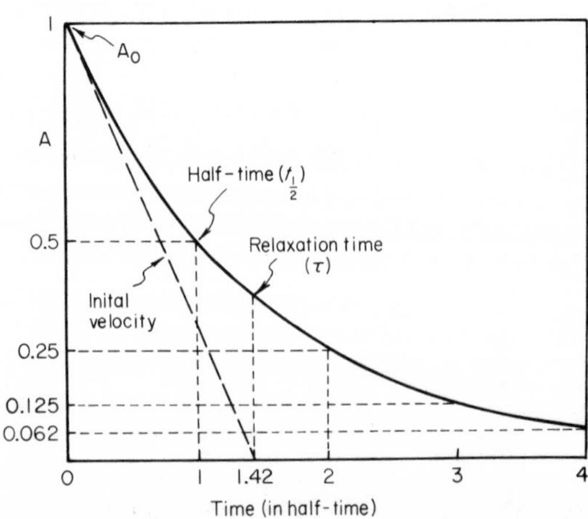


***Fig. 2-1.*** Plot of a first-order reaction.

always 0.3[A] per minute, i.e., 30%/min of the amount of A present at that moment.

Equation (2-1) describes the situation at any moment in Fig. 2-1, but we need an equation which sums up the total change in [A] in an interval of time ($t$). This (integrated) equation is

$$2.3 \log[A]_0/[A] = kt \tag{2-2}$$

where $[A]_0$ is the initial amount of A. When A is half gone, $2.3 \log(1/0.5) = kt = 0.7$ (actually 0.69). The time is the "half-time" or $t_{1/2}$, i.e.,

$$t_{1/2} = 0.7/k \tag{2-3}$$

For example, if $k = 0.3$/min, $t_{1/2} = 2.3$ min. Most people find the half-time of a reaction easier to visualize than the rate constant. [A somewhat more erudite term than the half-time is the "relaxation time," $\tau$, which is simply the reciprocal of the first-order rate constant (Fig. 2-1).]

Note that after 1, 2, 4, and 6 half-times, A will be 50%, 75%, 94%, and 98.5% gone; i.e., *a simple irreversible first-order reaction will be complete, for most practical purposes, in 5 or 6 half-times.* This applies to both true and pseudo-first-order reactions, and to enzymatic as well as nonenzymatic reactions.

## Reversible First-Order Reactions

If the reaction is reversible the situation is more complicated:

$$\mathrm{A} \underset{k_2}{\overset{k_1}{\rightleftharpoons}} \mathrm{B} \tag{2-4}$$

The net velocity is the difference between the two velocities ($v_1$ and $v_2$) in the forward and reverse direction. The velocity equation is

$$v = v_1 - v_2 = k_1[A] - k_2[B] \tag{2-5}$$

Thus the initial velocity (i.e., when $[B] = 0$) is the same as for an irreversible reaction, but the net velocity will diminish more rapidly and will become zero when $k_1[A] = k_2[B]$. When this happens the system is in equilibrium ($[A] = [A]_{eq}$, $[B] = [B]_{eq}$). It is clear that

$$\frac{[B]_{eq}}{[A]_{eq}} = \frac{k_1}{k_2} = K_{eq} \tag{2-6}$$

This describes the fundamental fact that the equilibrium constant is determined by the two rate constants. Since $[B] = [A]_0 - [A]$, [B] can be eliminated from Eq. (2-5) to give

$$v = (k_1 + k_2)[A] - k_2[A]_0 \tag{2-7}$$

This can be integrated, and by taking advantage of Eq. (2-6) the integrated equation can be put into a simple form:

$$2.3 \log \frac{[A]_0 - [A]_{eq}}{[A] - [A]_{eq}} = (k_1 + k_2)t \tag{2-8}$$

This is exactly analogous to Eq. (2-2). The half-time of *approach to equilibrium* is

$$t_{1/2} = \frac{0.7}{k_1 + k_2} \tag{2-9}$$

and the reaction will be 50%, 75%, and 98.5% as far as it will ever go in 1, 2, and 6 half-times. If, for example, $k_1 = 0.3$/min and $k_2 = 0.075$/min, $K_{eq} = 4$ [from Eq. (2-6)], i.e., equilibrium will occur when [A] has fallen to 20% of $[A]_0$. The half-time (i.e., when $[A] = 60\%$ of $[A]_0$) will be $0.7/0.375 = 1.9$ min. Note that the approach to equilibrium is *faster* than if the reaction had not been reversible and [A] had been approaching zero instead of 20%.

## Second-Order Reactions

Most analytical reactions have the form of first- or zero-order reactions, or something in between. However, we will encounter some analytical reactions which follow a second-order time course, and, moreover, a brief presentation of certain aspects of the second-order reaction may be helpful in understanding enzyme kinetics. Consider the irreversible reaction:

$$A + B \xrightarrow{k} C \tag{2-10}$$

The velocity equation is

$$v = k[A][B] \tag{2-11}$$

Thus the concentrations of both A and B affect the velocity, and the constant $k$ has a more complex meaning than in the case of a first-order reaction. It is numerically equal to the *instantaneous rate* expressed as the fraction of one reactant which is converted to product in *unit time* when the other reactant is present at *unit concentration*. For example, suppose $k = 0.3$ liter per mole per minute ($0.3\ M^{-1}\text{min}^{-1}$). This would mean that if [A] is 1 $M$, the instantaneous velocity will correspond to the disappearance of B at the rate of 30%/min. If [A] is only 0.01 $M$ the rate of disappearance of B will be only 0.3%/min. Thus a second-order rate constant has both a time and a concentration dimension.

The integrated form of Eq. (2-11), which would describe the time course of a second-order reaction, will not be presented. However, in the enzyme section below a graphic representation will be given for the time course when [A] is varied relative to [B].

If reaction (2-10) is reversible,

$$\mathrm{A} + \mathrm{B} \underset{k_2}{\overset{k_1}{\rightleftarrows}} \mathrm{C} \tag{2-12}$$

i.e., it is second order in one direction and first order in the other. The velocity equation is

$$v = k_1[\mathrm{A}][\mathrm{B}] - k_2[\mathrm{C}] \tag{2-13}$$

At equilibrium, $k_1[\mathrm{A}][\mathrm{B}] = k_2[\mathrm{C}]$ and

$$K_{eq} = \frac{[\mathrm{A}][\mathrm{B}]}{[\mathrm{C}]} = \frac{k_2}{k_1} \tag{2-14}$$

It is important to realize the difference between Eq. (2-14) and Eq. (2-6). $K_{eq}$ has a concentration dimension, and the position of equilibrium is no longer independent of absolute concentrations. For example, let $k_1 = 10\ M^{-1}\ \mathrm{min}^{-1}$ and $k_2 = 0.1\ \mathrm{min}^{-1}$. From Eq. (2-14), $K_{eq} = 0.01\ M = 10\ \mathrm{m}M$. At equilibrium, if $[\mathrm{A}] = [\mathrm{B}] = 0.1\ M$, $[\mathrm{C}] = 1\ M$, i.e., 91% of the total; if $[\mathrm{A}] = [\mathrm{B}] = 0.1\ \mathrm{m}M$, $[\mathrm{C}] = 0.001\ \mathrm{m}M$, i.e., only 1% of the total. (If the reaction is $\mathrm{A} + \mathrm{B} \rightleftarrows \mathrm{C} + \mathrm{D}$, the equilibrium constant has no concentration term and the position of equilibrium is, of course, independent of absolute concentrations.)

A shift in equilibrium with concentration occasionally has analytical significance. For example, glutamate can be oxidized enzymatically with $\mathrm{DPN}^+$ to form α-ketoglutarate and $\mathrm{NH}_3$. Two components react to form three, therefore the equilibrium position is concentration-dependent. In the millimolar concentration range the equilibrium is unfavorable, but in the micromolar range the reaction can easily be driven to essential completion.

## ENZYME KINETICS

The kinetics of catalyzed reactions are more complicated than those of spontaneous reactions, because the catalyst must of necessity enter into the reaction, usually by forming a transient complex with the reactant(s). The velocity of an enzymatic reaction will be determined by the amount of enzyme and by some function of the reactant (substrate) but, in contrast to a spontaneous reaction, the velocity will usually not be directly proportional to reactant concentration.

## One-step Reactions with One Substrate

In simple uncomplicated cases, the initial velocity of an enzyme reaction increases with substrate concentration in the manner shown in Fig. 2-2. At low substrate concentrations, the rate rises almost linearly with concentration. At higher concentrations, the increase gradually lessens, until the reaction is hardly accelerated by increasing substrate. Under such conditions, the substrate is said to have "saturated" the enzyme, and the rate observed is the "maximum velocity" or $V_{max}$ for that concentration of enzyme. The actual

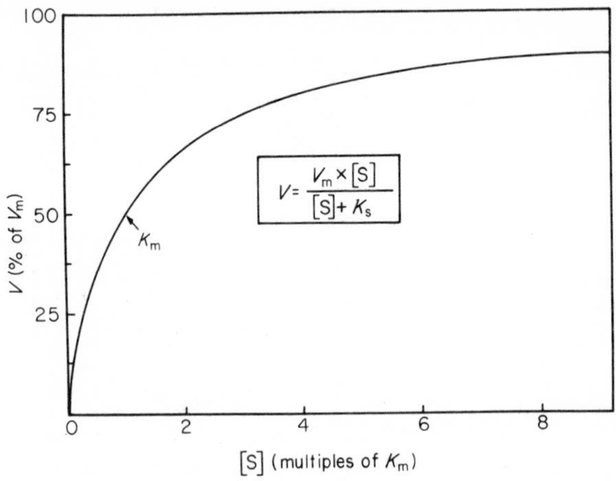


*Fig. 2-2.* Velocity of an enzyme reaction as a function of substrate concentration. This describes the situation for an enzyme with "normal" kintetics.

chain of events in an enzyme reaction may be very complex, but in the case of a single substrate, S, and an irreversible reaction which forms product, P, the situation can usually be explained for practical purposes by three reactions:

$$E + S \underset{k_2}{\overset{k_1}{\rightleftarrows}} ES \xrightarrow{k_3} P + E \tag{2-15}$$

where E is free enzyme and ES is a transient complex between enzyme and substrate. (It would be equally satisfactory to write "EP" instead of "ES.") From what has been said earlier, it will be evident that of the three reactions taking place, one is second order and two are first order. The third reaction ($ES \rightarrow P$) is the one ordinarily measured as the overall enzyme velocity (represented below as either $v$ or $v_3$).

Usually, at the start of such an enzyme reaction, S greatly exceeds the enzyme on a molar basis. In this case, reaction (2-15) can be considered to take place in three phases. In the first phase, ES is formed faster by reaction 1 than it breaks down by reaction 2 and 3; therefore, ES accumulates and E falls. This is the "pre-steady-state" phase which usually lasts only a matter of milliseconds. In the second phase, a "steady state" has been established in which reaction 1 is balanced by reactions 2 and 3, but S has not diminished appreciably. ES is therefore at its peak level and the net velocity (decrease in S or increase in P) is called "initial velocity." Finally, in the third (and usually longest) phase, S falls toward zero, during which time a steady state is maintained between E, S, and ES.

At all times the net velocity is proportional to [ES], i.e.,

$$v = k_3[\mathrm{ES}] \tag{2-16}$$

It will be seen from reaction (2-15) that as [S] is made larger and larger it will finally diminish [E] to a negligible value, i.e., the total enzyme, $E_T$, will be in the form ES and therefore the velocity will be maximal, thus,

$$V_{max} = k_3[\mathrm{E_T}] \tag{2-17}$$

When velocity is maximal it is independent of modest changes in S and is called "zero" order in S. (It is, in reality, at least a second-order reaction, and should properly be called *pseudo* zero order. It is difficult to conceive of a *true* zero-order reaction.)

At lower substrate levels, an equation is needed relating $v$ to S and $V_{max}$. The three separate velocities of reaction (2-15) are

$$v_1 = k_1[\mathrm{E}][\mathrm{S}],$$
$$v_2 = k_2[\mathrm{ES}],$$
$$v_3 = k_3[\mathrm{ES}]$$

When the steady state is reached, $v_1 = v_2 + v_3$, or

$$k_1[\mathrm{E}][\mathrm{S}] = k_2[\mathrm{ES}] + k_3[\mathrm{ES}]$$

Rearranging,

$$\frac{[\mathrm{E}][\mathrm{S}]}{[\mathrm{ES}]} = \frac{k_2 + k_3}{k_1} = K_m(\text{or } K_s) \qquad \text{or} \qquad \frac{[\mathrm{E}]}{[\mathrm{ES}]} = \frac{K_m}{[\mathrm{S}]} \tag{2-18}$$

$K_m$ (the "Michaelis constant") has the form of an equilibrium constant [cf. Eq. (2-14)]. It is a true equilibrium constant if $k_2$ is much larger than $k_3$. $K_m$ is a readily determinable constant, whereas, $k_1$ and $k_2$ are usually very difficult to evaluate. From Eq. (2-18), it is seen that when $[\mathrm{S}] = K_m$, [E] = [ES], i.e., half of the total enzyme will be present as ES. Since $v$ is proportional to [ES] [Eq. (2-16)], when $[\mathrm{S}] = K_m$, $v$ is half-maximal. Thus, $K_m$ is numerically equal to the substrate concentration which gives half-maximal velocity (see Fig. 2-2).

The desired equation can now be formulated. Rearranging Eq. (2-18), $[ES]K_m = [E][S]$. Substituting $[E_T] - [ES]$ for [E] and rearranging ($E_T$ = total enzyme),

$$[ES] = \frac{[E_T][S]}{[S] + K_m} \tag{2-19}$$

If both sides are multiplied by $k_3$, this becomes [from Eqs. (2-16) and (2-17)]

$$v = \frac{V_m[S]}{[S] + K_m} = \frac{V_m}{1 + K_m/[S]} \tag{2-20}$$

which is the Michaelis-Menten equation. This describes the hyperbolic curve plotted in Fig. 2-2. The reciprocal form of Eq. (2-20) is often more convenient because it describes a straight line:

$$\frac{1}{v} = \frac{1}{V_m}\left(1 + \frac{K_m}{[S]}\right) \tag{2-20a}$$

Equation (2-20a) is the basis of the widely used Lineweaver-Burk plot, or "reciprocal" plot, of $1/v$ against $1/[S]$ (Fig. 2-3). The intercept on the vertical axis is $1/V_m$, and the intercept on the horizontal axis is $-1/K_m$.

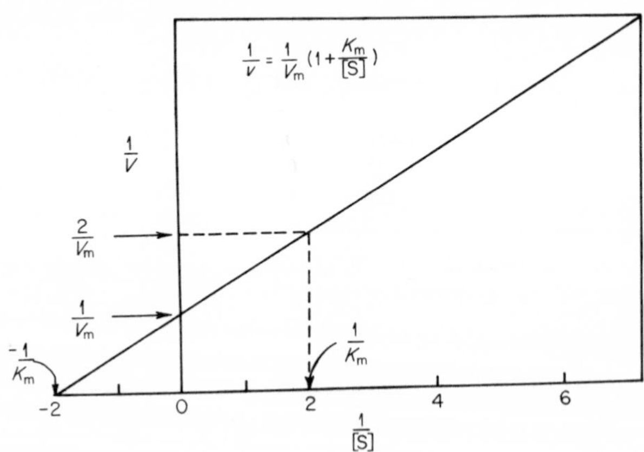


*Fig. 2-3.* Reciprocal plot of Lineweaver-Burk plot of velocity against substrate concentration for an enzyme with "normal" kinetics.

## First-Order Enzyme Reactions

When [S] is small compared to $K_m$, Eq. (2-20) reduces to

$$v = V_m[S]/K_m \tag{2-20b}$$

Since, with a given level of enzyme, $V_m$ and $K_m$ are both constants they can be replaced by their ratio, which we can denote by $k$, and call the "apparent first-order rate constant":

$$k = V_m/K_m \tag{2-21}$$

from which

$$v = k[S] \tag{2-1a}$$

This has the form of a first-order equation, and we speak in this case of the substrate concentration being at a "first-order level." Everything said above about nonenzymatic first-order reactions is applicable in this case. Equation (2-2) can be rewritten (S replacing A) as

$$t = \frac{2.3}{k} \log \frac{[S]_0}{[S]} \tag{2-22}$$

For the majority of analytical situations, $k$ is more important than either $V_{max}$ or $K_m$ alone. For example, in the hypothetical cases of Table 2-1,

**TABLE 2-1**

INFLUENCE OF MAXIMUM VELOCITY AND MICHAELIS CONSTANT ON APPARENT FIRST-ORDER RATE CONSTANT AND REACTION TIMES[a]

| Enzyme | $V_{max}$[b] ($\mu M$/min) | $K_m$ ($\mu M$) | $k$ ($min^{-1}$) | $t_{1/2}$ (min) | $t_{0.75}$ (min) | $t_{0.98}$ (min) |
|---|---|---|---|---|---|---|
| A | 700 | 1000 | 0.7 | 1 | 2 | 5.6 |
| A | 200 | 1000 | 0.2 | 3.5 | 7 | 20. |
| B | 50 | 50 | 1 | 0.7 | 1.4 | 3.9 |

[a] The reaction times apply to initial substrate levels well below the $K_m$.
[b] At the enzyme concentration used.

enzyme B at an activity ($V_{max}$) of 50 $\mu M$/min is more active toward low levels of substrate than enzyme A at an activity of 700 $\mu M$/min.

If a substance is to be measured enzymatically, and the concentration is well below the $K_m$, the time required, as shown earlier, is 5 or 6 half-times. The same time would be required whether $[S]_0$ is 1/10th or 1/1000 of $K_m$.

Note that in calculating $k$ it is necessary to put $V_m$ and $K_m$ into rational compatible terms. It is often convenient to express $K_m$ as $\mu M$ ($\mu$moles per liter), and to express $V_m$ as $\mu M$/min (i.e., $\mu$moles per liter per minute). In this case $k$ will be expressed as *per minute*. The advantage of this is that the international enzyme unit (U) is 1 $\mu$mole/min, therefore U/liter = $\mu M$/min. Stock enzyme solutions can be conveniently expressed as m$M$/min (U/ml).

### Mixed Zero-Order and First-Order Enzyme Reactions

If $[S]_0$ approaches $K_m$ or exceeds it, more time is required to complete the reaction than when $[S]_0$ is at a first-order level. The time required can be calculated from the integrated form of Eq. (2-20):

$$t = \frac{2.3K_m}{V_m} \log \frac{[S]_0}{[S]} + \frac{[P]}{V_m} \tag{2-23}$$

where P is the product formed (i.e., $S_0 - S$). If $k$ is again substituted for $V_m/K_m$:

$$t = \frac{2.3}{k} \log \frac{[S]_0}{[S]} + \frac{[P]}{V_m} \tag{2-24}$$

This may be seen to differ from the equation for a first-order reaction [Eq. (2-22)] only be the addition of the term $[P]/V_m$.

In other words, *to complete a one-step enzyme reaction, with any level of substrate, requires the time calculated for a first-order reaction plus the time that would be required if the entire reaction were to proceed at the* $V_m$.

For the reaction to be 98% complete (and ignoring the 2% difference between $[S]_0$ and [P]), Eq. (2-23) becomes (with sufficiently close approximation),

$$t_{0.98} = \frac{4}{k} + \frac{[S]_0}{V_m} \tag{2-25}$$

or since $k = 0.7/t_{1/2}$,

$$t_{0.98} = 6t_{1/2} + \frac{[S]_0}{V_m} \tag{2-26}$$

Consider as an example an enzyme with $K_m = 200\ \mu M$ and $V_m$ (at the level of enzyme used) $= 50\ \mu M$/min. This gives $k = 0.25$/min and $t_{1/2} = 3$ min. To give 98% conversion to product with initial substrate levels of 1 $\mu M$, 100 $\mu M$, and 1 m$M$ would require 18 min, (18 + 2) min, and (18 + 20) min, respectively. (See Table 2-2 for this and other examples.)

In a metabolite assay the problem is usually the converse of the above, i.e., how much enzyme should be used to complete a reaction in a given time. In this case it is simpler to replace $k$ by $V_m/K_m$ in Eq. (2-25), which then becomes

$$t_{0.98} = \frac{4K_m + [S]_0}{V_m} \quad \text{or} \quad V_m = \frac{4K_m + [S]_0}{t_{0.98}} \tag{2-27}$$

**TABLE 2-2**

EFFECT OF SUBSTRATE CONCENTRATION ON REACTION TIME[a]

| $V_{max}$ ($\mu M$/min) | $K_m$ ($\mu M$) | $[S]_0$ ($\mu M$) | $k$ ($min^{-1}$) | $t_{1/2}$ (min) | $t_{0.98}$ (min) |
|---|---|---|---|---|---|
| 50 | 200 | 1 | 0.25 | 3 | 18 |
| 50 | 200 | 100 | 0.25 | 4 | 20 |
| 50 | 200 | 1000 | 0.25 | 13 | 38 |
| 10 | 10 | 1 | 1 | 0.7 | 4 |
| 10 | 10 | 100 | 1 | 5.7 | 14 |
| 10 | 10 | 1000 | 1 | 51 | 104 |

[a] The half-times were calculated from Eq. (2-23); for example, in the third line, $t_{1/2} = 0.7/k + [P]/50 = 3 + 10 = 13$. The 98% times ($t_{0.98}$) were calculated in a similar manner (see text).

For example, it is desired to have a reaction (98%) complete in 10 min. The $K_m$ is 10 $\mu M$, and the highest concentration to be measured will be 100 $\mu M$ in the assay system. The enzyme stock is a 1% solution with activity of 8 units/mg, i.e., 80 $\mu$moles/min/ml, or 80,000 $\mu M$/min. How much of this stock will be needed? From Eq. (2-27), $V_m = (4 \times 10 + 100)/10 = 14$ $\mu M$/min. Therefore an enzyme concentration of 1 : 5000 should suffice. Figure 2-4 provides a graphic calculation of the enzyme requirement with different substrate levels and with enzymes having different Michaelis constants.

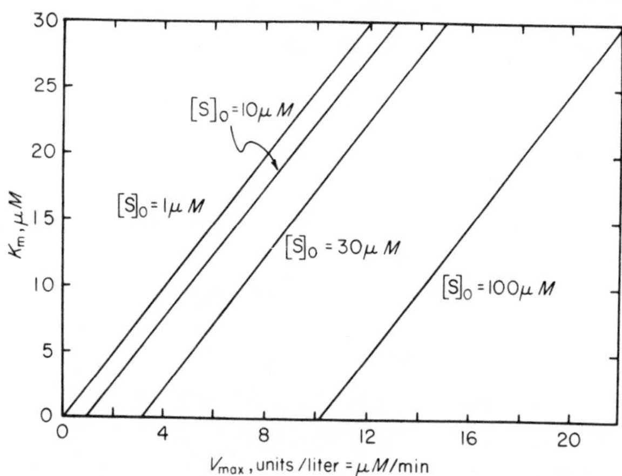


***Fig. 2-4.*** Enzyme required for an enzymatic reaction with normal kinetics to be 98% complete in 10 min. The lines are calculated from Eq. (2-27). For enzymes with $K_m$'s of 50, 100, 200, and 500 $\mu M$, the minimal enzyme requirement for $[S]_0 = 100$ $\mu M$ would be 30, 50, 90, and 210 U/liter, respectively, and only 10 U/liter less if $[S]_0 = 1$ $\mu M$.

When $K_M$ is substantially larger than $[S]_0$, Eq. (2-27) reduces to $V_m = 4 K_M/t$. Thus, if the $K_M$ is 200 $\mu M$, to complete the reaction (98%) in 20 min would require enzyme activity equal to 40 $\mu M$/min or 40 units per liter.

### One-Step Reactions with Two Substrates

The majority of analytical reactions to be considered will involve two substrates (A, B). In many cases, however, one of the reactants (B) will be present in substantial excess and will therefore not be greatly reduced during the reaction. This means that for practical purposes the kinetic situation will reduce to that for one substrate, except that the level of B will influence one or both of the kinetic parameters for A ($V_m$ and $K_A$). (For the case in which B as well as A is substantially reduced during the reaction see the next subsection.)

The mutual effects of two or more cosubstrates can be complex; nevertheless, if [B] is held constant, the reaction will usually obey Eq. (2-20) [and therefore Eqs. (2-22)–(2-26)], i.e.,

$$v = \frac{V_m'[A]}{[A] + K_m'}$$

where $V_m'$ and $K_m'$ only apply to the particular level of B. In designing or modifying an analytical method, it is useful to know how the concentration of B affects the kinetics of A in the analytical range, but it should not be necessary to go into a complete and perhaps complicated kinetic analysis. Without any attempt to be comprehensive, some of the common types of kinetic interactions between cosubstrates can be mentioned.

CASE 1: "RANDOM ORDER OF ADDITION"

In the simplest case, each substrate acts as though it combines independently with the enzyme.

$$\mathrm{E}\begin{cases} +\mathrm{A} \overset{K_A}{\rightleftharpoons} \mathrm{EA} + \mathrm{B} \overset{K_B}{\rightleftharpoons} \\ +\mathrm{B} \overset{K_B}{\rightleftharpoons} \mathrm{EB} + \mathrm{A} \overset{K_A}{\rightleftharpoons} \end{cases} \mathrm{EAB} \longrightarrow \mathrm{E} + \mathrm{P}$$

The net velocity is proportional to the amount of enzyme that is combined simultaneously with both substrates (EAB). The $K_m$ for each substrate ($K_A$ or $K_B$) is independent of the concentration of the other substrate, and the velocity equation is

$$v = \left(\frac{V_m[A]}{[A] + K_A}\right)\left(\frac{[B]}{[B] + K_B}\right) \tag{2-28}$$

which is simply Eq. (2-20) multiplied by a term applying to B. If [A] is much less than $K_A$, Eq. (2-28) reduces to

$$v = \frac{V_m[A]}{K_A}\left(\frac{[B]}{[B] + K_B}\right) = k_a[A]\left(\frac{[B]}{[B] + K_B}\right) \tag{2-29}$$

If [B] is also much less than $K_B$ this becomes

$$v = \frac{V_m[A][B]}{K_A K_B} = k_{ab}[A][B] \tag{2-30}$$

which is the same as the equation for a second-order nonenzymatic reaction [Eq. (2-11)]. Note that as in the nonenzymatic case, $k_{ab}$ is a second-order rate constant and therefore has both a time and concentration dimension (see examples in Table 2-3).

If [B] is constant, Eq. (2-28) reduces to

$$v = \frac{V_m'[A]}{[A] + K_A}$$

and Eqs. (2-29) and (2-30) reduce to $v = k_a'[A]$, where

$$k_a' = k_a\left(\frac{[B]}{[B] + K_B}\right) \tag{2-31}$$

Table 2-3 gives examples of the influence of Michaelis constants and concentrations of substrate B on the apparent first-order rate constant for substrate A. This table would be relevant for enzymatic analysis of substrate A.

### Case 2: "Cooperative Addition"

In a second two-substrate case, each substrate influences the $K_m$ for the other in a manner that can be represented by

$$\begin{array}{ccccc} & +A \xrightleftharpoons{K_A} & EA + B & \searrow K_B{}^A & \\ E & & & & EAB \longrightarrow E + P \\ & +B \xrightleftharpoons{K_B} & EB + A & \nearrow K_A{}^B & \end{array}$$

Here there are two Michaelis constants for each substrate, one ($K_A$ or $K_B$) which applies to the free enzyme and another ($K_A{}^B$ or $K_B{}^A$) which applies to the enzyme–second substrate combination. There are many instances in which the presence of one substrate will favor combination with the other

**TABLE 2-3**

EFFECT OF MICHAELIS CONSTANTS AND LEVELS OF ONE SUBSTRATE ON THE APPARENT FIRST-ORDER RATE CONSTANT OF THE OTHER SUBSTANCE[a]

| Enzyme | $V_{max}$ ($\mu M$/min) | $K_A$ ($\mu M$) | $K_B$ ($\mu M$) | $k_a$ ($min^{-1}$) | $k_{ab}$ ($\mu M^{-1}min^{-1}$) | [B] ($\mu M$) | $V'_{max}$ ($\mu M$/min) | $k_a'$ ($min^{-1}$) | $t_{0.98}$ (min) |
|---|---|---|---|---|---|---|---|---|---|
| I | 1000 | 500 | 200 | 2 | 0.01 | 5000 | 960 | 1.9 | 2.1 |
| | | | | | | 200 | 500 | 1 | 3.9 |
| | | | | | | 10 | 48 | 0.1 | 39. |
| II | 1000 | 500 | 20 | 2 | 0.1 | 5000 | 996 | 2.0 | 2.0 |
| | | | | | | 10 | 333 | 0.67 | 5.8 |
| III | 1000 | 50 | 200 | 20 | 0.1 | 5000 | 960 | 19.2 | 0.2 |
| | | | | | | 10 | 48 | 0.96 | 4.1 |

[a] This table applies to the case of a two-substrate reaction showing "random order of addition." $k_a$ is the apparent first-order rate constant ($V_{max}/K_A$) at infinite concentration of B; $k_a'$ is the apparent first-order rate constant at the particular concentration of B, and is defined by Eq. (2-31). The time required for 98% disappearance of A is designated "$t_{0.98}$."

(positive cooperation), i.e., $K_A{}^B < K_A$. If so, the effect must be mutual, i.e., $K_B{}^A < K_B$. In fact, whether the cooperation is positive or negative,

$$K_A/K_A{}^B = K_B/K_B{}^A \tag{2-32}$$

The velocity equation is

$$v = \frac{V_m[A][B]}{[A]([B] + K_B{}^A) + K_A{}^B([B] + K_B)} \tag{2-33}$$

If [B] is held constant this reduces to Eq. (2-20):

$$v = \frac{V_m'[A]}{[A] + K_A'}$$

where

$$V_m' = V_m\left(\frac{[B]}{[B] + K_B{}^A}\right) \tag{2-34}$$

and

$$K_A' = K_A{}^B\left(\frac{[B] + K_B}{[B] + K_B{}^A}\right) \tag{2-35}$$

At low levels of A the apparent first-order equation is

$$k_a' = \frac{V_m}{K_A{}^B}\left(\frac{[B]}{[B] + K_B}\right) \tag{2-36}$$

in analogy to Eq. (2-31) for the simpler case.

In the above equations, when [B] is much larger than either $K_B$ or $K_B{}^A$, $V_m' = V_m$, $K_A' = K_A{}^B$, and $k_a' = V_m/K_A{}^B$.

Consider an example in which $K_A = K_B = 1$ m$M$, $K_A{}^B = K_B{}^A = 0.1$ m$M$, and [B] is held constant at the levels shown in the tabulation below.

| [B] (m$M$) | $V_m'$ (m$M$/min) | $K_A'$ (m$M$) | $k_a'$ ($\text{min}^{-1}$) |
|---|---|---|---|
| Infinite | 1 | 0.1 | 10 |
| 1 | 0.91 | 0.18 | 5 |
| 0.1 | 0.5 | 0.55 | 0.9 |
| 0.01 | 0.09 | 0.92 | 0.1 |

Note that [B] has a favorable effect on both $V_m'$ and $K_A'$ and therefore an even more favorable effect on $k_a'$.

CASE 3: "ORDERED ADDITION"

A third type of two-substrate case is in reality an extreme example of the case just discussed in which $K_A{}^B$ is zero and $K_B$ is infinitely large.

$$E + A \underset{}{\overset{K_A}{\rightleftharpoons}} EA + B \underset{}{\overset{K_B{}^A}{\rightleftharpoons}} EAB \xrightarrow{k_5} E + P$$

This is called "ordered addition" of substrates since it describes the situation in which the second substrate can only combine with the enzyme if the first substrate is in place. For strictly ordered addition, "rapid equilibrium" kinetics, on which Eq. (2-33) is based, lead to erroneous results. Thus, according to rapid equilibrium kinetics, $K_A'$ becomes infinitely small as the concentration of B approaches infinity, whereas $K_A'$ in fact approaches $K_A$. The reason for the discrepancy is that the assumption of rapid equilibrium does not properly take into account the contribution of $k_5$ to the kinetic situation. Further consideration of this special case is beyond the scope of this book.

## Enzyme Reactions with Second-Order Kinetics

There are a number of analytical situations involving two substrates in which both substrates are initially below their respective Michaelis constants and in which the auxiliary substrate B falls substantially during the analytical reaction. This is often the case when B is DPNH or TPNH and the disappearance is measured directly in the fluorometer. It is not possible to use a large excess of the pyridine nucleotide because during the reaction the percentage change would be too small to measure accurately. In this situation, the reaction is no longer pseudo first order. The practical consequence is that the reaction velocity falls off faster than in a first-order reaction. As shown above, the velocity is described by the equation for a nonenzymatic second-order reaction:

$$v = k_{ab}[A][B] \tag{2-30}$$

The integrated form is somewhat complicated and awkward to use. Therefore a graphic representation is provided (Fig. 2-5). From this can be calculated the time required to use up a given percentage of substrate A when A is varied relative to B.

The graph is constructed with time recorded in multiples of the half-time which would be found when A is much smaller than B. When A is 20% of B, it requires 5.5 and 7 such half-times, respectively, to convert 95% and 98% of A to product (i.e., only about 25% longer than if A were much smaller than B). But when A is 90% of B, the graph shows it requires 15 and 25 half-times

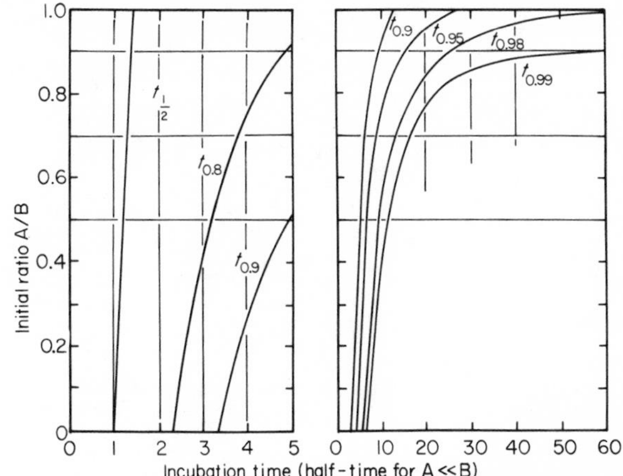


***Fig. 2-5.*** Incubation time required for an enzyme reaction between two substrates when both are present well below their respective Michaelis constants. The ratio of A to B is given on the vertical axis. The incubation time required to convert different fractions of A to product are indicated by the curves. Time is recorded on the horizontal axis as multiples of the half-time for the case when A is much smaller than B.

to convert 95% and 98% of A to product. When A equals B, it would require 70 such half-times (off the graph) to remove 98% of A!

This graph applies strictly only to the case of random addition with no cooperative effects between the two substrates. However, it will in fact apply well enough in most cases, since by definition both substrates are present at levels substantially below the Michaelis constants and cooperative effects will therefore usually be negligible.

## Two-Step Reactions

Many analytical enzyme reactions involve two or more steps. Usually all but the first step will be first order, therefore what follows will be limited to situations where both steps are first order or where the first step is zero order and subsequent steps are first order.

### Two-Step Reaction with First Step Zero Order, Second Step First Order

This is the usual situation in a two-step assay for enzyme activity. The enzyme to be measured catalyzes the first step and forms product, B, at a steady rate.

$$\mathrm{A} \xrightarrow{v_1} \mathrm{B} \xrightarrow{k_2} \mathrm{C} \tag{2-37}$$

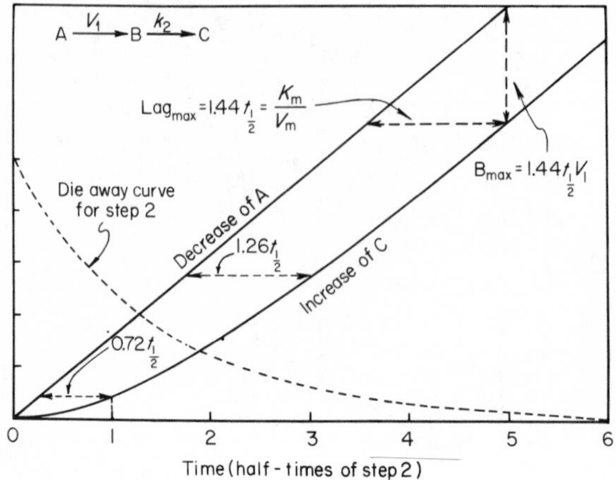


***Fig. 2-6.*** Lag in a two-step reaction in which the first step is zero order, the second step is first order. The vertical distance between the lines equals the amount of B accumulated, the horizontal distance between the lines equals the time lag between disappearance of A and appearance of C.

The problem is to determine what discrepancy there may be between the disappearance of A and the appearance of C, i.e., how fast and to what extent B accumulates (Fig. 2-6).

Since step 2 by definition is first order, $v_2 = k_2[\mathrm{B}]$, where, according to Eq. (2-21), $k_2 = V_{m2}/K_\mathrm{B}$. Ultimately, a steady state will be established, at which time $v_1 = v_2$. When this occurs,

$$v_1 = k_2[\mathrm{B}]_{max} \qquad \text{or} \qquad [\mathrm{B}]_{max} = v_1/k_2 \tag{2-38}$$

where $[\mathrm{B}]_{max}$ is the steady-state level of B.

Thus for the case of a zero-order reaction followed by a first-order reaction, $[\mathrm{B}]_{max}$ will be numerically equal to the velocity of the first reaction divided by the rate constant for the second reaction. For example, if the velocity of the first reaction is 1 $\mu M$/min, and the rate constant for the second reaction is 3/min, $[\mathrm{B}]_{max}$ will be 1 $\mu M/3 = 0.33\ \mu M$.

Since $k = 0.69/t_{1/2}$ [Eq. (2-3)], Eq. (2-38) can also be written as

$$[\mathrm{B}]_{max} = 1.44 t_{1/2} v_1 \tag{2-39}$$

where $t_{1/2}$ refers to the second step. This says that $[\mathrm{B}]_{max}$ represents the amount of A converted to B in 1.44 half-times for the second step.

It is often more useful to consider the lag time between the disappearance of A and the appearance of C. The maximum lag time (Fig. 2-6) is the time it takes to accumulate $[\mathrm{B}]_{max}$, or

$$\mathrm{lag}_{max} = \frac{[\mathrm{B}]_{max}}{v_1} = \frac{1}{k_2} = \frac{K_\mathrm{B}}{V_{m2}} = 1.44 t_{1/2} \tag{2-39a}$$

Notice that the lag time depends solely on the first-order rate constant of the *second step.*

The amount of B present before the steady state is reached is described by the integrated equation:

$$[\mathrm{B}] = (v_1/k_2)(1 - 0.5^{t/t_{1/2}}) \tag{2-40}$$

Accordingly when $t$ is equal to 1, 2, 3, etc. half-times (for step 2), [B] will be 1/2, 3/4, 7/8, etc., of the maximum, or what amounts to the same thing, the lag time will be 1/2, 3/4, 7/8, etc., of the maximum lag time (Fig. 2-6).

*Example.* $v_1 = 0.5\ \mu M/\mathrm{min}$, $t_{1/2}$ for the second step is 2 min. $[\mathrm{B}]_{\mathrm{max}} = 1.44 \times 2\ \mathrm{min} \times 0.5\ \mu M/\mathrm{min} = 1.44\ \mu M$. At 2, 4, 6, and 8 min, [B] will be 0.7, 1.1, 1.3, and 1.36 $\mu M$, respectively.

## Two-Step Reactions with Both Steps First Order

$$\mathrm{A} \xrightarrow{k_1} \mathrm{B} \xrightarrow{k_2} \mathrm{C} \tag{2-41}$$

This is a common situation in metabolic assays. It is useful to know the rate of appearance of C as a function of $k_1$ and $k_2$. When the first step is much slower than the second, the time lag curve is the same as that given for the previous case (step 1 zero order, step 2 first order). Departure from this simple situation is not serious until the rate of the first step approaches or exceeds the second. Therefore, in most analytical situations, the maximal time lag caused by the second step can be taken as 1.5 to 2 half-times (for the second step).

For those who may need more exact information, the following more detailed analysis is presented. We will approach this problem by regarding the time course of the first step as primary and asking what *time lag* in the overall reaction is caused by the second step. The kinetic equations are more complicated than when the first step is of zero order, but fortunately approximations and graphical representations will suffice for most purposes.

The appearance of C is the same as the disappearance of A + B. The integrated rate equation is

$$\frac{[\mathrm{A}] + [\mathrm{B}]}{[\mathrm{A}]_0} = \frac{k_2\,\mathrm{e}^{-k_1 t} - k_1\,\mathrm{e}^{-k_2 t}}{k_2 - k_1} \tag{2-42}$$

The equation for the disappearance of A is

$$[\mathrm{A}]/[\mathrm{A}]_0 = \mathrm{e}^{-k_1 t} \tag{2-2a}$$

This is the exponential form of Eq. (2-2).

The time lag in appearance of C can be calculated from the difference between these two equations. For example, let $k_1 = 0.5/\mathrm{min}$ and $k_2 = 1/\mathrm{min}$.

It is calculated from Eq. (2-2a) that the half-time for disappearance of A would be 1.39 min. It is calculated from Eq. (2-42) that the half-time for disappearance of (A + B) is 2.46 min. Therefore, after an elapsed time of 2.46 min the lag is 2.46 − 1.39 = 1.07 min.

Curves of time lags are given in Fig. 2-7 for $k_2$ held constant and $k_1$ varied

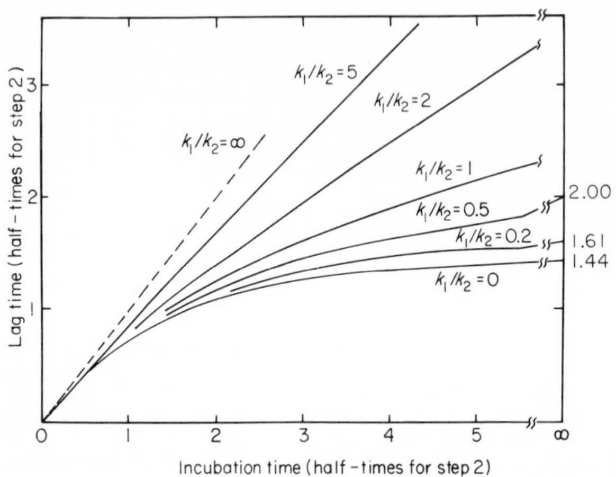


*Fig. 2-7.* Lag in a two-step reaction in which both steps are first order. Time is recorded in multiples of the half-time for the second step. Examples of use: Let $k_1/k_2 = 2$ and the step 2 half-time be 3 min. After incubation for 12 min (Four times the half-time for step 2), the lag in appearance of product is 7.5 min (2.5 half-times for step 2).

from 0.2 to 5 times $k_2$. When $k_1$ is smaller than $k_2$ the lag approaches a finite fraction of the half-time for step 2 (Table 2-4). If the ratio of $k_1$ to $k_2$ is made successively smaller, the lag curve approaches that obtained when the first step is zero order. That is, the maximum lag approaches 1.44 half-times (for step 2), and the lag is 1/2, 3/4, 7/8, etc., of the maximum after 1, 2, 3, etc., half-times (of step 2).

If the half-time for step 2 is known, the half-time and even the whole reaction curve for step 1 can be calculated with the aid of Table 2-4 and Fig. 2-7. As an example, suppose step 2 has a half-time of 2 min and the overall reaction has a half-time of 8 min (four times longer). Table 2-4 shows that the first step is about 0.4 times as fast as the second step ($k_1/k_2 = 0.4$). If the first step is known to be first order, this would indicate a half-time of 2.5 min. To see if the first step did in fact follow a first-order curve, its time curve could be calculated by deducting the lag times indicated in Fig. 2-7 along an interpolated curve for $k_1/k_2 = 0.4$. In the present example, readings made at 2, 4, and 6 min (1, 2, and 3 half-times) would be corrected by 0.75, 1.3, and 1.4 half-times, or 1.5, 2.6, and 2.8 min. This procedure would give a close

**TABLE 2-4**

LAG TIME FOR TWO SEQUENTIAL FIRST-ORDER REACTIONS[a]

| | Half-time | | Lag | 98% time | | Lag at | Maximum |
|---|---|---|---|---|---|---|---|
| $k_1/k_2$ | A decrease | C increase | at $t_{1/2}$ for C | A decrease | C increase | 98% time for C | lag time |
| 0.05 | 20 | 21.47 | 1.47 | 113 | 114 | 1.47 | 1.48 |
| 0.1 | 10 | 11.52 | 1.52 | 56.5 | 58 | 1.52 | 1.52 |
| 0.2 | 5 | 6.58 | 1.58 | 28.2 | 29.8 | 1.59 | 1.61 |
| 0.4 | 2.5 | 4.05 | 1.55 | 14.1 | 16.0 | 1.82 | 1.84 |
| 0.6 | 1.67 | 3.20 | 1.53 | 9.4 | 11.5 | 2.14 | 2.20 |
| 0.8 | 1.25 | 2.68 | 1.43 | 7.05 | 9.50 | 2.45 | 2.51 |
| 0.9 | 1.11 | 2.53 | 1.42 | 6.25 | 8.90 | 2.65 | 3.70 |
| 1.0 | 1.0 | 2.42 | 1.42 | 5.63 | 8.40 | 2.77 | [b] |
| 2.0 | 0.5 | 1.77 | 1.27 | 2.82 | 6.64 | 3.82 | [b] |
| 5.0 | 0.2 | 1.31 | 1.11 | 1.13 | 5.95 | 4.82 | [b] |
| 10.0 | 0.1 | 1.15 | 1.05 | 0.56 | 5.80 | 5.24 | [b] |

[a] The reaction sequence is $A \xrightarrow{k_1} B \xrightarrow{k_2} C$. The half-time for the second step is held constant at unity. The half-time of the first step is varied, as shown, from 0.1 to 20 times that of the second step. The times are given for 50% and 98% decrease in A, and 50% and 98% appearance of C. The times are recorded as multiples of the half-time of the second step.

[b] These lag times increase without theoretical limit.

approximation to the time curve for step 1, even if step 1 turned out not to be first order.

Another use for Table 2-4 and Fig. 2-7 would be to guide the choice of assay conditions. They allow prediction of amounts of enzymes needed to complete a reaction in a specified time. For example, from Table 2-4 it is easily seen that for the overall reaction to be 98% complete in about 10 min the half-times for step 1 and step 2, respectively, could be (in minutes) 1.7 and 0.17, 1.5 and 0.6, both 1.2, or 0.17 and 1.7. Note that the most efficient use of the enzymes occurs when the half-times are about equal and that interchange of the two half-times does not affect the overall time curve.

## Other Factors Affecting Enzyme Kinetics

The preceding has been devoted to the relationships of substrate concentrations to reaction velocity. It should be stressed that $V_m$, Michaelis constants, and equilibrium constants are all subject to the influence of temperature, pH, ionic strength, and the presence of activators and inhibitors. A brief comment on temperature and pH is in order.

### Temperature

Ordinarily $V_m$ is increased by increasing temperature until the point of enzyme instability is reached. Temperature coefficients vary greatly from one enzyme to another, and it is sometimes overlooked that enzyme activity does not fall to zero in an ice bath. Usually the $V_m$ at 0° is 3 to 10% of that at 38° and in the case of some enzymes may be as high as 20% or more.

Michaelis constants also would be expected to increase with temperature. Therefore, temperature may have a less favorable effect on rates at low substrate levels, i.e., in the usual situation encountered in analytical work. In some cases rates may actually decrease with temperature.

### pH Effects

The effect of pH on $V_m$ is, of course, extremely variable from enzyme to enzyme. For some enzymes the pH optimum is extremely sharp, for others very broad. The pH for maximum velocity does not necessarily coincide with the pH for minimum Michaelis constant. Consequently the pH optimum for an analytical reaction with low substrate level may be quite different from that for an enzyme assay carried out at near-saturating substrate concentration.

CHAPTER 3

# GLASSWARE

## CONSTRICTION PIPETTES

Constriction pipettes were originally described by Levy (1936) working in the laboratory of Linderstrøm-Lang. The large range of sizes in which they can be made (Fig. 3-1) and the ease of use makes them particularly valuable for microchemical analyses. In fact, for all of the analytical applications of the type considered in this book, the most important tool is the constriction pipette. The construction, calibration, and use of this instrument therefore deserve extended presentation.

At present, there are commercial sources for standard types and larger sizes of constriction pipettes (Appendix). Smaller sizes and pipettes of modified design for special purposes are not readily available.

### The Role of Surface Tension in Pipetting

Surface tension represents a force of great importance in handling small liquid volumes. It is usually useful, but is sometimes a nuisance. The purpose of this section is to show how surface tension governs the use of constriction pipettes and dictates their construction.

Gravity is the force responsible for emptying large pipettes and burettes. In contrast, delivery of a small constriction pipette is accomplished almost

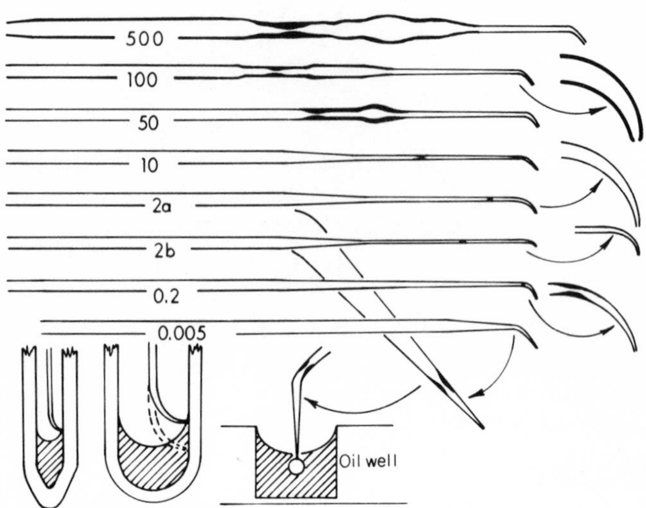


*Fig. 3-1.* Design of assorted types of constriction pipettes. The volumes are indicated in microliters. The upper five pipettes and the 0.2 μl pipette are for general use. The pipette labeled 2b is for use in a slender tube (2–3 mm bore). Note the finer tip and shorter bend as compared to 2a. This relationship between the tip and tube size is shown at the lower left. Note that ideally the pipette tip and meniscus have the same radius of curvature. A 0.005 μl oil well pipette is illustrated at the bottom. (The pipette with the tip actually in an oil well is of larger size.)

entirely by air pressure and the wall is swept clean by the opposing surface tension. A 10 μl pipette can be delivered equally well vertically, horizontally, or even upside down.

If it were not for surface tension a small pipette would deliver only a small fraction of its contents. Imagine a column of liquid standing in a tube of 1 or 2 mm bore, and then being forced to flow out slowly. The fastest flow will tend to occur down the middle of the tube. This, if unopposed, would result in a small channel down the middle, leaving a thick stagnant layer next to the wall. However, as flow begins, and the center of the meniscus tries to move faster than the periphery, surface tension exerts a powerful force to maintain the hemispherical shape of the surface (the shape with the smallest surface area). Thus, surface tension tries to force the liquid near the wall to keep up with that in the center. (With pipettes below 0.001 μl volume, the effect of surface tension is so great that it may be difficult to apply sufficient pressure by mouth to empty the pipette. In this case, the tremendous sweeping force is readily appreciated.)

For effective removal of liquid near the wall the flow must be slow enough to give surface tension time to overcome the viscosity of the liquid. A rule of thumb for flow rate was given by Peters and Van Slyke for larger pipettes and burettes (1932). This rule is that with aqueous fluids a linear speed of 1 cm/sec will leave behind a sufficiently thin film of liquid (1% or less of the contained volume) to permit good reproducibility.

We have made tests to see whether the rule applies to very small pipettes and to measure more exactly the relationship between flow rate and residual fluid. Measurements were made with tubing of five sizes ranging from 0.06 to 1.9 mm in bore. These would encompass the diameters of pipettes ranging from 0.003 to 200 $\mu$l. Within experimental limits the residual fluid for all sizes was found to be related to flow rate (cm/sec) by the same simple formula:

$$\text{Residual fluid} = 0.9\%\sqrt{\text{flow rate}}$$

Experience with pipettes down to 0.000,1 $\mu$l indicates that the formula applies even at this level.

The following tabulation gives some examples calculated from this formula.

| Linear rate (cm/sec) | Residual fluid (%) |
|---|---|
| 0.2 | 0.4 |
| 0.4 | 0.57 |
| 0.8 | 0.8 |
| 1.0 | 0.9 |
| 1.2 | 1.0 |
| 2.0 | 1.3 |
| 4.0 | 1.8 |
| 6.0 | 2.4 |

It will be seen that for rates near 1 cm/sec a 20% difference in rate makes only a 0.1% difference in the volume delivered. This fact helps explain why it is relatively easy to achieve a high degree of precision with even the smallest constriction pipettes.

It may seem surprising that the same rule governs the fraction of fluid left behind over the 10,000,000-fold range of useful constriction pipettes. For a 500 $\mu$l pipette, 1% of the volume represents a 7$\mu$ film on the wall. For a 0.000,1 $\mu$l pipette, 1% of the volume represents only a 0.05 $\mu$ film.

## Effect of Changes in Surface Tension and Viscosity

From what has been said it is clear that either a decrease in surface tension or an increase in viscosity of the liquid to be pipetted will increase the amount of residual fluid. The greatest difficulties arise if there is both low surface tension and high viscosity, as is true of certain organic liquids. In this case special pipettes may be needed which provide for very slow delivery.

Surface tension plays other roles in pipetting: (1) The effect of surface tension at the upper constriction makes it easy to fill pipettes with an exactly reproducible volume. (2) Surface tension acting at the constricted orifice at the tip of the pipette prevents air from escaping after the fluid volume has been delivered. Sudden escape of air would tend to spatter the sample. (3) Surface tension acting at the tip is also useful when drawing liquid into the pipette. If the tip accidentally gets out of the liquid while suction is being applied, air will not be drawn in unless the suction exceeds a critical value determined by the diameter and surface tension of the liquid. (4) Surface tension is relied on to keep liquid from running up the outside of a pipette during pipetting. When liquid runs out of a pipette tip hanging free in the air, the liquid clings to the tip until the weight of liquid overcomes surface tension and a drop falls. With pipettes of less than 10 $\mu$l the entire sample can collect on the outside without falling. If the shaft is fine enough, the liquid can even ascend up the shaft for a considerable distance. If, however, before beginning to expel the liquid, the pipette tip is touched to a larger surface (inside of a beaker or tube, etc.), surface tension will pull the emergent liquid to the larger surface, completely preventing liquid from running up the pipette shaft. The finer the pipette tip, the less the tendency for liquid to cling to the tip instead of the surface touched.

## Pipetting Technique

Constriction pipettes are capable of precision delivery ($\pm 0.3\%$ or better) in sizes ranging from the largest useful volumes (up to 1 ml) down to a volume of $10^{-7}$ ml or even less. It is difficult to make large errors in using a well-constructed 200 $\mu$l constriction pipette. With pipettes under 20 $\mu$l, however, good technique is necessary to achieve precision. It is believed, therefore, that attention to the "fine points" emphasized in this section may prove rewarding.

The pipettes are filled and emptied by mouth using very flexible rubber tubing of appropriate diameter attached to the pipette barrel. If the tubing is too stiff it will require undue force to manipulate the pipette and this will increase the danger of breakage. A suitable tubing is 3/32 inch bore with 1/32 inch wall. For working with pipettes under 20 $\mu$l a good light is recommended and, if necessary, eyeglasses to permit sharp vision at a distance of 10 to 12 inches. Eyeglasses with +2 or +3 diopters are very useful, particularly if one is a little presbyopic.

When ready to fill, the pipette tip is held against the wall of the vessel but is dipped only a minimum distance into the fluid. The bend of the tip keeps the shaft away from the vessel wall, preventing fluid from running up between shaft and wall. (If the pipette is accidentally dipped too far into the liquid, is should be withdrawn slowly; this permits surface tension to strip off excess liquid which would otherwise cling to the tip. It is clear that any material collecting on the tip will be rinsed off into the sample during delivery.) Fluid is now sucked in to just above the upper constriction and then driven back until the meniscus almost reaches the narrowest portion of the constriction. Before releasing or even decreasing the pressure, the tip must be withdrawn from the liquid surface, otherwise excess fluid would be drawn into the pipette by capillarity. With a slender 1 $\mu$l pipette this could result in disastrous errors.

To deliver the sample, the pipette tip is placed against the wall of the receiver. If liquid is already present, the tip is ordinarily touched to the wall through the edge of the meniscus but is never extended any further than necessary below the surface. The tip bend again functions to keep the shaft away from the wall. Pressure is now applied and increased until the critical pressure is reached at which delivery begins. For reasons already given, the delivery rate should be as nearly as possible the same each time the pipette is used. The easiest way to accomplish this is to maintain throughout delivery the pressure that is necessary to start delivery. If the pipette is properly made this will provide for a suitable flow rate. This requires a proper balance between the diameters of the upper and lower (tip) constrictions. The critical pressure is determined by the diameter of the upper constriction (and by the surface tension of the liquid). The resistance to delivery is determined by the diameter of the tip and shaft and by the viscosity of the liquid.

If at the end of delivery a very small, reproducible, amount of fluid remains in the tip it is ordinarily preferable not to try to blow this out. If it must be delivered, in order not to spatter, the tip is withdrawn and touched to the wall above the meniscus before blowing out. In any event, pressure must be maintained until the tip is removed from the meniscus, otherwise fluid will usually be drawn back into the pipette by capillarity. Even if pressure is maintained the tip should be lifted from the wall before withdrawal, because if this is not done liquid will follow the pipette tip up the receiver wall.

## Special Considerations with Small Pipettes and Small Tubes

When pipetting into or out of a slender tube there exists considerable danger fluid will inadvertently run up between the pipette and the wall. This is likely to spoil the analysis. Such accidents are avoided in the first place by using a well-made pipette with a bend at the tip which is equal to the radius of the tube (see section on pipette making). Pipetting into small tubes is also made easier if the tip, as it touches the inside of the tube, is pointed toward the pipetter. This permits exact positioning of the shaft in the middle of the tube. Otherwise it would be difficult to judge the position of the shaft in relation to the wall; if it should twist sideways a run-up is sure to occur. Manipulation of the pipette is facilitated if the hands are turned, so that the little fingers touch each other. This will also help prevent breakage of pipettes, which if designed for small bore tubes, are of necessity somewhat fragile.

To make a clean withdrawal of a sample from a small bore tube, the pipette tube is inserted into the meniscus, rather than into the body of the fluid, and is then moved down the side of the tube to keep pace with the meniscus. Similarly, if a substantial volume of fluid is added into a slender tube, delivery is started at the bottom or at the surface of any fluid already present, and the pipette is drawn up along the side of the tube, keeping the tip always in the meniscus.

## Transferring the Whole Sample ("Total Transfer")

In classical volumetric analysis, when a sample is to be transferred from one vessel to another, it is poured or pipetted into the second vessel, the rinsings are added to it, and the whole is diluted to volume. With the small volumes dealt with here, this is impractical. One satisfactory procedure is to transfer an exact aliquot, representing the bulk of the sample, to the new vessel.

An alternative procedure, which is sometimes advantageous, is to transfer the whole sample, or actually about 99% of the whole sample. The technique is to draw up the fluid using a constriction pipette that is *somewhat smaller than the volume to be transferred* (and which need not be calibrated). As the last of the fluid is being drawn in, the suction is reduced. As stated earlier, due to the effect of surface tension at the constricted orifice, the flow will stop automatically when the last of the sample is in the pipette. Slight negative pressure is maintained until the pipette is out of the tube and against the wall of the receiving vessel, where it is delivered at the appropriate rate for that pipette. The upper constriction, which is surrounded by fluid when the pipette is full, serves a safety function. In case air is accidentally sucked in, this constriction acts as a brake to prevent fluid from travelling far up the pipette.

The total transfer described is at least as reproducible as any other pipetting. Because the fluid moves slowly down the walls of the original vessel, the volume left behind is reproducible and equal to less than 1% of the whole. This clean removal is due to the same sweeping action of surface tension which applies to pipette deliveries. There is one precaution to be mentioned. The tubes should not be mixed or agitated immediately before the transfer. Liquid splashed on the walls must drain down under the force of gravity; it is not subject to the sweeping action of surface tension. Whereas a column of liquid 2 cm high in the bottom of a tube can be withdrawn to the extent of 99% or more in 2 or 3 sec, if a sample has been splashed up the walls for 2 cm it will take several minutes before 99% has drained back. (In special cases, for example after heating, droplets may collect on the wall and not drain down at all. In this case they can be brought down by centrifugation.)

Total transfer is preferable to transferring an aliquot if there is any danger that there has been evaporation, or if for reasons of convenience the samples differ somewhat in volume. For samples under 100 μl total transfer is at least as fast as taking an aliquot.

## Common Pipetting Errors

1. Using a bad pipette (see Making Constriction Pipettes)
2. Using stiff rubber tubing
3. Dipping pipette too deep into liquid of reagent or sample
4. When filling, not removing tip from meniscus before releasing pressure
5. During delivery, increasing or decreasing pressure from that used to start delivery
6. Blowing air after the sample

## Rinsing Pipettes

Since a pipette constructed and used properly leaves after delivery a residual fluid volume of 1% or less, it is usually satisfactory (in fact preferable) not to clean or rinse the pipette between samples in a series. If an error of 2% (from failure to rinse) is acceptable, it is clear that only when the previous sample had been more than three times *stronger* than the sample concerned would an unacceptable error arise. (Because in this case the pipette will be filled with 1% of concentration 3 = 0.03, plus 99% of concentration 1 = 0.99; the sum is concentration 1.02) Rinsing is nevertheless desirable before pipetting blanks, since the carry-over from a sample would have a large relative effect, and any error in the blank affects the evaluation of every other sample. When rinsing is believed necessary the rinsing fluid (ordinarily water) should be delivered slowly to leave a minimum of liquid in the pipette. Cleaning during the analysis, other than by rinsing with water, is not recommended unless the pipette is not delivering properly because of accumulation of protein or lipid.

## Making Constriction Pipettes

### Properties of a Good Pipette

Precision and convenience are enhanced by proper construction. The constriction should be narrow enough and abrupt enough so that the meniscus will stop at an easily reproducible point. The constriction should be wide enough so that is is not easily obstructed and an excessive amount of force is not needed to drive the meniscus past the constriction. The region around the constriction should be such that once delivery is started the orifice remains free until delivery is complete. This condition can usually be attained by narrowing the tubing above and below the constriction. This prevents fluid from collecting in the neighborhood of the constriction. If, instead, the constriction is made abruptly in the middle of a wide tube (Fig. 3-2), fluid will not be swept away cleanly during the delivery and will drain back into the constriction to give a jerky and less precise delivery.

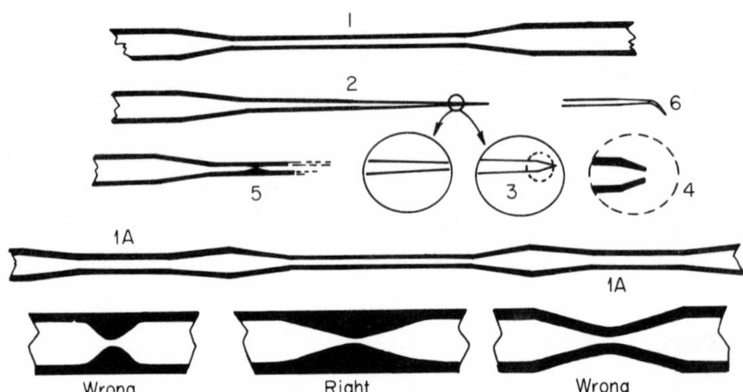


***Fig. 3-2.*** Steps in making a pipette. Two pipettes are made from each piece of tubing. Steps 1–6 are shown for making a pipette of 10 $\mu$l or less. Step 1A shows the extra step for large pipettes in order to narrow the tubing above and below the future constriction. At the bottom are shown one correct and two faulty constrictions; that on the right is too fragile, that on the left would trap fluid and below the constriction and give jerky delivery.

The tip itself should be related to the size of the pipette. It should taper abruptly to a sufficiently fine point so that the amount of adherent liquid will be small in comparison to the delivered volume. Most important of all, the diameter of the upper constriction and the size of the tip orifice need to be balanced against each other to provide a suitable rate of delivery, and the upper constriction must be slightly larger than the lower constriction (i.e., less force ought to be needed to start delivery than to blow air out of the tip after delivery).

The shape of the bend at the pipette tip is not critical except for pipettes to be used with small tubes. The ideal bend is an approximation of a quarter circle. In slender tubes the meniscus is hemispherical. When the quarter circle of the pipette tip crosses a meniscus of the same radius, the shaft and fluid are at right angles to each other. This minimizes any tendency for fluid to run up the shaft (Fig. 3-1). The optimal bend at the tip will place the shaft in the center of the tube (Fig. 3-1) where it will be least likely to trap fluid against the wall. As smaller and smaller tubes are used this problem of preventing fluid from running up between shaft and wall becomes more and more difficult. A practical limit is reached with tubes of about 1.5 mm bore. With tubes in this range the danger of a "run-up" is minimized if the pipette shaft is made more slender than usual for the last few mm, and if the shaft in turn tapers at the very tip to a somewhat finer point than usual.

### Construction

When making constriction pipettes for routine use, Pyrex tubing is recommended because it is stronger than soft glass. For even greater durability, quartz tubing may be used, and it is especially desirable for pipettes of 3 $\mu$l or less.

For most of the glassblowing described, tinted eyeglasses should be worn. Ordinary sun glasses are adequate when working with Pyrex, but special dark glasses are needed when using quartz (Appendix). For making and bending the tips less light is produced and glasses may be unnecessary even with quartz.

As in all glassblowing, success is increased by using a flame of correct size and temperature. Two burners are ordinarily required, a multiflame burner or blowpipe torch for large and middle-sized flames and a special microtorch (Appendix).

Tubing of 4 or 5 mm outer diameter is cut into 20 to 25 cm lengths, each of which will make two pipettes. (The 4 mm tubing is suitable for pipettes of 150 $\mu$l or less, the 5 mm for 200 to 500 $\mu$l pipettes.) For pipettes of 10 $\mu$l and larger, the tubing in the region of each of the future constrictions is thickened (for strength) and narrowed for a length of 10 or 20 mm to a bore of 0.5 to 2 mm, depending on the intended volume (Fig. 3-2). This is accomplished by rotating the tube in a broad oxygen–gas or oxygen–air–gas flame with care not to pull the ends apart. The distance between the two thickened regions is determined by the desired pipette size.

The portion of tubing between the two future constrictions is heated, usually with little or no thickening, and drawn down to a diameter of 1.5 to 3 mm. The tubing is always removed from the flame to make the pull. The center portion is narrowed further to a diameter of 0.5–0.8 mm and the tubing finally separated at the center.

For pipettes in the 1–10 μl range it is satisfactory to slightly thicken the original tubing in the center and pull it down without any special thickening in the region of the future constrictions.

The actual tip of each pipette is made in a small cool flame, provided by the microtorch. Air rather than oxygen may be preferable for smaller Pyrex pipettes, to avoid collapsing the wall. A small area of glass is heated, and the glass pulled to give a sharp taper (Fig. 3-2). This provides a fine tip without undue fragility. With a diamond point, the tip is cut off at a point judged suitable for the volume of the pipette concerned. The upper end of the pipette is fire-polished.

The exact position for the constriction is determined in the case of 1–500 μl pipettes, by drawing into the pipette a measured volume of water. A graduated pipette of suitable size (0.1, 0.2, or 1 ml) is partially filled with water and attached almost horizontally to a ring stand (Fig. 3-3). The tip of

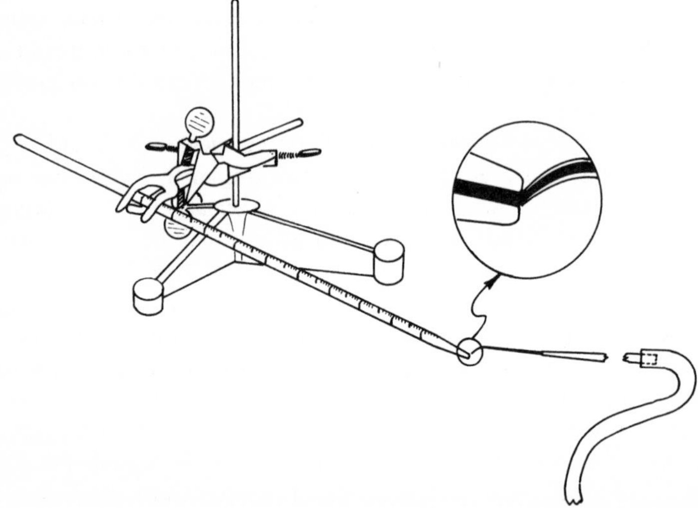

***Fig. 3-3.*** Drawing in a measured volume of water to determine the constriction site. The same arrangement is used for calibrating the finished pipette (see text).

the new pipette is touched to the tip of the graduated pipette, and the desired volume of water is drawn into the thickened (constriction) area. A mark is made with a wax pencil at the point indicated, and a temporary label with the approximate volume is attached to the barrel of the pipette.

The constriction is put in with the same microburner used to pull the tip, using a small hot oxygen flame. It is well to experiment with burner tip sizes, gas, and oxygen supply until the optimal flame for this purpose is obtained. The constriction is placed slightly above the pencil mark to compensate for shrinkage just below the constriction. The pipette is heated, first in the cooler part of the flame, and gradually in the hotter part of the flame as the glass thickens. The pipette is rotated or rocked in the flame to ensure uniform thickening. It is necessary to avoid pushing or pulling in the process; this can be done if the left hand holds the pipette whereas the right hand merely cradles the smaller segment. It is easier to make the constriction if one's hands are supported at the proper height above the flame by resting on wooden blocks. This also helps to control lateral motion of the pipette in the flame, which could unduly lengthen the constricted region. When the glass is cool, the constriction is tested by drawing water into the pipette. The constriction can be made smaller by further heating but cannot be made larger. The delivery rate is adjusted at this stage by either fire polishing the tip if delivery is too fast, or by cutting it off slightly if delivery is too slow. Narrowing the constriction will also accelerate delivery, by increasing the force needed to start delivery. Fire polishing of the tip is done at the edge of a very cool flame. If fire polishing is carried too far, the pipette can usually be rescued by pulling out a new tip.

The final step is to bend the tip. A very cool flame is used, and the tip of the pipette gradually brought into the flame while pressure is applied to the tip with a flat piece of metal, such as a flattened nail or, better, a piece of platinum. No more heat than necessary should be applied for the bending in order not to narrow or seal off the tip.

### Quartz Pipettes

The directions given above are for pipettes made from Pyrex. If quartz tubing is used, a hotter flame is required, and the working temperature range is narrower so that the quartz must be pulled more quickly when it is out of the flame. Because quartz is a better conductor of heat than Pyrex, it is not necessary to rotate the tubing as much for the initial pull or constriction.

### Oil Well Pipettes

Pipettes for use in oil wells are provided with a very slender tip, with a 45° bend 5 or 10 mm from the end, but usually without a bend below this (Fig. 3-1 and Chapter 13). The slender tip is to minimize the amount of adherent oil and to make it easier to strip off the aqueous volume when the pipette is withdrawn through the oil–air interface. Because of the fragility of the slender tip, quartz is usually preferable to glass.

### Smaller Pipettes

In making pipettes of less than 1 $\mu$l, it is also desirable, as said earlier, to use quartz instead of glass, not only because it is stronger, but because it is actually easier to handle in construction of these smaller pipettes. It is also desirable to use magnification. Most suitable is a wide angle (dissecting) microscope with a long working distance and a zoom lens. For most purposes a total magnification of from 5 × to 15 × will be adequate.

The smaller the pipette the smaller the flame that will be needed. A flame with an exceedingly small hot zone can be achieved by using a torch tip with an orifice of 0.3 to 0.5 mm and by increasing the ratio of oxygen to gas until it is just below the point of extinguishing the flame. We have found an oxygen–gas mixture to be preferable to an oxygen–hydrogen mixture. With hydrogen the flame appears to be just as small but the hot zone is not so well localized. Whatever the flame size, the effective width can be narrowed by working on the edge of the flame.

When making very small pipettes it is desirable to keep the bore as wide as possible by reducing the length as well as the bore. If the bore is too narrow, surface tension makes delivery difficult. This is accentuated by the fact that the smaller the bore of the pipette the smaller must be the constriction. This results from the need to provide a sharp difference between the pressure needed to move the liquid down to the constriction and that needed to make the constriction "let go." In designing small pipettes a rough guide is to keep the ratio between the bore at its largest point and the length between constrictions in the range from 1:25 to 1:50. This means that the distance between upper constriction and tip for a 0.001 μl pipette may be only 1 or 2 mm.

Because the smallest pipettes are very short it is usually convenient to change the order of construction from that given for larger pipettes. The tubing (quartz) is pulled down to a suitable diameter for the constriction and cut off 8 or 10 mm below the site of the constriction. The constriction is now made with the free end as a handle which can be supported as necessary by another piece of quartz. If the free end is too short to pull with the fingers, a longer piece of fine quartz is now attached, the flame is enlarged, and the body and tip are now pulled out. In contrast to the situation with larger pipettes, the pulling can often best be done directly in the cooler part of the flame.

**Calibration of Pipettes**

Three ways of calibrating constriction pipettes are offered: gravimetric, colorimetric, and volumetric. For volumes of 20 μl and over the gravimetric method is the most accurate (0.1 % to 0.3 %). The volumetric method is by far the most rapid and is accurate to 0.5–1 % down to perhaps 5 μl. The colorimetric method is accurate to 0.5 % or better down to 0.002 μl. It is usually desirable to have a selection of larger pipettes calibrated accurately by the gravimetric method, whereas, the bulk of the pipettes in the 5 to 50 μl range will be sufficiently accurate if calibrated volumetrically. Under 5 μl the colorimetric method is required.

Although is it desirable to have available accurately calibrated pipettes, in most analyses this is not essential, since the pipette volumes will cancel out of the final calculation if standard solutions and samples are handled with the same pipettes.

The calibration of pipettes is one of the best ways to test pipetting technique. Failure to obtain satisfactory checks means faulty technique or bad pipettes. Since the gravimetric method is the most precise, it offers the most critical test, but only for larger pipettes. Also, there is one fault that will not be detected gravimetrically: that of leaving too much fluid on the outside of the pipette, since this would be expected to evaporate before delivery and not appear in the weight. Colorimetric calibration will detect this fault because the tip is washed off in the diluent. Colorimetric calibration is also more likely to reveal bad technique since it can be used with the smallest pipettes, for which pipetting faults can cause very serious errors.

## Gravimetric Calibration

Constriction pipettes are calibrated by delivery of water into a vessel already containing a little water. The increase in weight of the vessel is measured on an analytical balance. A narrow-mouthed vial without stopper and with a capacity not more than 20 times the volume of the pipette is recommended. The pipette to be calibrated is delivered 3 or 4 times into the vessel and the increase in weight recorded for each delivery. The pipette should be carefully touched to the wall before starting delivery just as in ordinary use. Care should be taken not to blow air after the sample since this would drive *moist* air out of the calibration vessel. The vessel should be handled as little as possible with the fingers. In order to correct for possible evaporation, sham deliveries are made and any weight change in the vessel used as a correction for calibration. The successive weights of the water delivered by the pipette should agree within $\pm 0.1$–$\pm 0.3\%$. The volume is calculated from the weight and the apparent specific gravity of water in air at the temperature at which the calibration is done (Amounts of water which appear to weigh 1 mg in air at 21°, 25°, and 29° are equivalent to 1.003 $\mu$l, 1.004 $\mu$l, and 1.005 $\mu$l, respectively.)

## Colorimetric Calibration

The pipette to be calibrated is used to deliver a volume of concentrated *p*-nitrophenol into a much larger accurately measured volume of diluent. The resulting solution is read at 400 nm in a spectrophotometer and the reading is compared with readings of dilutions made with previously standardized pipettes. (*p*-Nitrophenol is chosen because it is highly colored but unlike many colored materials is not strongly adsorbed to surfaces. Full color is obtained at any pH above 9, but above pH 11 oxidation becomes noticeable.)

For pipettes in the 0.5–50 $\mu l$ range the stock solution is 0.5% *p*-nitrophenol in 0.05 *M* $Na_2CO_3$. It is carefully protected from evaporation throughout the period of calibration. The diluent is 25 m*M* $NaHCO_3$–25 m*M* $Na_2CO_3$ (pH 10). The *p*-nitrophenol is pipetted into an approximately 1000-fold larger volume of diluent measured with accurate volumetric pipettes. It is recommended that the constriction pipettes be calibrated in duplicate or triplicate and the diluting volumetric pipettes be themselves calibrated for this purpose.

Two standard solutions are made with 100 ml of diluent in a volumetric flask. To one is added 100 $\mu l$ and to the other 120 $\mu l$ of the *p*-nitrophenol *after* the flasks have been diluted to 100 ml. The volumes of *p*-nitrophenol, although they need not be exactly 100 and 120 $\mu l$, must be accurately known and are added with constriction pipettes standardized gravimetrically.

The standards are read in a spectrophotometer, with appropriate correction for the readings of the cuvettes with carbonate buffer alone. From the corrected readings and the dilution of the standards a factor is obtained for the other pipettes. For example, suppose the two standard volumes of *p*-nitrophenol were 100.6 and 120.4 $\mu l$ of *p*-nitrophenol and the corrected spectrophotometer readings were 0.506 and 0.608.

$$\text{Factor 1} = \frac{100.6\ \mu l}{100\ \text{ml} \times 0.506} = 1.988\ \mu l/\text{ml}$$

$$\text{Factor 2} = \frac{120.4\ \mu l}{100\ \text{ml} \times 0.608} = 1.980\ \mu l/\text{ml}$$

The factor to be used would be the average of the two standard calculations, or 1.984. The use of two standards provides a check on the proportionality of the spectrophotometer or the possibility that one of the standardizing pipettes is in error.

The use of the factor is illustrated by the following example: A pipette that was previously estimated to be about 3 $\mu l$ is delivered into exactly 3.01 ml of carbonate buffer. The corrected triplicate readings are 0.515, 0.513, and 0.512 for an average of 0.513. The calculation is

$$3.01\ \text{ml} \times 0.513 \times 1.984\ \mu l/\text{ml} = 3.06\ \mu l$$

Calibration within 0.3% is easily attainable.

For pipettes larger than 50 $\mu l$ the same principle is followed but weaker *p*-nitrophenol (0.05%) is used and the dilutions are approximately 1 to 100 instead of 1 to 1000 in order to keep the total volumes within convenient limits.

For pipettes smaller than 0.5 $\mu$l the *p*-nitrophenol concentration can be increased to 5% which is prepared in 0.4 *M* KOH (to increase the solubility). The dilution is increased to 1 to 10,000 for pipettes in the 0.05–0.5 $\mu$l range (total volume 0.5–5 ml). Still smaller pipettes can be calibrated by increasing the dilution to 1 to 25,000 and if necessary using microspectrophotometer cells. Thus, 0.002 $\mu$l of 5% *p*-nitrophenol diluted with 50 $\mu$l of carbonate solution will give an optical density of nearly 0.3, which permits 0.5% precision.

Although it is beyond the scope of this section, still smaller pipettes can be calibrated fluorometrically with strong quinine solutions. Others have found it convenient to calibrate pipettes with a highly radioactive solution pipetted directly onto the planchet of the counter.

### Calibration by Volume

Pipettes ranging from 5 to 200 $\mu$l can be rapidly calibrated using a series of graduated pipettes. The graduated pipettes are selected for uniform bore, determined by withdrawing successive aliquots of water with a constriction pipette previously calibrated by weight or color. If the pipette is not exactly uniform, it can nevertheless serve if it is itself first calibrated. A pipette of 0.1 ml volume, with 1000 subdivisions per milliliter is of suitable size for calibrating constriction pipettes of 5–50 $\mu$l. A similar 0.2 ml pipette is appropriate for constriction pipettes of 30–150 $\mu$l. For larger pipettes, a 1 ml graduated pipette with 100 divisions per milliliter can be used.

The recommended procedure is as follows. The graduated pipette is mounted on a ring stand as in Fig. 3-3 and filled with water. The angle is adjusted so that water will not spontaneously flow out, but will fill the pipette exactly to the tip. The constriction pipette to be calibrated is filled from a small beaker and then delivered back into the beaker at its proper delivery rate. (This creates a space in the pipette exactly equal to the volume that it ordinarily delivers.) The tip of this pipette is now carefully inserted at an angle into the tip of the graduated pipette and filled at a uniform rate. This rate should be such that the meniscus moves down the graduated pipette not faster than 1 cm /sec. (See section on the role of surface tension in pipetting.) The rate of movement of the meniscus of the graduated pipette should be the same for all pipettes calibrated and should be the same when the graduated pipette is itself calibrated or checked for uniformity.

By observing these precautions and by careful estimation of the meniscus position without parallax, readings can be made reproducibly to $\pm 0.1$ $\mu$l. With pipettes of less than 20 $\mu$l, precision can be increased by repeating the withdrawal two to four times between readings, thus bringing the total volume change up to at least 20 $\mu$l. Significant evaporation must either be ruled out or corrected for on the basis of sham pipettings.

## Cleaning and Storage of Constriction Pipettes

After use, if a pipette is to be employed subsequently with a similar solution, there seems to be little reason for rigorous cleaning; a simple rinse with water should suffice. Otherwise, thorough cleaning is desirable. The following procedure is recommended for routine purposes.

The narrow constrictions and fine tips of the pipettes preclude the usual procedure of pipette cleaning. Instead, each pipette is cleaned individually by drawing and expelling a succession of cleaning agents into the pipette. If many pipettes are to be cleaned, the process can be facilitated by using the vacuum line in the laboratory. A suction flask is interposed between the pipette and line to trap the cleaning fluids. Pressure tubing is used between the line and flask and pipette. A short section of rubber tubing of the type used for pipetting is joined to the pressure tubing with a section of glass tubing. The end of the pipette is inserted in the tubing and the pipettes filled with suction. The cleaning fluids must fill the pipette both below and above the constriction. In some instances complete filling is made easier if the negative pressure is reduced by pinching the rubber tubing. The cleaning solutions can be kept in wide-mouth glass stoppered 60 ml bottles. The pipette is first rinsed in distilled water to remove any residual material, and the following solutions drawn through in sequence:

1. 0.1 *N* NaOH
2. Deionized (or glass-distilled) water (bottle No. 1)
3. Fuming $HNO_3$
4. Deionized water, at least three rinses (bottle No. 2)
5. Deionized water, at least three rinses (bottle No. 3)
6. Acetone

The acetone is used to dry the pipette for storage but will probably leave material on the inner walls of the pipette. Before use, the pipette should be rinsed with deionized water, and, if possible, with the reagent to be used.

Because the constrictions and tips are very small, the orifices may become plugged during use if certain precautions are not taken. Pipettes that have been used for protein solutions should be rinsed two or three times with water immediately after use. Any concentrated solution is likely to clog the pipette if left to dry on the walls around the tip and constriction. If the pipette is to be left overnight the last traces of fluid are removed from the tip by sucking in air after rinsing. The danger of leaving material at the orifice is thus minimized.

Pipettes with very small constrictions (those of less than 1 $\mu$l volume or oil well pipettes) are only occasionally cleaned by this routine, because of the danger of plugging with particles in the cleaning fluids. Instead they are rinsed in and out successivly with the above fluids by suction from the mouth. Each fluid is drawn up a short distance above the constriction. The water rinses are drawn up further than the alkali or acid, to be sure the latter are completely removed.

If a pipette becomes plugged, the first step is to determine whether it is at the tip, the constriction, or both. The pipette is first dipped into water. If the tip is open, water will usually enter to a slight degree due to capillarity. To test the constriction, water is introduced above the constriction with a small plastic wash bottle or another pipette. The fluid is shaken down to the constriction and pressure applied by mouth through rubber tubing. If the constriction is open, the pressure will force some water through. If it is not open, the water may dissolve the material if it is allowed to remain in the pipette a few minutes. It is useful to know what substance is responsible for plugging the pipette. If the material is a protein, it is better to fill the pipette with 0.1 *N* NaOH than with acid. In either case, the cautious application of heat may hasten solution of the plug. This can be done by using a match flame or immersing the pipette in hot water (60°). Once the constriction is open, the pipette can be filled from above by shaking it like a thermometer. If the tip is closed the same procedure is followed. The fluid is allowed to remain in contact with the plugged orifice and can be very gently heated if necessary. Once the openings are clear, the pipette can be cleaned in the standard way.

Occasionally it may be found that the pipettes do not deliver properly after cleaning, and bubbles form on the walls when the pipette is filled. This usually means that the pipette is not thoroughly clean. If the material last used was protein, the pipette should be filled with 0.1 *N* NaOH and gently heated with a match flame or by immersing it in hot water (60°). It is helpful in many instances to fill the pipette with fuming nitric acid and allow it to remain for several minutes. Such procedures are followed by routine cleaning.

If the pipette is dirty on the outside, it can be immersed as far as necessary in the cleaning solutions. The pipette should not be handled on the outside below the constriction and particularly not near the tip. There is a temptation to warm the pipette by rubbing it with one's fingers and thus hasten drying. Such practices must be scrupulously eschewed to avoid contaminating the outside of the pipette.

The pipettes are stored in small cabinets when not in use to protect them from dust and fumes. A convenient cabinet can be made by using small metal four-drawer tool cabinets ($10 \times 10 \times 11\frac{1}{2}$ inches) available at most hardware stores. The metal partitions are replaced with two parallel wooden strips, about 2 inches wide, notched to hold the pipettes in a horizontal position. During active use, pipettes are conveniently kept on the table top in grooved wooden racks (Fig. 3-4).

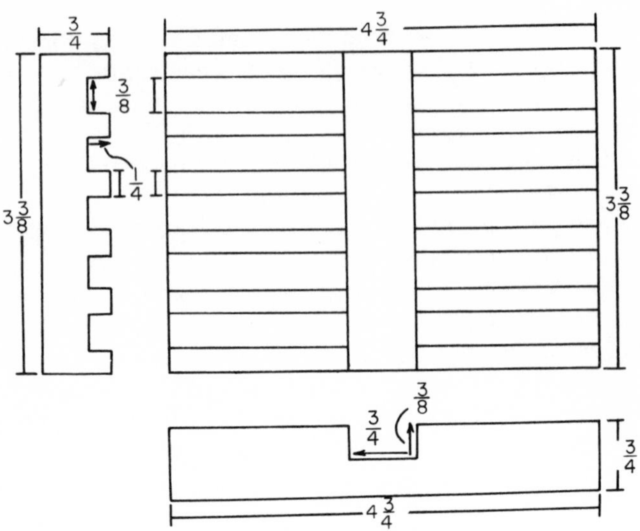


*Fig. 3-4.* Wooden rack for holding pipettes on the desk top. (Measurements given are in inches.)

## TUBES

### Fluorometer Tubes

The tubes used for fluorometric assays are ordinary 3 ml (10 × 75 mm) Pyrex tubes. They are selected for uniformity of diameter by choosing tubes which slip easily into the holder in the fluorometer but with a minimum clearance. They are also selected for freedom from scratches and are discarded when scratched or roughened from usage. This is particularly important for analyses at high fluorometer sensitivities. Special racks for these tubes are made from Lucite (Plexiglas) to minimize scratching, which can occur when metal racks are used. The diagram in Figure 3-5 gives the dimension and construction details of these racks. Racks holding 50 tubes are a convenient size; with 10 holes per row it is easy to keep track of samples without numbering the tubes. A volume of 1 ml fills these tubes sufficiently for reading in the Farrand or Turner fluorometer. Rapid easy mixing becomes difficult with volumes much greater than about 1.2 ml. For reactions involving two steps, the first step can often be carried out in the fluorometer tube, if the volume is in the 50–200 μl range. Before reading, 1 ml of reagent or diluent is added.

### Tubes for 25–200 μl Volumes

For reactions that are carried out in a volume of 25–200 μl, commercial culture tubes, 7 × 70 mm, or 6 × 50 mm are suitable. In this case, since the tubes are not used for fluorometric or optical measurement, metal racks can be used, and are preferable because of greater heat stability. The dimensions and construction details are given in Fig. 3-6.

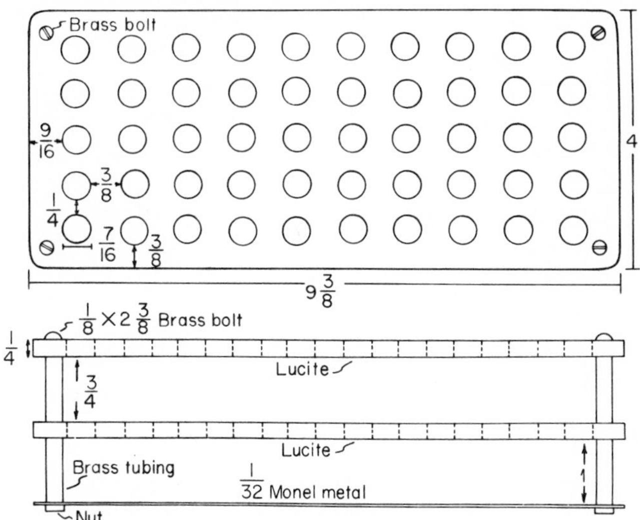


***Fig. 3-5.*** Rack for holding 10 × 75 mm tubes. Plexiglas is used to prevent scratching fluorometer tubes. (Measurements given are in inches.)

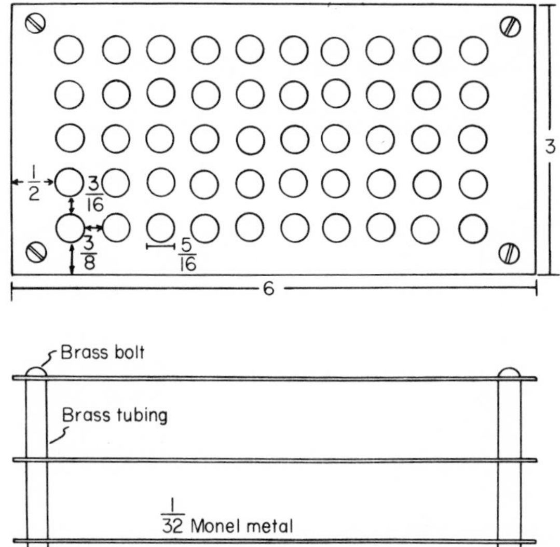


***Fig. 3-6.*** Rack for holding 7 × 70 mm culture tubes. Racks with the same dimensions are used for smaller tubes except that the diameter of the holes is reduced to $\frac{7}{32}$ inches. (Measurements given in inches.)

### Tubes for Volumes of Less than 25 $\mu$l

The tubes to be described are easily made from Pyrex tubing. Assays performed in 5–100 $\mu$l volumes can be conveniently carried out in 5 × 60 mm tubes of 3 mm i.d. When using volumes of 2–20 $\mu$l, 4 × 50 mm tubes (2.0–2.5 mm i.d.) are recommended, and 3 × 50 mm tubes (1.6–1.8 mm i.d.) are useful for 0.5–10 $\mu$l volumes. The upper limit in each case is the volume which fills the tube for a third of its length, since this is about the limit for satisfactory mixing. The *lower* limit is set by the tendency to evaporate and spread over the surface.

Ordinary constriction pipettes can usually be used with tubes down to 3 mm i.d.; special pipettes are necessary for delivery into the smaller tubes (see pipette section).

Metal racks, constructed as detailed in Fig. 3-6, are used for these tubes, with suitable modification of the openings according to the size of tube. In each case, racks holding 50 tubes are convenient.

### How to Make Micro Test Tubes

Tubing of the desired diameter and bore is cut into lengths twice that of the finished tube. Particularly in the case of smaller tubes, it is worthwhile to select tubing with inner diameters falling at least within the i.d. ranges shown above. The tubing is heated in the middle of its length by rotating in a small hot oxygen–gas flame. When the glass is sufficiently soft, the tubing is removed from the flame and pulled to form a short, double, conical taper (Fig. 3-7). Each taper is then cut off at the proper point with the edge of the

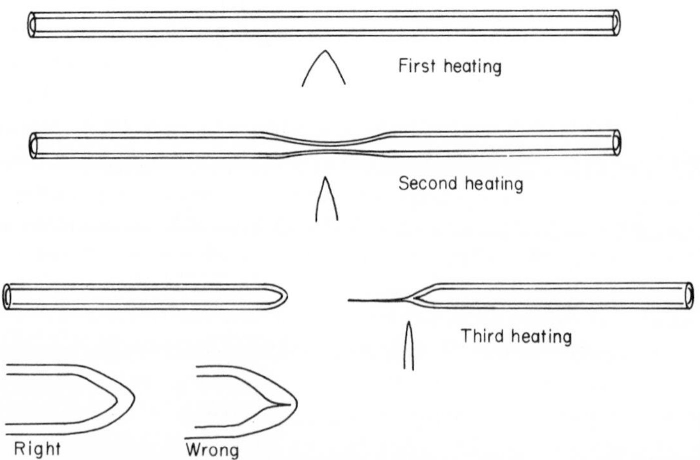


*Fig. 3-7.* Construction of small tubes. The faulty tip contains a funnel which will trap material. It is usually the result of the use of a flame that is not hot enough (see text).

hot flame to form two small test tubes (Fig. 3-7). The degree and uniformity of the first heating will determine the length and uniformity of the taper. The taper must be short enough to permit pipettes to go all the way to the bottom of the tube for delivery. When the tubes are cut in two by the flame there is danger that hairline "funnels" may be formed (Fig. 3-7). This can be prevented by using a very small hot flame so as to heat the glass to a fully molten state. The finished tubes should be inspected under a dissecting microscope, and those tubes having defects discarded. Tubes having funnels are impossible to clean, and materials which could disturb measurements would be carried over from one analysis to another. The open ends of the tubes should be lightly fire-polished. It is desirable to make tubes that are uniform in length to facilitate cleaning.

## Cleaning and Storage of Tubes

Rigorous cleaning of glassware and proper storage thereafter are rewarding in terms of increased precision and avoidance of analytical failures, and need not add materially to the total time per analysis.

It is recommended that all tubes be cleaned with hot 50% $HNO_3$, by the following special procedures adapted to the size of the tube. The person washing tubes should be protected with glasses and acid-resistant gloves. The need for the differences in procedure comes from the fact that cleaning fluids will not easily flow in or out of tubes of less than 4 or 5 mm i.d.

### LARGER TUBES (10 × 75 mm, 7 × 70 mm, AND 6 × 50 mm)

The tubes are placed in 600 ml beakers fitted with a stainless steel screen (Fig. 3-8). The beaker is completely filled with tubes to maintain alignment. The tubes are filled by pouring fluid into the beaker, shaking the beaker if necessary to eliminate air bubbles. The tubes are emptied by inverting the tubes against the stainless steel screen and pouring the liquid into a sink (or bottle if the liquid is to be saved). The tubes are shaken vigorously with the stainless steel screen in place to remove as much of the liquid as possible. An effective cleaning protocol for these larger tubes is as follows.

1. Rinse tubes twice with water to remove gross particles and left over fluids.
2. Fill tubes with 0.1 *N* NaOH and add enough extra liquid to cover them. Place the beaker of tubes in a stainless steel pan half full of water and boil for 5 or 10 min.
3. Allow tubes to cool for safety in handling, empty, and rinse with distilled water. The NaOH can be saved and used several times before discarding if the tubes are not too dirty.

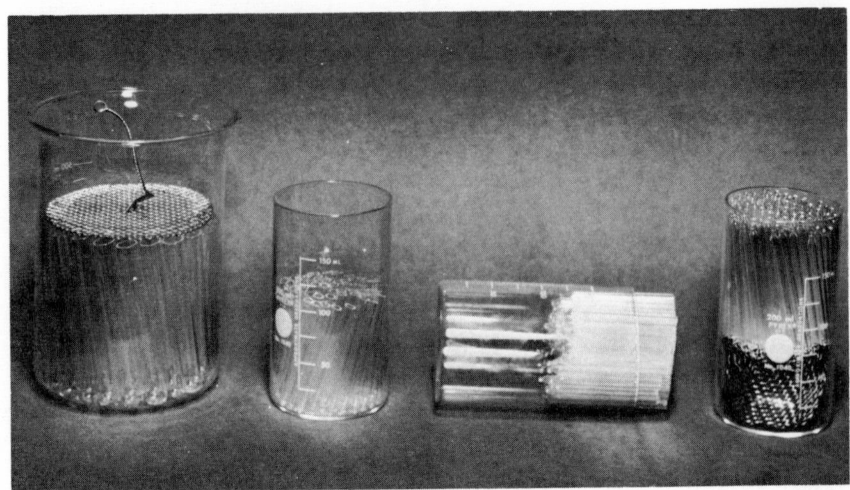

*Fig. 3-8.* Cleaning tubes. In the case of the large tubes on the left, the screen makes it easy to drain or shake out most of the liquid. (The screen is not left in the beaker during the other cleaning steps.) Three steps in the cleaning of smaller tubes are shown. First a beaker full of tubes ready to fill. Next the tubes are shown projecting from one beaker ready to transfer upside down into a second beaker. Finally, they are shown inverted against a wire screen insert ready to empty by centrifuging.

4. Fill tubes with 50% $HNO_3$, cover them with the liquid, and boil 5 or 10 min as with NaOH, and cool before handling. The $HNO_3$ can be reused many times before discarding.
5. Rinse twice with distilled water.
6. Fill tubes with distilled water and boil 5 or 10 min.
7. Rinse twice with distilled water.
8. Rinse once with glass-distilled or deionized water and shake as dry as possible.
9. Cover beakers with foil, perforate foil to permit water to evaporate, and dry tubes in a clean oven at 100°.
10. The tubes can be stored in the beakers after covering them with fresh foil, or removed to large glass-lidded jars.

### Fluorometer Tubes

Tubes used in the fluorometer are cleaned as described, except that the NaOH step is omitted. The base tends to etch the glassware and therefore increase the blank reading. In addition, fluorometer tubes are air-dried because heating also tends to increase the blank readings in the fluorometer. To air-dry, the fluorometer tubes are inverted over a stainless steel screen which is elevated above the table or bench surface to permit circulation of air. When working at highest sensitivity, the tubes may with advantage be rerinsed just before use with glass-distilled or deionized water. The air-dried tubes appear to gradually accumulate a film of some kind from the air in the laboratory which can increase the fluorescent blank as much as severalfold. It is possible to eliminate this film by rinsing the tubes and aspirating the residual water with plastic tubing. The tubes are dried on the outside with a lint-free towel. The blank fluorescence reading of tubes filled with clean water need not exceed the equivalent of $10^{-7}$ *M* DPNH (TPNH).

MICROTUBES

Tubes with inner diameters of less than 4 mm are filled and emptied by centrifugation. This is a slower process than that described for larger tubes, but since the tubes are more completely emptied by centrifugation fewer rinses are necessary. Moreover, because of their smaller size, many more tubes are contained in each beaker so that hundreds of tubes can be cleaned at a time. The following procedure is recommended.

The tubes are collected open end up in a 200 ml flat-bottomed Pyrex beaker (tall form), which has been cut off at a height of $3\frac{3}{4}$ inches (about $\frac{1}{2}$ inch below the original rim, Fig. 3-8). It saves time to clean a number of beakers of tubes at once. Two sets of the cut-off beakers are required during the cleaning process. The second set of beakers used for emptying the tubes are fitted with stainless steel inserts made of perforated metal or screen that permit the fluid emptied from the tubes to collect in the bottom of the beaker (Fig. 3-8). The cleaning process is much the same as for regular tubes but the transfer of tubes requires a little practice and will be described in detail.

The tubes are covered with the fluid with which they are to be filled and centrifuged for 2 or 3 min at low speed in a swinging-bucket centrifuge head. The slow speed (not over 1200 rpm) is to prevent breaking the beakers. The tubes should be inspected for air bubbles after centrifugation to make sure that all tubes are completely filled. The excess fluid is poured off and the tubes transferred to a second beaker with the perforated stainless steel insert. A flat surface, such as a clean petri dish, is held against the top of the beaker and the tubes inverted against it. The tubes are then eased out to about half their length, using the petri dish to maintain them in position so that the tubes are projected at a uniform length from the beaker. It is important that all the tubes extend exactly the same distance out of the beaker (Fig. 3-8), or the transfer may fail. The beaker with the screen insert is then used to receive the tubes. The projecting tubes are placed inside the mouth of the receiving beaker, until the second beaker meets the first. The first beaker is removed, and the tubes pushed down into the second beaker so that the open ends touch the screen. A small clean Erlenmeyer flask (50 ml), which has the same diameter at the bottom as the inside of the 200 ml beaker is convenient for this purpose. The tubes are emptied by centrifugation in the same manner as for the filling process. The centrifuge should be allowed to stop without applying the brake because otherwise the fluid emptied from the tubes may splash back and be caught in the open ends of the tubes. The fluid collected below the screen is removed by inserting a polyethylene tube between the tubes and aspirating the liquid. The tubes are then returned *en masse* to the original beaker in the manner described above for the original transfer. At the final step, the tubes are centrifuged somewhat longer than usual (20 min) to remove as much water as possible, and the tubes dried in an oven or at room temperature as the needs dictate. If an oven is used, it should be reserved for glassware.

The following steps are recommended:

1. Rinse once with water.
2. Fill the tubes with 0.1 $N$ NaOH and heat in a pan of boiling water for 5 or 10 min.
3. Rinse once with water.
4. Fill the tubes with 50% $HNO_3$ and heat in boiling water for 5 or 10 min.
5. Rinse twice with glass-distilled or deionized water. Heat the tubes for 5 or 10 min in the second change of water.
6. Rinse twice more with glass-distilled or deionized water. The last time spin in the centrifuge about 20 min to remove as much water as possible.
7. Dry at room temperature or in an oven free of contaminating vapors. The tubes can be stored in the beakers covered with foil, or in glass jars. In either case, cover in such a way as to protect from dust and fumes.

A final precaution is worth mentioning. Even though the screen support is made of stainless steel, it is not completely resistant to acid. It should be removed before heating the tubes at any step and should not be allowed to sit for any prolonged period in the cleaning or rinsing fluids, even at room temperature, and especially not in the $HNO_3$. Stainless steel contains some copper, and when the screen has not been removed before heating in $HNO_3$, $Cu(NO_3)_2$ has been found deposited on and in the tubes.

### Modified Cleaning Method for Microtubes

A modification of the two-beaker transfer method has been designed which makes the manipulation of the tubes easier. This technique requires greater caution to avoid contamination of the tubes in the process of cleaning, because there is more likelihood of the liquids splashing back. However, since this second method is faster and can with care result in satisfactory cleaning, a description is warranted.

The same 200 ml beakers with the lip cut off and the perforated stainless steel supports are used, but the tubes are never transferred from the original

***Fig. 3-9.*** Alternate procedure for cleaning small tubes. On left is beaker of tubes with perforated steel insert in place. In the middle is polyethylene cup. On the right the beaker of tubes with insert is inverted into polyethylene cup which has been cut away to permit better visibility.

beaker. The tubes are filled as described, and the stainless steel support is placed on top of the tubes (Fig. 3-9). The beaker is then inverted into a plastic cup, made from a 250 ml polyethylene centrifuge cup cut off at a height of $3\frac{1}{4}$ inches (Fig. 3-9), and centrifuged. The beaker, together with the tubes, is removed very carefully so the fluid now in the bottom of the cup is not tipped up into the tubes. The stainless steel support is removed, the tubes filled, and the cleaning process carried out in the steps described. If a large volume of fluid is in the bottom of the plastic cup and there is danger of tilting it up into the tubes, the fluid can be aspirated with fine polyethylene tubing inserted between the beaker and plastic cup while the beaker is still inverted.

CHAPTER 4

# TYPICAL FLUOROMETRIC PROCEDURES FOR METABOLITE ASSAYS

An almost unlimited number of substances can be measured at a variety of levels by fluorometric analysis with pyridine nucleotides. However, there are certain principles and limitations that are common to all such analyses. To illustrate these principles and show how a given basic method can be adapted to the assay of different amounts of material, a few typical procedures are described in some detail.

In each case, the conditions recommended for a specific measurement are those believed to be most favorable under average circumstances. Frequently, however, other conditions of assay are perfectly satisfactory when required by special situations. An understanding of the basic elements in an assay permits an easy modification of the procedure when special circumstances demand.

## A ONE-STEP METABOLITE ASSAY WITH $TPN^+$

The measurement of glucose-6-P requires only one enzyme, glucose-6-P dehydrogenase, for which $TPN^+$ is the coenzyme. Glucose-6-P is oxidized to 6-P-gluconolactone and the $TPN^+$ is reduced to TPNH. The lactone is unstable and breaks down spontaneously (but not rapidly) to 6-P-gluconate.

$$\text{Glucose-6-P} + \text{TPN}^+ \rightleftharpoons \text{6-P-gluconolactone} + \text{TPNH} + \text{H}^+ \quad (4\text{-}1)$$

$$\text{6-P-Gluconolactone} + \text{H}_2\text{O} \longrightarrow \text{6-P-gluconate} \quad (4\text{-}2)$$

Under favorable conditions, the formation of TPNH is stoichiometric with the oxidation of glucose-6-P.

The choice of pH for the analysis of glucose-6-P is influenced by the kinetics of glucose-6-P dehydrogenase, the equilibrium constant of reaction (4-1), and the rate of reaction (4-2). The Michaelis constants for both substrates increase with increase in pH, i.e., become less favorable. Although the maximum velocity also rises with pH (to a peak at about pH 9), the increase is not enough to compensate. As a result, the first-order rate constant is maximal at about pH 7 [see Chapter 2, Eq. (2-21)]. pH 7 would therefore be chosen for glucose-6-P measurement if it were not for the fact that the equilibrium constant for reaction (4-1) is unfavorable at this pH (about unity).

Therefore, without a very large excess of $TPN^+$, the reaction would not be complete until the lactone broke down [reaction (4-2)]. The spontaneous hydrolysis of 6-P-gluconolactone has a half-time of at least 30 min at pH 7. A compromise is reached by choosing pH 8. Here the equilibrium constant for reaction (4-1) is much more favorable than at pH 7 and the rate of reaction (4-2) is increased eight-fold (half-time about 4 min).

(In spite of this conclusion as to the best pH, there are situations in which a different pH is chosen for the reaction. Thus, when the reaction is used as an auxiliary to the measurement of glycogen, pH 7 is chosen for the sake of the phosphorylase and P-glucomutase reactions. In this case the unfavorable equilibrium is handled by using a large excess of $TPN^+$ and compensation for any incompleteness of the reaction is made with suitable standards.)

### *Side Reactions*

As has been mentioned earlier, any enzyme reaction is exposed to the danger of side reactions of one kind or another. These may occur as the result of contamination of the enzyme with other enzymes or be due to a second activity of the enzyme itself. In the case of glucose-6-P dehydrogenase, the enzyme itself is capable of slow reaction with free glucose. The Michaelis constant for glucose is very high, but if, as is often the case, glucose-6-P is to be measured in the presence of a much greater concentration of glucose, this slight lack of specificity can be a problem. The apparent first-order rate constant for glucose-6-P is a million times greater than for glucose. Nevertheless, if a quantitative measure of 1 $\mu M$ glucose-6-P is required in the presence of 50 m$M$ glucose (as may be the case in an assay for glucokinase) the amount of enzyme will need to be kept to a minimum.

Two side reactions due to contaminating enzymes may be mentioned. Glucose-6-P dehydrogenase may contain a TPNH oxidase. If the presence of this enzyme is not recognized, some of the TPNH produced can be reoxidized and give a low result. The use of proper standards will control this problem, and it can be minimized by avoiding the use of more enzyme than necessary.

A more subtle problem is presented by the presence of glutathione reductase, either in the glucose-6-P dehydrogenase, or in one of the other enzymes of a more complex system. Here, there is no problem unless oxidized glutathione is also present, in which case some of the TPNH will be reoxidized. Many tissue extracts contain oxidized glutathione. Since the standard solutions would not contain glutathione, standards would yield the proper amount of TPNH, whereas the tissue samples would give low results. A careful time curve with tissue extracts, if carried well beyond the expected end point, will detect this difficulty (a decrease in fluorescence after it reaches a peak). The problem can be cured by introducing 0.1 m$M$ dithiothreitol into the reagent. This keeps any glutathione that may be present in the reduced state. Since a very low level of dehydrogenase is required for glucose-6-P analysis, the danger is remote in this particular case, but it may arise when, as in the $P_i$ assay below, a much higher level of dehydrogenase is required.

One other attribute of the glucose-6-P dehydrogenase reaction should be mentioned. This is its susceptibility to strong competitive inhibition from a number of anions, e.g., $P_i$, trichloroacetate and especially sulfate. Since the enzyme is ordinarily prepared as a suspension in strong $(NH_4)_2SO_4$, sulfate inhibition cannot be ignored. To make matters worse, the inhibition increases with the square of the sulfate concentration (50% inhibition at 10 m$M$, 90% with 18 m$M$ levels at pH 8.)

*Kinetic Situation* (For more detail see Chapter 2)

At pH 8 the Michaelis constants for glucose-6-P and $TPN^+$ are approximately 20 $\mu M$ and 2 $\mu M$, respectively. There is little or no effect of one substrate on the Michaelis constant of the other. With most of the adaptations to be described, $TPN^+$ concentrations will be kept well above the Michaelis constant, therefore the velocity will be a function only of the concentrations of enzyme and glucose-6-P.

Let us assume that it is desirable to complete the enzyme reaction in 10 min. This means that when glucose-6-P levels are such as to give approximately first-order kinetics (5 $\mu M$ or less) the half-time must be 1.5 min or less. As shown in Chapter 2,

$$V_{max} = 0.7\ K_m/t_{1/2}$$

or in this case,

$$V_{max} = 0.7 \times 20\ \mu M/1.5\ \text{min} = 10\ \mu M/\text{min}$$

(i.e., 10 $\mu$moles per liter per minute or 10 U/liter). It is well to increase this to 15 or 20 U/liter to provide a margin of safety.

When glucose-6-P levels are very low, it may be useful to lower the $TPN^+$ concentration (see below). If $TPN^+$ is decreased to 0.5 $\mu M$ (one-fourth of the $K_m$) this will reduce the velocity to one-fifth of what it would be with high $TPN^+$ ($v/V_m = [S]/([S] + K_m)$. Consequently, the enzyme concentration must be increased five-fold to at least 50 U/liter.

When glucose-6-P levels are increased above the first-order zone, then the amount of enzyme must also be increased. The *extra* enzyme needed is simply that which would be required if the enzyme were operating on *all* of the glucose-6-P at the $V_{max}$. For example, if the initial glucose-6-P level is 200 $\mu M$ (0.2 m$M$), the *extra* enzyme for a 10 min analysis time would be 20 $\mu M$/min or 20 U/liter, or a total of 30–40 U/liter.

## Spectrophotometric Procedure

### A. Spectrophotometric Analysis of Glucose-6-P (1–5 $\times 10^{-8}$ Mole)

This procedure is that recommended for standardizing stock glucose-6-P solutions, but it is also satisfactory for analyzing unknown samples when the concentrations are sufficiently high.

*Reagent*. One hundred m$M$ Tris-HCl (pH 8.1), 0.5 m$M$ $TPN^+$.

Dilute glucose-6-P dehydrogenase to about 10 U/ml (10,000 $\mu M$/min) with 20 m$M$ Tris-HCl (pH 8.1) containing 0.02% bovine plasma albumin. (This is about a 25 $\mu$g/ml concentration of crystalline glucose-6-P dehydrogenase.)

To 500 $\mu$l of reagent in the cuvette is added 2–5 $\mu$l of 10 m$M$ glucose-6-P (giving a concentration of 40–100 $\mu M$). After mixing, a reading is taken at 340 nm, and 2 $\mu$l of the diluted enzyme are added (giving an activity of 40 $\mu M$/min). The reaction is followed until completed. The concentration of glucose-6-P is calculated from the increase in optical density corrected for any change in the blank (glucose-6-P omitted). (See General Standardization Protocol, Chapter 1, for calculation.) Glucose-6-P analyzed in this way can be diluted quantitatively to various lower concentrations to serve as standards for the fluorometric analyses.

With the ordinary spectrophotometer this method provides satisfactory results with concentrations of glucose-6-P from 20 to 200 $\mu M$.

## Five Fluorometric Procedures for the Measurement of Glucose-6-P

The range of glucose-6-P levels which can be analyzed satisfactorily by direct fluorometry is from 10 to 0.2 $\mu M$, or with care somewhat below 0.2 $\mu M$. Throughout this range the reaction is first order for glucose-6-P, or nearly so.

At the higher levels of glucose-6-P, only enough $TPN^+$ need be used to be sure it is in sufficient excess of the glucose-6-P to compensate for the equilibrium with 6-P-gluconate. A three- or four-fold excess of $TPN^+$ is adequate. At the lowest levels of glucose-6-P, the choice of $TPN^+$ level will be dictated by the experimental needs. If the $TPN^+$ does not contribute to the blank to a disturbing degree, enough can be added to saturate the enzyme. If experimental conditions require that $TPN^+$ be kept at a very low concentration, it can be reduced and the enzyme suitably increased, as discussed earlier.

*General Procedure*

In all of the fluorometric assays readings are made in 3 ml (10 × 75 mm) tubes in final volumes of slightly more than 1 ml. It is standard practice before analyzing a large number of samples to validate the reagent by testing one blank, one standard, and one sample. An instrumental setting is chosen such that the readings will be nearly full scale. The reaction time curve for standard and sample should be identical. It is especially important to check the time course of the reaction when new reagents or a new enzyme preparation are used. The time curve is a test of the specificity of the method. It is particularly important to make sure that there is no drift upward or downward with samples after the standard curve indicates that the reaction should be over. Such drifts would indicate side reactions, which if undetected would lead to erroneous results. It is advised that the same medium as that used for preparing the sample, and equal to the sample in volume, be added to the blanks and standards. This checks on the possible presence of interfering substances in the reagents and makes the rate comparisons more reliable. Duplicate analyses of blanks and standards are made; if there are a large number of measurements, a set of blanks and standards are included at the end as well as the beginning. Whenever possible, measurements of samples are also made in duplicate.

### B. Analysis of Glucose-6-P at 5 μM Concentration ($5 \times 10^{-9}$ Mole)

*Reagent.* Fifty millimolar Tris-HCl (pH 8.1), 50 μ*M* $TPN^+$.

Dilute glucose-6-P dehydrogenase to 10 U/ml in 20 m*M* Tris-HCl (pH 8.1), containing 0.02% bovine serum albumin.

One milliliter of the reagent is pipetted into the appropriate number of tubes. Duplicate standards are provided at each of two levels (e.g., 2 and 5 μl of 1 m*M* glucose-6-P). Aliquots of the samples are added which will give concentrations in the range of 2 to 5 μ*M*. The same volume of sample medium is added to the blanks and standards. The tubes are mixed and read at the previously determined appropriate instrumental setting. Glucose-6-P dehydrogenase, 2 μl, is added and a final reading is made based on the time course for the test standard and sample. The expected $t_{1/2}$ is 1–1.5 min, and the reaction should be complete in 10 min. In some cases it may pay to take second readings on some or all of the samples to make sure the reaction is finished. Calculations are based on the glucose-6-P standards.

## C. Direct Analysis of Glucose-6-P at 0.2 μM Concentration ($2 \times 10^{-10}$ Mole)

*Reagent.* Fifty millimolar Tris-HCl (pH 8.1), 5 μ*M* $TPN^+$. The glucose-6-P dehydrogenase diluted stock is the same as in B.

The fluorescent blank of the reagent is decreased by reducing the concentration of Tris-HCl and $TPN^+$. In addition, freshly rinsed fluorometer tubes should be used (see Chapter 3). The reagent blank should not exceed the equivalent of 0.2 μ*M* TPNH.

If it has not already been done, the turntable of the fluorometer is fixed rigidly in position so that only one tube holder is used. Each tube is carefully placed in an identical position for each reading by means of a reference scratch mark. These changes increase the precision of reading and therefore minimize the consequence of relatively high blank readings.

The analyses are conducted as for the 5 μ*M* level. Standards contain 2 μl of 0.1 m*M* glucose-6-P, prepared from the 10 m*M* stock. After the initial reading, 3 μl of the enzyme are added. The enzyme is increased slightly over the preceding test to compensate for the reduction in $TPN^+$. The time of the final readings is based on the tests of blank, standard, and sample. The expected half-time is 1–1.5 min, and the reaction should be complete in 10 min.

## D. Two-step Analysis of Glucose-6-P at the Level of $2 \times 10^{-10}$ Mole

By conducting an analysis in two steps it is possible to achieve a number of favorable effects. Some of these result from the fact that the first step can be conducted in a reduced volume. This reduces difficulties arising from fluorescent components or impurities in the reagent, or of contamination of the reagent with the substance to be measured. Contamination is a particularly difficult problem when measuring ubiquitous materials such as glucose, lactate, or phosphate, which may be hard to keep out of reagents and glassware. Even though the first step is conducted in a small volume, and not in a fluorometer tube, it is usually possible to validate this first step by conducting a preliminary test in 1 ml directly in the fluorometer. For this purpose, although larger volumes of reagents are used, all the *concentrations* are maintained equal to those to be used in the actual assay. A preliminary test of this kind is highly desirable, especially when many samples are concerned. It can increase precision by indicating the optimal reaction time, and it may prevent the tragedy of a set of valuable samples ruined by an enzyme failure.

*Step 1*

The reagent is 50 m$M$ Tris-HCl (pH 8.1), 20 $\mu M$ $TPN^+$, and 0.02% bovine serum albumin. Dilute glucose-6-P dehydrogenase to 10 U/ml in 0.02 $M$ Tris-HCl, pH 8.1, containing 0.02% bovine plasma albumin. The albumin is added to the reagent to protect the enzyme from surface denaturation. This is no problem in 1 ml in the fluorometer tubes, but becomes increasingly significant as the volume of reagent is reduced. In the present case, the first step reaction is planned for a volume of 100 $\mu$l; the glucose-6-P standards in the test will be 2 $\mu M$. Consequently, the preliminary validation of the reagent in 1 ml is made with 2 $\mu M$ glucose-6-P. The calculated amount of enzyme necessary for completion in 10 min is 2 $\mu$l/ml. (With a large series of samples, a longer incubation time may be preferable, in which case the amount of enzyme is reduced.)

Assuming the preliminary trial is satisfactory, enzyme is added, just prior to use, in the same proportion to sufficient reagent for the entire set of analyses. Aliquots of 100 $\mu$l are pipetted into 7 × 70 mm tubes. Standards are prepared by adding 2 $\mu$l of 0.1 m$M$ glucose-6-P. Samples are added in suitable volume (1 to 25 $\mu$l). (The volumes of reagent and concentrations of the stock standards are adjusted so that final volumes of standards and samples are alike.) The tubes are incubated at room temperature until the reaction is complete, and the reaction is stopped by heating the tubes 2 min in boiling water.

*Step 2*

The reagent is carbonate buffer, pH 10 (10 m$M$ $Na_2CO_3$, 10 m$M$ $NaHCO_3$). One milliliter of the carbonate buffer is pipetted into freshly rinsed fluorometer tubes. Before adding the samples, a reading is taken at an instrumental setting such that when standards and samples from Step 1 are subsequently added, the reading will approach full scale. The whole sample or a major aliquot from Step 1 is transferred to the carbonate buffer. The tubes are mixed and read again in the identical position and at the same setting as before.

## E. Analysis of Glucose-6-P at the Level of $2 \times 10^{-11}$ Mole

It is not ordinarily possible to measure accurately $10^{-11}$ mole of TPNH by its native fluorescence in 1 ml. However, if the fluorescence of the TPNH is developed with strong alkali and peroxide, the 10-fold increment in fluorescence makes reproducible measurements possible. In order to measure the TPNH in this way, the $TPN^+$ must be first destroyed in weak alkali. A buffer is used in preference to weak base to reduce the fluorescence developed from $TPN^+$ (see Chapter 1).

*Step 1*

The Step 1 reagent and enzyme are prepared exactly as for the preceding two-step analysis at the $2 \times 10^{-10}$ mole level. The reagent is pipetted in 10 $\mu$l volumes into tubes $5 \times 60$ mm (3 mm i.d.). Standards of 2 $\mu$l of 10 $\mu M$ glucose-6-P and samples in the same volume are added. The tubes are incubated at room temperature until the reaction is complete. (The time is checked, as before, by a preliminary incubation in a 1 ml volume in the fluorometer.)

*Step 2: Destruction of TPN⁺ in Weak Alkali*

The reagent is phosphate buffer, pH 12 (250 m$M$ $Na_3PO_4$ : 250 m$M$ $K_2HPO_4$). The pH, concentration, and volume of the phosphate buffer are selected with regard to the first step reagent. After addition of this buffer the pH should be near 12 and well-buffered to ensure destruction of the $TPN^+$ with minimal induction of fluorescence. In this case, the phosphate buffer is of sufficient concentration so that when added in a volume equal to that of the Step 1 reagent, the pH will be little affected.

To each sample is added 10 $\mu$l of the phosphate buffer, and the tubes are thoroughly mixed, so that no part of the $TPN^+$ in the reagent fails to be made alkaline. The tubes are then heated at 60° for 15 min. The TPNH is unaffected by this treatment, and can be allowed to remain at this alkaline pH at room temperature for many hours.

With $10^{-11}$ mole of glucose-6-P, the $TPN^+$ is in 20-fold excess. If 1% of the $TPN^+$ escapes destruction a 20% error would result. If difficulty on this score is encountered the $TPN^+$ concentration can be reduced from 20 $\mu M$ to perhaps 5 $\mu M$.

*Step 3: Development of Fluorescence in Strong Alkali*

The reagent is 0.03% $H_2O_2$ in 6 $N$ NaOH. The NaOH should be prepared at least a day or two in advance and placed in sunlight (Chapter 1). The illumination reduces the fluorescence of the alkali which after treatment need not exceed that of $H_2O$. $H_2O_2$ is added to the NaOH within an hour of use. The stock 30% $H_2O_2$ is diluted to 3% in water, and this is added to the NaOH in the ratio of 1 ml per 100 ml.

One milliliter of the NaOH : $H_2O_2$ is pipetted into freshly rinsed fluorometer tubes. Before use, the volumetric pipettes should be rinsed with water, and then with the NaOH : $H_2O_2$ solution. (To increase precision, before adding the samples the tubes can be read in the fluorometer at a setting such that the subsequent readings of standards and samples will approach full scale.) The whole sample (or a 10 $\mu$l aliquot) is then pipetted from the reaction tubes and added to the NaOH : $H_2O_2$ with immediate and thorough mixing. The fluorescence is developed by heating 15 min at 60°, 60 min at 38°, or standing

2 hr at room temperature (23°–25°). If the tubes are heated, they must be cooled exactly to room temperature before reading. This can be done quickly in a water bath, after which the tubes are dried with a lint-free towel and transferred to a dry rack. Once the samples have been added to the alkali the tubes are shielded from direct illumination because the fluorescent product is light-sensitive. In addition, the readings should be made in the fluorometer with exciting light reduced to the point at which fading is not a problem. (See Chapter 1 if this causes difficulty.) The tubes are read in the same position and at the same setting as for the initial reading. Calculations are based on the glucose-6-P standards.

## F. Analysis of Glucose-6-P at the $10^{-13}$ Mole Level

The measurement of $10^{-13}$ mole of glucose-6-P requires special modifications. The volume of reagent must be reduced still further to decrease the blank and to keep a sufficiently high concentration of glucose-6-P. A concentration of 0.2 $\mu M$ or greater during the enzyme reaction is usually satisfactory.

The reduction in volume makes evaporation a serious problem; therefore the analysis is performed under oil (see Chapter 13).

Because the amount of product of the reaction is not sufficient to read directly in the fluorometer, the TPNH generated is amplified by an enzymatic cycling procedure (see Chapter 8).

The need for measurement of as little as $10^{-13}$ mole of glucose-6-P has been encountered in studies with frozen-dried histological material. The protocol is intended for such analyses, but could be adapted for other purposes.

### *Step 1*

Volumes of 0.15 $\mu$l of 15 m*M* HCl are pipetted under the oil in the oil wells for samples and blanks, and 0.15 $\mu$l volumes of 0.5 to 1 $\mu M$ glucose-6-P in 15 m*M* HCl serve in the same way for standards. The dry tissue samples are introduced into the acid drop through the oil (see Chapter 13). The oil well rack containing the blanks, standards, and samples is heated 20 min in an 80° oven. This destroys tissue enzymes which would interfere with the analysis.

### *Step 2*

The reagent is 200 m*M* Tris-HCl (pH 8.1), 0.1% bovine plasma albumin, 20 $\mu M$ $TPN^+$, and 0.03 U/ml of glucose-6-P dehydrogenase.

After the rack is cooled to room temperature from Step 1, 0.15 $\mu$l of Step 2 reagent is added to each sample and allowed to incubate for 30 min. The concentration of Tris-HCl buffer is increased above that prescribed for larger samples to compensate for the acid in the first step and to ensure adequate buffering in the very small volume. The surface-to-volume ratio is very high during the enzyme action. Therefore, to ensure stability of the analyzing enzyme, the level of bovine plasma albumin is increased over that recommended for analyses in larger volumes.

*Step 3*

After incubation, 0.25 $\mu$l of 0.2 *N* NaOH is added to each droplet and the rack is heated 20 min at 80° to destroy excess $TPN^+$. Direct fluorescent blanks are not troublesome in this analysis, therefore NaOH can be used instead of the phosphate buffer used in E. The more alkaline pH is used to provide unequivocal destruction of the $TPN^+$. Any residual $TPN^+$ would be indistinguishable from TPNH in the cycling system, and thus increase the blank.

*Step 4*

The cycling reagent consists of 100 m*M* Tris-HCl (pH 8.1), 5 m*M* $\alpha$-ketoglutarate, 1 m*M* glucose-6-P, 100 $\mu M$ ADP, 25 m*M* ammonium acetate, 0.2 mg/ml of beef liver glutamate dehydrogenase, and 7 U/ml of glucose-6-P dehydrogenase (see Chapter 8).

The rack is cooled after the NaOH heat step and 10 $\mu$l of cycling reagent are added to each sample. Since cycling will occur at a substantial rate at room temperature, the addition is made rapidly to prevent a significant cycling differential between the first and last samples. After incubation at 38° for 1 hr, the cycling is stopped by heating 10 min at 100°. The enzymes when heated form a precipitate that would make pipetting difficult. Therefore, to dissolve the precipitate, to each droplet is added 1 $\mu$l of 2 *N* NaOH and the rack is heated again for 10 min at 100°.

*Step 5*

The reagent is 20 m*M* Tris-HCl (pH 8.1), 40 $\mu M$ $TPN^+$, 100 $\mu M$ EDTA, 25 m*M* ammonium acetate, and 0.5 $\mu$g/ml of crystalline yeast 6-P-gluconate dehydrogenase. One milliliter volumes of the reagent are pipetted into fluorometer tubes, and 8 $\mu$l aliquots of the samples in the oil wells are transferred to the fluorometer tubes and mixed. 6-P-Gluconate standards, in the expected range of the analysis (1–2 $\mu M$), are included. After incubation for 30 min at room temperature the tubes are read in the fluorometer. Calculations are based on the original glucose-6-P standards.

## A ONE-STEP METABOLITE ASSAY WITH DPNH

The measurement of α-ketoglutarate is accomplished in one enzymatic step by reductive amination with glutamate dehydrogenase in the presence of ammonia and DPNH (Kornberg and Pricer, 1951a).

$$\alpha\text{-Ketoglutarate} + NH_4^+ + \text{DPNH} + H^+ \rightleftharpoons \text{glutamate} + \text{DPN}^+ \quad (4\text{-}3)$$

In this reaction, as in any metabolic assay with DPNH, either the amount of DPNH which disappears or the $DPN^+$ which appears can be measured. The disappearance of DPNH can be followed directly by the decrease in absorption at a wavelength of 340 nm, or by the decrease in fluorescence. The appearance of $DPN^+$ can be measured by first destroying excess DPNH with acid and then measuring the $DPN^+$ either by the fluorescent product that can be formed in alkali, or by enzymatic cycling.

Measurement of the disappearance of DPNH has advantages and disadvantages. It permits direct observation of the rate of the reaction and the easy detection of difficulties and troubles from side reactions. On the negative side, it is not practical to use a large excess of DPNH; therefore the range of sample size is more limited than if the $DPN^+$ is to be measured.

In any reaction involving disappearance of DPNH or TPNH, it must be made certain that the reduced nucleotide is not completely used up. Both in the spectrophotometer, and especially in the fluorometer at high sensitivity, it is necessary to determine what reading corresponds to complete oxidation of the nucleotide (the " bottom "). It is not safe to assume that this is equal to the reading of the reagent alone, since DPNH and TPNH solutions usually have impurities which absorb at 340 nm or which are fluorescent. This is particularly true of old stored solutions. And the problem is increased by the fluorescence of the sample, which adds to the "bottom" fluorescence. When in doubt, the safest move is to either add more reduced nucleotide to sample and blank and see if the increment remains unchanged or to add standard α-ketoglutarate solution to the sample to see if a further drop in absorption or fluorescence occurs. (In the latter case, if no drop occurs, that particular sample is lost, but a fallacious result is detected.) When analyzing samples which vary widely in content of the metabolite, it may be desirable to use initial DPNH (TPNH) concentrations which are appropriate for weaker specimens, and plan to add accurately measured increments of reduced nucleotides to stronger specimens as needed.

The kinetic characteristics of glutamic dehydrogenase, the equilibrium constant for this reversible reaction, and the stability of DPNH influence the choice of analytical conditions.

*Equilibrium Constant*

As reaction (4-3) indicates, the equilibrium position becomes more favorable for α-ketoglutarate measurement as the pH is decreased. According to the data of Olson and Anfinsen (1953), if $(H^+)$ is held constant at pH 7

$$K_{eq} = \frac{[\alpha KG][NH_4^+][DPNH]}{[Glut][DPN^+]} = 0.45 \times 10^{-6}\ M = 0.45\ \mu M \qquad (4\text{-}4)$$

(where αKG is α-ketoglutarate, and Glut is glutamate). If instead $[H^+]$ is held constant at pH 8, $K'_{eq} = 4.5\ \mu M$.

Since $(NH_4^+)$ is ordinarily used at millimolar concentrations, the equilibrium position is very favorable at both pH 7 and pH 8 for reduction of α-ketoglutarate. Therefore the fact that this is a reversible reaction would not be of analytical concern if it were not that glutamate levels in tissues are very high, and $DPN^+$ levels may be at least as great as those of α-ketoglutarate. Therefore, further discussion of the equilibrium situation is in order.

A useful way to arrange the equilibrium equation is

$$\frac{[\alpha KG]}{[Glut]} = \frac{K'_{eq}[DPN^+]}{[NH_4^+][DPNH]} \qquad (4\text{-}5)$$

At pH 8, if all concentrations are micromolar, and $[NH_4^+]$ is set at 10,000 $\mu M$,

$$\frac{[\alpha KG]}{[Glut]} = \frac{4.5[DPN^+]}{10{,}000\ [DPNH]} = \frac{[DPN^+]}{2200\ [DPNH]} \qquad (4\text{-}6)$$

At pH 7,

$$\frac{[\alpha KG]}{[Glut]} = \frac{0.45[DPN^+]}{10{,}000[DPNH]} = \frac{[DPN^+]}{22{,}000[DPNH]} \qquad (4\text{-}7)$$

From Eqs. (4-6) and (4-7) it is simple to calculate how nearly complete the analytical reaction will be. Take as an example a tissue that contains per kilogram 0.1 mmole of α-ketoglutarate, 10 mmoles of glutamate, and 0.2 mmole of $DPN^+$. (Endogeneous DPNH is destroyed in making an acid tissue extract.) Let us suppose that during the enzymatic assay the overall tissue dilution is 200-fold, and a 50% excess of DPNH is provided. The following tabulation shows the initial and the equilibrium concentrations that would result during the assay. It also shows what would happen if the pH were shifted from 8 to 7.

| | Initial ($\mu M$) | Final (pH 8) ($\mu M$) | Final (pH 7) ($\mu M$) |
|---|---|---|---|
| α-Ketoglutarate | 0.5 | 0.093 | 0.013 |
| Glutamate | 50 | 50.4 | 50.5 |
| $DPN^+$ | 1.0 | 1.41 | 1.49 |
| DPNH | 0.75 | 0.345 | 0.263 |
| ΔDPNH | — | 0.405 (81%) | 0.487 (97%) |

At pH 8 the reaction would be only 81% complete, whereas at pH 7 it would be 97% complete. The equilibrium is even more favorable below pH 7, but DPNH instability is a deterrent. Another way to favor the reaction would be to increase $NH_4^+$ concentration to 25 m*M* or even 50 m*M*. Levels above 50 m*M*, although favorable to the equilibrium, would slow the rate.

*Reaction kinetics*

Another, and usually more important, reason for keeping the pH low is that it has a very favorable effect on the reaction rate at low substrate levels. This is not due to any effect on the $V_{max}$, which is no greater at pH 7 than at pH 8 (45 μmoles/mg/min). The favorable effect of low pH is on the Michaelis constants for both α-ketoglutarate and DPNH. The $K_m$ for α-ketoglutarate is nearly 10 times smaller at pH 7 than at pH 8. Although the reaction is even faster below pH 7, the advantage is not great enough to compensate for the beginning instability of DPNH.

In addition to pH, the type of buffer and its concentration have profound influence on the Michaelis constants for both substrates. With DPNH and α-ketoglutarate at low levels (5 μ*M* and 3 μ*M*, respectively) comparisons were made at pH 7 between a number of buffers and buffer strengths. With 20 m*M* imidazole-acetate buffer taken as 100%, rates were 50% with 100 m*M* imidazole-acetate, 20% with 100 m*M* imidazole-HCl, and 10% with 100 m*M* phosphate buffers. As in the case of pH, the $V_{max}$ values were only moderately affected by the changes in buffer and buffer concentration. As a result of these tests imidazole-acetate is recommended at pH 6.9 and a concentration of 40 m*M*. A lower concentration would give a somewhat faster rate but the higher buffering capacity is ordinarily desirable.

When measuring low levels of α-ketoglutarate directly in the fluorometer, the DPNH levels, as has been said, are of necessity also low. This means that both substrates may be well below their respective Michaelis constants (approximately 100 μ*M* for α-ketoglutarate and 4 μ*M* for DPNH under recommended conditions). In this event the reaction becomes truly second order. This has two consequences. For a given percentage excess of DPNH more enzyme will be needed as the concentration of DPNH is decreased. In addition, with a given initial DPNH concentration, the time for completion will vary with the amount of α-ketoglutarate present. This general situation has been described fully in Chapter 2, but a specific illustration is included here. Figure 4-1 shows the course of the reaction if the initial α-ketoglutarate level is either 40% or 80% of the initial DPNH level. The half-time is only 16% longer at the higher level but the time for 98% completion of the reaction is more than double. A first-order reaction with the same initial velocity is shown for comparison. Whereas the first-order reaction is 98% complete in 5.6 half-times, the second-order reactions, to be 98% complete, require 7.3

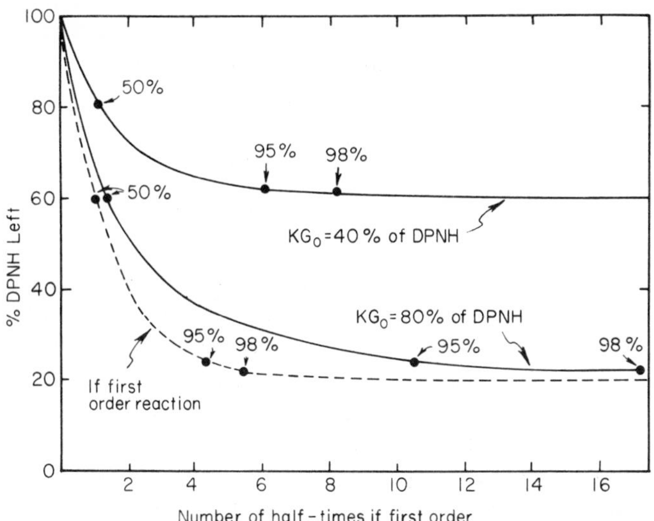


***Fig. 4-1.*** Time course of reaction between DPNH and α-ketoglutarate when both are well below their respective Michaelis constants. Curves are shown for initial α-ketoglutarate concentrations ($KG_0$) that are either 40% or 80% of initial DPNH concentrations.

and 13 half-times at the 40% and 80% levels respectively. There is obviously a need for caution when the analytical reaction is second order and most of the pyridine nucleotide is used up.

## Spectrophotometric Procedure

### A. Spectrophotometric Analysis of α-Ketoglutarate (1 to $4 \times 10^{-8}$ Mole)

*Reagent.* Imidazole-acetate buffer, pH 6.9 (20 m*M* imidazole base: 20 m*M* imidazole acetate), 25 m*M* ammonium acetate, 100 μ*M* ADP, and 100 μ*M* DPNH.

Dilute beef liver glutamate dehydrogenase to 1 mg/ml in 20 m*M* Tris-HCl (pH 8.1) containing 100 μ*M* ADP. A stock solution of 100m*M* α-ketoglutaric acid is prepared by dissolving 14.6 mg/ml of water. The sodium salt may be used, but the free acid is more stable. From the stock solution a 10 m*M* standard solution is prepared.

Five hundred microliters of reagent are added to each of three cuvettes, and 2 μl of 10 m*M* α-ketoglutaric acid are added to two of them. The contents are mixed, and after making preliminary readings at a wavelength of 340 nm, 2μl of the diluted enzyme are added. The reaction is followed until completed, approximately 10 min. The concentration of α-ketoglutarate is calculated from the average decrease in optical density corrected for any change in the blank. The α-ketoglutarate solution analyzed in this way can serve as a standard for fluorometric analyses.

**Four Fluorometric Procedures for α-Ketoglutarate**

B. ANALYSIS OF α-KETOGLUTARATE AT 5 μM CONCENTRATION ($5 \times 10^{-9}$ MOLE)

The reagents are the same as in A except for reduction of DPNH to 10 *μM*. The reagent volume is 1 ml. Duplicate standards are provided at each of two levels (e.g., 3 and 6 μl of 1 m*M* α-ketoglutaric acid). Samples are added with volumes chosen to give expected final concentrations in the 2–8 *μM* range. Initial readings are made at an instrumental sensitivity that will provide nearly full-scale deflections. Glutamate dehydrogenase (5 μl) is added to each tube and mixed well. A final reading is made after an interval based on the time course of a preliminary test of a standard and sample. The expected half-time is 1 or 2 min; final readings may be made after 10 half-times.

C. ANALYSIS OF α-KETOGLUTARATE AT 0.2 μM CONCENTRATION BY NATIVE FLUORESCENCE ($2 \times 10^{-10}$ MOLE)

With α-ketoglutarate at lower levels than in the previous example, the DPNH concentration in the reagent is varied accordingly to provide an excess of 20–80%. Since the Michaelis constant for DPNH is about 4 *μM*, the amount of enzyme and/or time must be increased as DPNH is decreased. For example, with DPNH levels of 2 *μM* the amount of enzyme can be doubled and the incubation time kept the same. In the following example with 0.5 *μM* DPNH both the enzyme concentration and the incubation time are doubled.

*Reagent.* This is the same as for the 5 *μM* concentration, except that the DPNH is reduced to 0.5 *μM*, and if the imidazole is too fluorescent its concentration may be cut in half.

All of the precautions regarding tubes and fixing of the turntable given for the 0.2 *μM* level of glucose-6-P apply here. Standards are 1 and 2 μl of 0.1 m*M* α-ketoglutaric acid. The procedure is the same as for the 5 *μM* level, except that the glutamic dehydrogenase is increased to 10 μl of a 1 mg/ml solution and the necessary incubation time will be 20 or 30 min, depending on how much of the DPNH is used up.

There is little to be gained by measuring the residual DPNH in two steps by dilution after incubation (as described for measuring the TPNH generated in the glucose-6-P analysis). If blanks are troublesome, or the results erratic, it is recommended instead that sensitivity be increased 10-fold and the useful range extended by forming the alkaline fluorescent product of the $DPN^+$ product. This can be accomplished by "scaling up" the next protocol.

### D. Analysis of α-Ketoglutaric Acid by Fluorescence in Strong Alkali at the $2 \times 10^{-11}$ Mole Level

As discussed in Chapter 1, when DPNH is present at high dilution, particularly in a small volume, it is susceptible to spontaneous partial oxidation to $DPN^+$. This could result in a troublesome or even intolerable blank in an analysis of this indirect type. Such oxidation can be largely controlled with ascorbic acid, which is therefore included at 2 m*M* concentration in the reagent below. (If ascorbic acid is present at a concentration of more than 0.02 m*M* at the final strong alkali step, there will be a decrease in fluorescence development from $DPN^+$. This decrease can be prevented by adding 0.03% $H_2O_2$ to the 6 *N* NaOH).

$DPN^+$ is the substance finally measured. Therefore, if the sample to be analyzed contains a significant amount of $DPN^+$ this must either be eliminated or evaluated by means of a sample blank. ($DPN^+$ levels in animal tissues are of the same order of magnitude as those of α-ketoglutarate.) $DPN^+$ can be destroyed by heating the sample in alkali; 10 min at 60° at pH 11 or above would suffice to destroy $DPN^+$ (Chapter 1) without loss of α-ketoglutarate. Alternatively a suitable blank can be obtained by substituting a reagent with glutamic dehydrogenase omitted.

*Reagent*. This is the same as for the assay at the 5 μ*M* level except that the DPNH is reduced to 5 μ*M*, ascorbic acid is added just before use (10 μl of 200 m*M*/ml of reagent), and bovine plasma albumin is added to give 0.02% concentration. In this case, on the day used the DPNH stock solution should be heated 20 or 30 min at 60° to destroy any $DPN^+$ that may be present. Within an hour of use, the enzyme is added to give a concentration of 5 μg/ml.

*Step 1*

Ten microliters of the reagent are pipetted into 5 × 60 mm tubes (3 mm i.d.). Standards of 1 and 2 μl of 10 μ*M* α-ketoglutaric acid are added. The samples should not be larger than 2 μl in volume. The tubes are incubated at room temperature until the reaction is complete. The anticipated time of incubation is 10 or 20 min but should be checked by preliminary tests of blanks, standards, and if possible samples. These trials can be carried out in 1 ml using the same concentration of reactants as in the smaller volume.

*Step 2: Destruction of DPNH in Acid*

After incubation, excess DPNH in the reagent is destroyed by adding 10 μl of 0.3 *N* HCl with thorough mixing. The acid excess over the imidazole buffer is a little more than 0.1 *N*, giving a pH of about 1, consequently the destruction of DPNH is almost immediate (see Chapter 1). It will be noted that a greater relative excess of DPNH is permitted by this indirect assay than in the direct assay in the fluorometer.

*Step 3: Development of Fluorescence in Strong Alkali*

The reagent is 6 *N* NaOH. The NaOH should be prepared as described for the glucose-6-P assay. Except for the omission of $H_2O_2$ and shorter incubation time, the last step in the analysis is handled in the same way as for $2 \times 10^{-11}$ mole glucose-6-P.

The tubes, each containing 1 ml of NaOH, are read, the samples are transferred from the small tubes to the fluorometer tubes, and immediately mixed. [They can either be transferred *in toto* (Chapter 3) or a 10 $\mu$l aliquot can be transferred. In the latter event all samples must have the same total volume before taking the aliquot.] The fluorescence is developed by incubation for either 10 min at 60°, 30 min at 38°, or an hour at room temperature (25°). The precautions cited in the strong alkali method for glucose-6-P also apply here.

If difficulty is encountered due to fading during the reading, Step 3 can be changed as follows: The NaOH is made 0.03% in $H_2O_2$ and the volume is reduced to 100 $\mu$l. The samples from Step 2 are added and the tubes are heated for 10 min at 60°. After cooling, 1 ml of $H_2O$ is added and the tubes are read with much less danger of fading (Chapter 1). The $H_2O_2$ prevents a reduction by ascorbic acid of fluorescence development (see above).

## E. Analysis of α-Ketoglutaric Acid at the $10^{-13}$ Mole Level By Enzymatic Cycling

As with glucose-6-P measured by cycling, the volume of reagent is reduced to increase the concentration of α-ketoglutrate to a level that permits accurate measurement. In the case of analyses involving DPNH oxidation, maintenance of a suitably high concentration of the substance to be measured is particularly important. Even with ascorbic acid present some DPNH may be spontaneously oxidized to $DPN^+$ if its concentration is very low and the volume is very small. The method is designed for analysis of frozen-dried tissue from histological sections and is carried out in oil wells. Native tissue enzymes are destroyed, as in the glucose-6-P method, except that alkali instead of acid is used. The change is to permit destruction of $DPN^+$ (see above) as well as the enzymes.

*Step 1: Destruction of Native Tissue Enzymes with Alkali*

A volume of 0.05 $\mu$l of 0.05 *N* NaOH is introduced into each oil well under the oil for samples and blanks. Standards are provided by 0.05 $\mu$l volumes of 1 to 4 $\mu M$ α-ketoglutaric acid in 0.05 *N* NaOH (i.e., 0.5 to $2 \times 10^{-13}$ moles each). The tissue samples are introduced into the acid through the oil. The racks are heated in an 80° oven for 20 min.

*Step 2: Reductive Amination of α-Ketoglutaric Acid*

The reagent is imidazole-acetate buffer, pH 6.6 (50 m*M* imidazole base, 150 m*M* imidazole acetate), 50 m*M* ammonium acetate, 0.2 m*M* ADP, 0.05% bovine plasma albumin, 10 μ*M* DPNH, 4 m*M* ascorbic acid, and 20 μg/ml of glutamate dehydrogenase. This reagent should be prepared and kept in ice until just before use to protect the DPNH at this slightly acid pH.

The rack is cooled to room temperature, then 0.05 μl of Step 2 reagent is added and incubated for 20 min at room temperature. The reagent is double strength to compensate for the dilution from the first step. The acid:base composition of the buffer is also adjusted so that when the buffer is mixed with the alkali the pH becomes 6.9. The higher final buffer concentration than in previous examples requires an increase in enzyme concentration for the kinetic reasons given earlier.

*Step 3: Destruction of Excess DPNH*

To each sample is added 0.1 μl of 0.4 *N* HCl.

*Step 4: Enzymatic Cycling*

The reagent is Tris-HCl buffer, pH 8.4 (70 m*M* Tris base, 30 m*M* Tris-HCl), 100 m*M* sodium lactate, 0.3 m*M* ADP, 5 m*M* α-ketoglutarate, 50 m*M* ammonium acetate, 400 μg/ml of glutamate dehydrogenase, and 50 μg/ml beef heart lactate dehydrogenase (see Chapter 8).

To each sample is added 5 μl of cycling reagent. This is added rapidly to all samples for the reason given in the cycling method for glucose-6-P. The rack is placed at 38° for an hour, then heated at 100° for 10 min.

*Step 5: Measurement of Pyruvate*

The reagent is imidazole-HCl buffer, pH 6.4 (70 m*M* imidazole base, 330 m*M* imidazole-HCl), 0.6 m*M* DPNH, and 1.5 μg/ml of beef heart lactate dehydrogenase.

The enzyme is added to ice-cold reagent within an hour of use, and the DPNH within 15 min. The rack is allowed to cool to room temperature after Step 4; 5 μl of Step 5 reagent are added to each sample and incubated for 15 min.

*Step 6: Destruction of Excess DPNH*

To each sample is added 1 μl of 5 *N* HCl.

*Step 7: Development of $DPN^+$ Fluorescence*

An aliquot of 8 μl from each sample is transferred to 1 ml of 6 *N* NaOH in fluorometer tubes, with immediate and thorough mixing. The fluorescence is developed by heating 10 min at 60°; the tubes are cooled and read in the fluorometer.

## A MULTISTEP METABOLITE ASSAY (FOR INORGANIC PHOSPHATE)

Most substances of biological interest which do not react enzymatically with pyridine nucleotides can, nevertheless, be analyzed in a pyridine nucleotide system. Such an analysis requires one or more additional enzymes. Accurate measurement depends on the proper functioning of each step in the multiple enzyme sequence. As in the one-step method, the reagent must be validated before use, and in this case it is wise to check each step.

The enzymatic analysis of inorganic phosphate is based on a sequence of three reactions which are carried out simultaneously (Fawaz *et al.*, 1966).

(1) Glycogen + $P_i$ ⟶ glucose-1-P

(2) Glucose-1-P ⟶ glucose-6-P

(3) Glucose-6-P + $TPN^+$ ⟶ 6-P-gluconolactone + TPNH

Great care must be taken to avoid $P_i$ contamination of the reagents if the analytical blank is to be kept low. The stock muscle phosphorylase *a* suspension [for reaction (1)] is dialyzed overnight at 4° against 1000 volumes of 20 m*M* imidazole: HCl (pH 7.0) containing 1 m*M* EDTA, 0.5 m*M* dithiothreitol, and 0.01m*M* 5′-AMP. The stock P-glucomutase and glucose-6-P dehydrogenase suspensions in $(NH_4)_2SO_4$ are centrifuged, and the supernatant fluids removed. The enzymes are washed three times with 2 volumes of $(NH_4)_2SO_4$ of the same concentration as the original solution. Each enzyme is finally suspended in the original volume of $(NH_4)_2SO_4$. [Phosphorylase becomes inactivated in the absence of sulfhydryl compounds, but can be reactivated by adding dithiothreitol (0.5 m*M*) or mercaptoethanol (5 m*M*) and warming to 38° for a few minutes.] An 8% solution of rabbit liver glycogen (500 m*M* expressed as glucosyl units) is dialyzed overnight against 200 volumes of 0.025 *M* acetate buffer (pH 4.6). Some $TPN^+$ preparations have been found to contain disturbing amounts of $P_i$, but it has been possible to select lots sufficiently free of $P_i$ for analytical purposes.

The imidazole used for making the buffer should be selected for a low fluorescent blank or treated with charcoal. The buffer and pH are a compromise from among the conditions best for each enzyme. Thus, phosphorylase is 25% more active in glycylglycine than in imidazole, but P-glucomutase is three times more active in imidazole. Phosphorylase is slightly more active at pH 6.6 than at pH 7.0, but this would be less favorable for glucose-6-P dehydrogenase (and would decrease TPNH stability). 5′-AMP is included for its favorable effects on the kinetics of phosphorylase (see below). The 5′-AMP is kept low, however, because high levels may cause a progressively increasing blank. This is the result of release of $P_i$ by certain phosphorylase preparations and is presumably due to a contaminating phosphatase. The glycogen level recommended is 20 times the Michaelis constant, which is about 0.25 m*M* (calculated as glucosyl units) at the low levels of $P_i$ measured, and is therefore not critical.

As discussed earlier, glutathione reductase is a common contaminant of glucose-6-P dehydrogenase, which in combination with oxidized glutathione from tissues can reoxidize TPNH. Dithiothreitol is added to prevent this by keeping glutathione in the reduced state. (It also ensures the maintenance of phosphorylase in the reduced state.)

The assay is designed for measurement of $P_i$ in the presence of organic phosphate metabolites, including ATP. EDTA is added in excess over $Mg^{2+}$ concentration in order to prevent possible ATPase action. Commercial preparations of phosphorylase have been found to be seriously contaminated with ATPase. In addition, ATPase is difficult to destroy completely in tissue samples. The excess of EDTA over $Mg^{2+}$ lowers the free $Mg^{2+}$ below that required for ATPase action and yet leaves a sufficient level for P-glucomutase to act. The mutase rate is inhibited 80 to 90% be the EDTA, but this is easily taken care of by increasing the concentration of P-glucomutase in the reagent.

*Kinetic Considerations*

This method represents a more complex situation kinetically than the previous one-step systems.

The most critical enzyme in the assay is glycogen phosphorylase *a*. Its kinetics are somewhat complicated in that the two substrates, glycogen and $P_i$, show strong cooperativity and the Michaelis constants are affected (favorably) by the presence of 5′-AMP. Glycogen can be used at a high level, and $P_i$ will always be at a relatively low level. Consequently the significant Michaelis constants are $K_P{}^G$, i.e., that for $P_i$ with glycogen saturating, and $K_G$, i.e.,that for glycogen in the absence of $P_i$ (see Chapter 2). In the presence of 5′-AMP these constants for muscle phosphorylase *a* are, respectively, 1 m*M* and 0.25 m*M* (in glucosyl units). (In the absence of 5′-AMP they are both much higher, 7 and 1 m*M*, respectively.) The high $K_m$ for $P_i$ (even with AMP present) sets a high requirement for phosphorylase, but means that the same enzyme concentration will suffice for 1 μ*M* or 200 μ*M* $P_i$ concentrations. For a half-time of 2 min,

$$V_{max} = \frac{0.7 \times 1\ \text{m}M}{2\ \text{min}} = 0.35\ \text{m}M/\text{min} \qquad (350\ \mu M/\text{min or } 350\ \text{U/liter})$$

The specific activity of phosphorylase *a* in the assay direction is about 15 μmoles/mg/min. Therefore a concentration of 25 mg/liter (25 μg/ml) should suffice. To compensate for lack of purity or deterioration of the enzyme, twice this amount is recommended.

Much lower concentrations of the other two enzymes are sufficient. P-Glucomutase has a high specific activity and low Michaelis constant (5 $\mu M$) and would be used in very small amount if it were not for the handicap imposed by the low level of free $Mg^{2+}$. The low Michaelis constant means that substantially more enzyme is required at the spectrophotometric level of assay than at fluorometer levels.

Glucose-6-P dehydrogenase is used in higher concentrations than in the method for glucose-6-P and at least a four- or five-fold excess of $TPN^+$ is provided. At pH 7 the dehydrogenase activity is less than at pH 8, and there is greater slowing as the reaction proceeds, due to the less favorable equilibrium position for the reaction (see glucose-6-P methods) and the fact that TPNH inhibits more strongly at this pH. Keeping the dehydrogenase reaction relatively fast also compensates a little for the delay occasioned by the other two enzymes.

Both auxiliary enzymes are inhibited by sulfate; this is particularly true of the dehydrogenase, which is 50% and 80% inhibited by 5 m$M$ and 16 m$M$ sulfate, respectively. Therefore an increase in the amounts of the auxiliary enzymes beyond those recommended might actually decrease the rate of reaction, unless most of the $(NH_4)_2SO_4$ used in the suspending medium is removed by centrifugation and resuspension in sulfate-free buffer.

The kinetics of three-step reactions were not discussed in Chapter 2, but with the glucose-6-P dehydrogenase step kept relatively fast, the present case approximates the two-step situation in which both steps are first order. Under the conditions recommended for fluorometric assay the conversion of glucose-6-P to 6-P-gluconolactone has a half-time of about 0.5 min. According to Chapter 2, if the first (phosphorylase) step has a half-time of 2 min, and the second step a half-time of 0.5 min, the overall half-time would be 2.8 min and the 98% time 12 min. The reaction times for glucose-6-P as well as for glucose-1-P and $P_i$ should be checked before analyzing a series of samples. The possibility of TPNH oxidation by glutathione reductase or TPNH dehydrogenases, the TPNH inhibition of glucose-6-P dehydrogenase, and the equilibrium position of the glucose-6-P reaction, all contribute to a possible negative error in analysis of $P_i$. For these reasons, when greatest precision is needed, calculations should be based on carefully prepared $P_i$ standards made from a primary standard of $KH_2PO_4$.

### Spectrophotometric Procedure

#### A. Spectrophotometric Analysis of $P_i$ (1 to $5 \times 10^{-8}$ Mole)

*Reagent*. Imidazole-HCl buffer, 0.05$M$ (pH 7.0) containing 0.8 mg/ml of glycogen (5 m$M$ in glucosyl units), 1.0 m$M$ $TPN^+$, 10 $\mu M$ 5′-AMP, 0.5 $\mu M$ glucose-1, 6-$P_2$, 1 m$M$ EDTA, 0.5 m$M$ magnesium acetate, 0.5 m$M$ dithiothreitol, 0.3 U/ml of glucose-6-P dehydrogenase, and 10 $\mu$g/ml of P-glucomutase.

Five hundred microliters of reagent are pipetted into each of three cuvettes, and 5 $\mu$l of a 5 m$M$ $P_i$ standard are added to two of them. After reading at 340 nm, 50 $\mu$g of phosphorylase *a* (2 $\mu$l of a 2.5% solution) are added. Readings are made at intervals until the reaction is complete, usually 10 to 15 min. For the reasons given, calculations are based on the $P_i$ standards, not the extinction coefficient of TPNH, although the difference should not be more than a few percent.

## Four Fluorometric Procedures for the Measurement of $P_i$

The lower range of concentrations of $P_i$ analyzed by fluorometric procedures permits the use of lower levels of P-glucomutase. This enzyme has a Michaelis constant for glucose-1-P of about 3 $\mu M$. It is also possible to reduce the levels of both phosphorylase and glucose-6-P dehydrogenase, since a somewhat longer assay time in the fluorometer is satisfactory. (In the spectrophotometer a long assay time is a disadvantage unless a great many cuvettes are available.)

Basically the fluorometric procedures are like those described for measurement of glucose-6-P. There is one difference in that $P_i$ is a very common substance and it is therefore much more difficult to avoid contamination. The measurement of $P_i$, therefore, provides a lesson in the assay of other common substances which present similar problems (lactate, ammonia, glucose, and in the modern laboratory even $DPN^+$). If the $P_i$ method is to be used at full sensitivity, all traces of $P_i$ must be removed from glassware. The fluorometer tubes are cleaned as described in Chapter 4 and rerinsed just before use. Other glassware, especially if washed with detergent, is repeatedly rinsed before use.

### B1. Direct Analysis of $P_i$ at 5 $\mu$M Concentration ($5 \times 10^{-9}$ Mole)

*Reagent.* This is similar to that for the spectrophotometric analyses except that the $TPN^+$ is reduced to 30 $\mu M$ and only 3 $\mu$g/ml of P-glucomutase and 0.12 U/ml of glucose-6-P dehydrogenase are required.

The reagent is pipetted in 1 ml volumes into fluorometer tubes. The blank solutions, standards, and samples are added and preliminary readings made. The phosphorylase, 2 $\mu$l of a 2.5% solution, is added to each tube and a final reading made after a predetermined interval which is usually 20 to 30 min.

If the sample to be analyzed is negligibly fluorescent and contains no significant amounts of glucose-6-P or glucose-1-P, there is some advantage in adding phosphorylase to the original reagent. This permits any $P_i$ in the reagent or contaminating the tube to react ahead of time. After allowing time for this blank reaction, an initial reading is made and the reaction is started by addition of the sample.

### B2. Direct Analysis of $P_i$ at Concentrations of 1 $\mu$M or Less

As in the case of glucose-6-P, $P_i$ can be measured in one step at concentrations below 1 $\mu M$. Suggestions for reducing the fluorescence blank and increasing reproducibility were given in the procedure for measuring glucose-6-P at a concentration of 0.2 $\mu M$. These are recommended in this case as well. $TPN^+$ should probably not be reduced below 5 $\mu M$; no other changes are necessary. The problem with $P_i$, however, is that it is difficult to reduce contamination from all sources much below 1 $\mu M$ ($10^{-9}$ mole/ml). Therefore, with samples containing less than $10^{-9}$ mole of $P_i$ the following two-step procedure is preferred.

## C. Two-step Analysis of $P_i$ at the $2 \times 10^{-10}$ Mole Level by Native Fluorescence

By carrying out the analysis in two steps, the blank can be greatly reduced. In the first step the reaction is carried only as far as glucose-6-P. Since $TPN^+$ and glucose-6-P dehydrogenase are not required, any contamination from these components is eliminated. More important is the fact that the volume is reduced during the first step, which is the only time $P_i$ can react. Thus $P_i$ contamination is reduced in proportion to the volume reduction.

This illustrates a not uncommon situation in which difficulties may be avoided by conducting a multienzyme reaction sequence in two steps (see also Chapter 6).

*Step 1*

The reagent is the same as for the preceding assay except that $TPN^+$ and glucose-6-P dehydrogenase are omitted and bovine plasma albumin is added to a concentration of 0.02%. The phosphorylase and P-glucomutase are added just before use. The complete reagent is pipetted in 10 $\mu$l volumes into 5 × 60 mm (3 mm i.d.) test tubes. Prior to addition the rack of tubes is placed in an ice bath, and blanks, standards, and samples are added in approximately the same volumes. Appropriate standards are 2 $\mu$l of 0.1 m$M$ $P_i$. The rack is transferred to a 25° water bath for 45 min, then heated 2 min at 100°.

*Step 2*

The reagent contains 0.02 $M$ Tris-HCl buffer (pH 8.0) and 30 $\mu M$ $TPN^+$. Glucose-6-P dehydrogenase is diluted to a concentration of 8 U/ml with 20 m$M$ Tris HCl-buffer (pH 8.0) containing 0.02% bovine plasma albumin. The total fluorescence blank of the Step 2 reagent should not exceed an equivalent of 0.2 $\mu M$ TPNH.

The reagent is pipetted in 1 ml volumes into fluorometer tubes and to each is added a large aliquot or the whole sample from the first step. The tubes are read at a setting such that full scale is equivalent to 0.5 to 1 $\mu M$ TPNH depending on the expected size of the sample. To each tube are added 2 $\mu$l of the diluted glucose-6-P dehydrogenase, and the tubes read again when the reaction is complete (about 10 min).

The procedure given calls for a 10 $\mu$l reagent volume at the first step with samples of the order of 2 $\mu$l or less (i.e., samples with $P_i$ at concentrations of the order of 100 $\mu M$ or more). If the $P_i$ should be present at greater dilution it would be necessary to use a larger volume of sample and the reagent volume would also need to be increased. The first step could easily be carried out in 100 $\mu$l without a prohibitively large increase in blank. Alternatively, 5 $\mu$l of double strength reagent could be used with 5 $\mu$l of sample, *etc.*

### D. Analysis of $P_i$ at the $2 \times 10^{-11}$ Mole Level with Fluorescence Developed in Strong Alkali

As described for the case of glucose-6-P, sensitivity can be increased 10-fold by destroying excess $TPN^+$ and converting the TPNH to the strongly fluorescent alkaline product. It is unnecessary to repeat the details which were given for glucose-6-P. However, there are a few points to be made which are relevant to this more complicated situation.

Because of the extra steps required, it is desirable to keep the rest of the analysis as simple as possible. Therefore, ordinarily it would be recommended that the three enzyme steps be carried out simultaneously as given for the 5 $\mu M$ level, rather than separating them as in the previous section.

If necessary, however, the extra step could be introduced. In this case step 2 would be conducted not in 1 ml as in the previous section, but in perhaps 20 $\mu$l, by adding 10 $\mu$l of a suitably altered Step 2 reagent; i.e., it would be suitable to double the $TPN^+$ and to incorporate the glucose-6-P dehydrogenase in the reagent at a concentration of at least 0.02 U/ml.

### E. Analysis of $P_i$ by Enzymatic Cycling ($10^{-13}$ Mole)

The procedure for measurement of $10^{-13}$ mole of $P_i$ will be presented in a form appropriate for the analysis of frozen-dried tissue sections.

*Step 1*

The reagent, 0.02 $N$ NaOH, is pipetted under oil in 0.025 $\mu$l volume. Standards consist of an equal volume of NaOH containing $P_i$ at a concentrating 4 $\mu M$. The tissue sections are added through the oil into contact with the NaOH. The rack of samples is heated 20 min at 80° to destroy native tissue enzymes.

*Step 2*

The reagent is imidazole-HCl buffer, pH 6.5 (30 m*M* imidazole base, 70 m*M* imidazole-HCl), 10 m*M* glycogen, 60 $\mu M$ $TPN^+$, 20 $\mu M$ 5′-AMP, 0.5 $\mu M$ glucose-1,6-$P_2$, 0.05% bovine plasma albumin, 2 m*M* EDTA, 1 m*M* magnesium acetate, 1 m*M* dithiothretiol, 0.6 U/ml glucose-6-P dehydrogenase, 6 $\mu$g/ml P-glucomutase, and 100 $\mu$g/ml of muscle phosphorylase *a*. The enzymes are added to the reagent just before use.

To the cooled samples after Step 1 are added 0.025 $\mu$l volumes of the reagent and the rack is incubated at 25° for 45 min.

*Step 3*

To each sample is added 0.15 $\mu$l of 0.2 *N* NaOH and the rack is heated for 20 min at 80° to destroy the $TPN^+$. The rest of the procedure is exactly as described for glucose-6-P at the $10^{-13}$ mole level.

CHAPTER 5

# MEASUREMENT OF ENZYME ACTIVITIES WITH PYRIDINE NUCLEOTIDES

This chapter contains typical methods for measuring three different enzyme activities, in each case at a wide range of sensitivity levels. The purpose is similar to that of the previous chapter, that is, to illustrate the principles for adapting a given method to a particular sensitivity requirement.

Enzyme measurements can be made with DPN and TPN in much the same way as metabolite measurements. It is, of course, necessary to control more carefully the reaction time and temperature. It may also be necessary to control more carefully the reagent composition, although in designing an enzyme method conditions will usually be selected such that modest changes in component concentrations and pH do not significantly affect enzyme velocities.

If the enzyme to be measured is not pyridine nucleotide-dependent, it may be coupled to an auxiliary enzyme which is so dependent, just as in a multistep metabolite assay. In this case the auxiliary enzyme reaction(s) must either be made so fast that it does not affect the result, or it must be controlled and capable of evaluation in calculating the velocity of the primary enzyme step.

In general, the purpose of measuring an enzyme activity is either to study the kinetic properties of the enzyme or to determine the amount of enzyme present. We shall concentrate on the second objective; however, in working out an enzyme assay it is essential to learn something of its kinetic properties.

There is one great problem in the measurement of enzymes that is not encountered in metabolite assays. This is that an enzyme from one source may have very different properties from an enzyme of the same type from another source. This is often true of enzymes from different organs of the same animal. And there are, of course, many well-known examples of a multiplicity of enzymes of the same type even within a given tissue. The differences among enzymes of a given specificity may be in regard to any of the parameters important for an assay: solubility, stability, pH optimum, kinetic constants, temperature coefficient.

This variability in properties means that there can be no guarantee that an enzyme assay worked out for one particular biological source will be suitable for the same type of enzyme from another source. This increases the demands on the analyst and his skill in adapting procedures to special needs.

## Advantages of High Dilution

The high sensitivity possible with fluorometric methods permits measurements at unusually high tissue dilutions. This usually decreases or eliminates problems from side reactions caused by the presence of other enzymes or of other tissue components. Assays can often be made on whole homogenates in spite of turbidity which would cause great difficulty at lower dilutions.

## Direct Versus Indirect Procedures

As in the case of metabolites, enzyme assays may be conducted either directly in the fluorometer or spectrophotomer, or indirectly with two or more analytical steps.

The *direct assay* has advantages and disadvantages. It permits a reaction to be followed continuously to determine whether it is linear. There may be, for example, a lag period due to insufficient amounts of auxiliary enzymes; in a direct assay it is easily possible to discount a preliminary period while the auxiliary enzymes are catching up.

The direct assay gives an immediate answer. It is therefore much more convenient for the study of a reaction. The effects of changes in the reagent can be determined at once. Additions can even be made in the middle of the reaction.

*Indirect assays*, however, offer advantages when large numbers of analyses are to be carried out. Temperature control is usually easier, because the first step can be made in a water bath. A large number of analyses can be carried out simultaneously for a fixed period of time, after which the subsequent steps can be made, usually without exact control of time.

Indirect assays may also offer advantages for situations with troublesome side reactions. Take as an example the measurement of pyruvate kinase:

$$\text{ADP} + \text{P-pyruvate} \longrightarrow \text{ATP} + \text{pyruvate} \qquad (5\text{-}1)$$

$$\text{Pyruvate} + \text{DPNH} \longrightarrow \text{lactate} + \text{DPN}^+ \qquad (5\text{-}2)$$

If reactions (5-1) and (5-2) are combined in a direct assay there will be in crude homogenates an unrelated oxidation of DPNH. This can be avoided by carrying out the reaction in two steps. The first reaction is stopped with heat or acid to kill the disturbing enzymes, after which pyruvate is measured by addition of a reagent containing DPNH and lactic dehydrogenase.

CALCULATION OF ENZYME ACTIVITY

Finding a satisfactory basis for calculation of enzyme activity has been a troublesome problem. Particularly in the past, highly arbitrary bases were often used. A unit of activity might be defined as an optical density change of 0.01/min in a volume of 3 ml, or the formation of 1 mg of product in 7 min. It is urged that whenever possible enzyme activity be reported on a rational mole–time basis. For tissue analyses there is merit in reporting results in terms of millimoles or micromoles per kilogram per minute (or per hour) since this permits direct comparison between the enzyme activity and the tissue concentration of its substrate(s) or product(s), which are usually reported on a kilogram basis. There are, of course, situations where a dry weight or protein basis is necessary or preferable.

## A ONE-STEP ENZYME ASSAY WITH $TPN^+$ (ISOCITRIC DEHYDROGENASE)

$$\text{Isocitrate} + \text{TPN}^+ \xrightarrow{Mn^{2+}} \alpha\text{-ketoglutarate} + CO_2 + \text{TPNH} + H^+ \qquad (5\text{-}3)$$

Procedures will be described for direct spectrophotometric analysis, and for direct and indirect fluorometric analysis at varying levels of sensitivity.

In hog heart, which has been well studied, the activity in fresh tissue is about 15 mmoles/kg/min at 25° and 30 mmoles/kg/min at 38°. The Michaelis constants for $TPN^+$, isocitrate, and $Mn^{2+}$ are 2 $\mu M$, 3 $\mu M$, and 10 $\mu M$, respectively. $Mg^{2+}$ can be used in place of $Mn^{2+}$; however, the maximum activity is somewhat lower. The pH optimum is 7.4. The procedures are given for an enzyme with these properties.

### Spectrophotometric Procedure

A. SPECTROPHOTOMETRIC MEASUREMENT OF ISOCITRIC DEHYDROGENASE ($1$ TO $5 \times 10^{-9}$ MOLE/MIN)

*Reagent.* Tris-HCl buffer, pH 7.5 (10 m$M$ Tris base, 40 m$M$ Tris-HCl), containing 500 $\mu M$ $TPN^+$, 100 $\mu M$ $MnCl_2$, 500 $\mu M$ isocitrate, and 0.02%

bovine plasma albumin. The tissue is homogenized in 10–20 volumes of 20 m$M$ Tris-HCl buffer, pH 7.4, and is assayed without removing any of the insoluble material.

Five hundred microliters of the reagent are pipetted into two cuvettes. The third cuvette is filled with reagent from which isocitrate is omitted. A reading is taken at 340 nm and an amount of enzyme solution is added to each of the three cuvettes to give the desired rate. An optical density change of 0.01–0.06/min is a convenient rate. This is equivalent to roughly 2–10 $\mu M$/min ($\mu$moles per liter per minute). In the case of pig heart this would correspond to a total tissue dilution of 1:8000–1:1500, or 1.2–7 $\mu$l of 1:20 heart homogenate per 500 $\mu$l of reagent.

The reaction is followed at intervals of a minute or two. The rate should be linear with time and amount of enzyme. Unless this has been thoroughly tested, it is recommended that duplicates be provided by using two levels of the homogenate. The rates are corrected for any change in reading of the blank cell.

The temperature is either kept constant at some standard temperature, or the temperature is recorded and rates corrected to a standard temperature on the basis of an experimentally determined temperature coefficient.

The calculation of enzyme activity is based on the rate of optical density change, the molar extinction coefficient of TPNH, and the total tissue dilution. If the result is to be expressed as mmoles $kg^{-1}$ $min^{-1}$, the calculation is ($1000 \times \Delta$O.D./min/6270) $\times$ total tissue dilution. For example, if 2 $\mu$l of a 1:20 homogenate are used in a total volume of 502 $\mu$l, the total dilution would be $251 \times 20 = 5020$-fold. If the $\Delta$O.D/min is 0.0175, the enzyme activity would be $0.0175 \times 1000 \times 5020/6270 = 14.0$ mmoles $kg^{-1}min^{-1}$.

## Three Fluorometric Procedures

### B. Direct Fluorometric Analysis of Isocitric Dehydrogenase (2 to $100 \times 10^{-11}$ Mole/Min)

*Reagent.* This is the same as for spectrophotometric analysis except that $TPN^+$ and isocitrate concentrations are both reduced to 50 $\mu M$. The reduction in $TPN^+$ is made because it usually contains fluorescent impurities; the reduction in isocitrate is made because of the potential risk of contamination with traces of substrates for other enzymes.

*Conduct of the Assay.*

One milliliter volumes of the reagent are pipetted into 3 ml fluorometer tubes. The fluorometer sensitivity is set so that full scale reading (100 divisions) is equivalent to about 10 $\mu M$ TPNH. (The exact equivalence of the scale setting is determined by adding a small volume of a TPNH solution that has been standardized in the spectrophotometer.) A reading is made and an amount of enzyme added to give the desired velocity. A convenient rate is 2–10 divisions/min (0.2–1 $\mu M$/min). This would correspond to a total dilution in the case of pig heart of 75,000–15,000-fold, or the equivalent of 15 to 70 $\mu$g of heart tissue per tube (1.5 $\mu$l of 1:100–1:20 tissue homogenate). The tube should be read three to six times at suitable intervals according to the rate of reaction. This will determine if, as expected, the reaction is linear with time. A number of samples can be analyzed simultaneously (with staggered readings) if the rates are not too fast. In untried situations tissue blanks should be included, using reagent from which isocitrate has been omitted.

When necessary the sensitivity of the test can be easily increased 10-fold or more by simply increasing the fluorometer sensitivity. The same reagent can be used with correspondingly less enzyme. For example, to obtain a rate of 2 divisions/min at a fluorometer setting of 0.01 $\mu M$ TPNH per division, would require only 1.3 $\mu$g of heart tissue. The ability to assay enzyme activity at such high tissue dilutions frequently makes purification of the enzyme unnecessary for kinetic and other studies.

### C. Indirect Fluorometric Measurement of Isocitric Dehydrogenase ($3$ to $30 \times 10^{-11}$ Mole/Min)

Here the objective is convenience, rather than special sensitivity, which will come later.

*Reagent.* This is the same as for the spectrophotometric assay.

Volumes of 100 $\mu$l are introduced into fluorometer tubes in an ice bath. To each is added an amount of the enzyme preparation with activity equivalent to 30–300 pmoles/min at 38°. This is the activity of 1–10 $\mu$g of heart or 1–10 $\mu$l of a 1:1000 dilution of heart tissue. (Any tissue dilution greater than 1:100 should be made in buffer containing at least 0.02% bovine plasma albumin.) Tubes of reagent for blanks and standards (e.g., 5 $\mu$l of approximately 1 m$M$ standardized TPNH) are included.

The rack of tubes is transferred to a 38° water bath for exactly 30 min then returned to the ice bath. To each tube is added 1 ml of carbonate buffer, pH 10 (25 m$M$ $Na_2CO_3$, 25 m$M$ $NaHCO_3$), containing 1 m$M$ EDTA. The tubes are transferred to a pan of water at room temperature, wiped dry, and read at a suitable fluorometer setting. Calculations are based on readings of the TPNH standards.

The EDTA by complexing with $Mn^{2+}$ serves two functions. It stops the reaction, and it prevents $Mn^{2+}$ from enhancing the fluorescence of TPNH at this alkaline pH (Chapter 1).

In previously untried situations, tissue blanks should be included, as in the direct assay.

## D. Indirect Fluorometric Measurement of Isocitric Dehydrogenase ($0.3$ to $10 \times 10^{-12}$ Mole/Min)

Analyses at these high levels of sensitivity are required ordinarily only for biochemical investigations with very small amounts of tissue (histochemical studies, measurements on single ova, etc.). Special techniques may be required for preparation of the sample and for bringing the sample into the reagent (see Chapter 13). Here it will be assumed that the sample has been suitably prepared and is added directly to the reagent. To produce a rate of $0.3 \times 10^{-12}$ mole/min at 38° would require only 0.01 $\mu$g of a tissue as rich in isocitric dehydrogenase as heart.

*Step 1*

*Reagent.* This is the same as for the direct fluorometric analysis, i.e., $TPN^+$ and isocitrate are each present at 50 $\mu M$ concentration.

Each sample is added to 10 $\mu$l of reagent in a $5 \times 60$ mm (3 mm i.d.) tube which is immediately placed in a rack in an ice bath. Included are blanks and standards (1 $\mu$l of 50–500 $\mu M$ TPNH, depending on the sensitivity required). Care is taken not to drag any of the reagent up the walls of the tube, where it might fail to mix with the alkaline buffer in the second step.

The rack of tubes is incubated in a water bath at 38° for 60 min, after which the rack is transferred back to the ice bath.

*Step 2 Destruction of* $TPN^+$ *in Weak Alkali*

*Reagent.* Phosphate buffer, pH 12 (0.25 $M$ $Na_3PO_4$:0.25 $M$ $K_2HPO_4$).

Ten microliters of the phosphate buffer is added to each tube with thorough mixing. The tubes are placed in a water bath at 60° for 15 min. It is essential that none of the reagent escapes the alkalinization, otherwise some of the excess $TPN^+$ will fail to be destroyed and will cause large and erratic blanks.

*Step 3 Development of Fluorescence in Strong Alkali*

*Reagent.* NaOH, 6 $N$, with 0.03% $H_2O_2$ (Chapter 1).

Each sample is transferred to 1 ml of the 6 $N$ NaOH–$H_2O_2$ in a 3 ml fluorometer tube and immediately mixed. After heating for 15 min at 60°, the tubes are cooled and read.

Inasmuch as the final product is present at lower concentration than in the previous protocol, stronger exciting light may be required in the fluorometer. If, because of this, fading causes difficulty (see Chapter 1) the procedure can be modified: The volume of NaOH:$H_2O_2$ is reduced to 200 $\mu$l. After heating at 60°, 1 ml of $H_2O$ is added to each tube. By this change, the rate of fading in light is reduced 20–25-fold.

## A ONE-STEP ENZYME ASSAY WITH DPNH (GLYCERO-P DEHYDROGENASE)

$$\text{Dihydroxyacetone-P} + \text{DPNH} + \text{H}^+ \rightleftarrows \text{glycero-P} + \text{DPN}^+ \qquad (5\text{-}4)$$

The equilibrium constant is $7 \times 10^{-12}$ *M*. Since $H^+$ is one of the reactants, the pH affects the equilibrium between the other four components. At pH 7,

$$\frac{K_{eq}}{[H^+]} = K_{eq}' = \frac{[\text{DHAP}][\text{DPNH}]}{[\text{GOP}][\text{DPN}^+]} = 7 \times 10^{-5} \qquad (5\text{-}5)$$

At pH 10, $K'_{eq} = 7 \times 10^{-2}$. Because of the position of equilibrium it is usually most satisfactory to measure the reaction in the direction of $DPN^+$ formation, although if necessary it could be measured in the opposite direction, particularly if the pH is raised to 9 or 10.

The following applies to the enzyme in mouse brain, but is probably suitable for other tissues if adjustment is made for differences in enzyme content and possible differences in kinetic constants. Under the conditions of assay the enzyme in mouse brain has Michaelis constants for dihydroxyacetone-P and DPNH of 40 $\mu M$ and 2 $\mu M$, respectively. $V_{max}$ is about 1.6 mmoles/kg$^{-1}$min$^{-1}$ at 25° and twice this at 38°.

In general, assays based on *direct* measurement of DPNH (or TPNH) disappearance are less convenient than those based on DPNH (or TPNH) formation because absorption or fluorescence decreases, rather than increases, during the assay. Consequently, there is less latitude in regard to the amount of enzyme that can be measured with a given initial coenzyme concentration. Moreover, the fact that in a direct assay a measurable fraction of DPNH must be used up means that the rate may fall off during the assay. This would occur unless the DPNH level is high enough to nearly saturate the enzyme. A related consideration limits the advantage to be gained (in a one-step direct assay) from the high sensitivity of the fluorometer; an increase in sensitivity requires a decrease in initial DPNH concentration, and therefore a decrease in velocity unless the Michaelis constant for DPNH is very low (see also below).

### Spectrophotometric Procedure

A. Spectrophotometric Analysis of Glycero-P Dehydrogenase (1 to $5 \times 10^{-9}$ Mole/Min)

*Reagent.* Imidazole-HCl buffer, pH 7.2 (35 m*M* imidazole base, 15 m*M* imidazole HCl), 1 m*M* dihydroxyacetone-P, 100 $\mu M$ DPNH, 5 m*M* sodium amytal, and 0.02% bovine plasma albumin. An homogenate of mouse brain is made with 9 volumes of 40 m$M$ phosphate buffer (pH 7) and subsequent dilutions are made in the same buffer containing 0.02% bovine plasma albumin.

Five hundred microliters of the reagent are pipetted into each of two cuvettes. Reagent without dihydroxyacetone-P, but containing DPNH, is used in the third cuvette for a tissue blank. A reading is taken at 340 nm, and 10 $\mu$l of the brain homogenate are added to each cuvette. The amount of tissue added is calculated to give a rate of about 0.2 mmole liter$^{-1}$ hr$^{-1}$, or an optical density change of 0.02/min. A tissue blank is particularly necessary when, as in this case, the amount of homogenate is such as to produce significant turbidity. Moreover, tissues contain enzymes capable of oxidizing DPNH in the absence of substrate. This oxidation is largely, but not completely, controlled by the addition of amytal.

Readings are taken at several intervals to determine if the rate is linear with time. The calculations are made as for the isocitric dehydrogenase assay.

## Three Fluorometric Procedures

### B. Direct Fluorometric Analysis of Glycero-P Dehydrogenase (2 to $100 \times 10^{-11}$ Mole/Min)

The reagent is the same as for the spectrophotometric analyses, except that the DPNH must be reduced in concentration. The upper limit of DPNH that can be measured and still obtain linearity is 10 $\mu M$. Up to a point it is an advantage to reduce DPNH. As DPNH is decreased, a given amount of enzyme will cause an increased *percentage* change in the reading, until the DPNH falls so low that the reaction becomes first order. As long as DPNH is near saturating levels, any decrease in its concentration will result in a pure gain in sensitivity. Ultimately, however, sensitivity will be increased at the expense of linearity with respect to time. The following tabulation illustrates this in the case of mouse brain glycero-P dehydrogenase ($K_{DPNH} = 2\ \mu M$, $V_m = 1.6$ mmoles/kg$^{-1}$min$^{-1}$).

| Initial DPNH ($\mu M$) | Tissue concentration ($\mu g$/ml) | $t_{0.1}$ (min) | $\frac{t_{0.5}}{t_{0.1}}$ | $\frac{t_{0.9}}{t_{0.1}}$ |
|---|---|---|---|---|
| 100 | 2100 | 3 | 5.0 | 9.2 |
| 10 | 248 | 3 | 5.2 | 11.1 |
| 1 | 64 | 3 | 6.1 | 17.6 |
| 0.1 | 45 | 3 | 6.5 | 21.2 |

(The times required for decreases of 10%, 50%, and 90% in initial DPNH are indicated by $t_{0.1}$, $t_{0.5}$, and $t_{0.9}$.) The concentration of tissue is shown which would give a value for $t_{0.1}$ of 3 min at each of the four DPNH levels. With 100 $\mu M$ DPNH, as suggested for the spectrophotometric assay, the reaction is nearly linear with time (for strict linearity the last two columns would be 5 and 9, respectively). With a reduction of DPNH to 1 $\mu M$ there is a 30-fold increase in effective sensitivity, but linearity has suffered somewhat. (Notice that there is little further advantage in reducing DPNH below 1 $\mu M$).

### *Conduct of the Assay*

Reagent (1 ml) with the selected DPNH concentration is placed in the fluorometer and the sensitivity adjusted to give 80–100% full scale reading. The sample is added and four to six readings are made, preferably covering the time period until 50 to 75% of the DPNH is gone. (A number of samples may be assayed simultaneously with staggered readings.) Unless experience has shown it to be unnecessary, tissue blanks are measured over the same time period, with reagent having the same DPNH concentration but no dihydroxyacetone-P.

### *Calculation*

If the reaction is linear or almost linear, or the initial velocity can be accurately measured, calculation is straightforward. Example: The reagent contained 4 $\mu M$ DPNH. To 1 ml of this reagent was added 2 $\mu$l of 1:20 homogenate (100 $\mu$g) giving a total dilution of 10,000-fold. The initial velocity (corrected for tissue blank velocity, if any) was 0.12 $\mu M$/min, corresponding to an activity in the tissue of $0.12 \times 10{,}000 = 1200$ $\mu$moles $\mathrm{kg^{-1}min^{-1}}$ (1.2 mmoles $\mathrm{kg^{-1}min^{-1}}$). Since the initial DPNH level is only twice the $K_m$, the initial velocity is only 67% of the $V_m$ (since $v = V_m[S]/([S] + K_m)$). Therefore, $V_m$ for the tissue is 1.8 mmoles $\mathrm{kg^{-1}min^{-1}}$.

An alternative calculation procedure, useful if fall-off is marked, is to determine the half-time (graphically or by direct observation), and from this to calculate what the velocity would be with a saturating level of DPNH. This calculation can be made with the formula

$$V_{max} = \frac{0.7\, K_m + [S]_0/2}{t_{1/2}} \qquad (5\text{-}6)$$

The equation is derived from Eq. (2-23). $[S]_0$ is the initial DPNH level. Example: To 1 ml of reagent containing 1 $\mu M$ of DPNH is added 2 $\mu$l of 1:20 homogenate. The half-time is found to be 10 min. $V_{max}$ (in the fluorometer tube) $= (0.7 \times 2\mu M + 1\mu M/2)/10$ min $= 0.19$ $\mu M$/min. The total tissue dilution is 1:10,000. Therefore, $V_{max}$ in the tissue is 1.9 mmoles $\mathrm{kg^{-1}min^{-1}}$.

### C. Indirect Measurement of Glycero-P dehydrogenase ($3–30 \times 10^{-12}$ Mole/Min)

*Reagent.* This is the same as that for the spectrophotometric assay except for reduction of DPNH to 50 $\mu M$ and the incorporation of 20 m$M$ nicotinamide and 2 m$M$ ascorbic acid. The nicotinamide is added to protect $DPN^+$ from destruction by the enzyme DPNase. The ascorbic acid is to protect DPNH from spontaneous oxidation in the reduced volume (see Chapters 1 and 4).

*Step 1*

One hundred microliter volumes are introduced into fluorometer tubes in an ice bath. To each is added an amount of enzyme preparation with activity equivalent to 3–30 pmoles/min at 38°. This is the activity of 1–10 $\mu$g of mouse brain (1–10 $\mu$l of a 1–1000 dilution). Tubes of reagent for blanks and $DPN^+$ standards are included. The strength of the standard is chosen according to the expected activity range, e.g., 5 $\mu$l of 200 $\mu M$ $DPN^+$ ($10^{-9}$ mole) would be appropriate for activities corresponding to 5–10 $\mu$g of brain. The rack of tubes is transferred to a 38° water bath for exactly 30 min, then returned to the ice bath.

*Step 2 Destruction of Excess DPNH*

Ten microliters of 1 $N$ HCl are added to each tube with thorough mixing, after which the tubes can be removed from the ice bath.

*Step 3 Development of Fluorescence*

To each tube is added with a syringe pipette 1 ml of 6 $N$ NaOH: 0.03% $H_2O_2$, after which the contents of each are immediately and thoroughly mixed. The tubes are heated 10 min at 60°, cooled, and read in the fluorometer. Instrumental sensitivity is adjusted to give generous scale readings for the samples analyzed. Precautions are taken, as given in Chapter 1, to avoid destruction of the fluorescent product by light.

*Calculation*

Because the DPNH concentration is far above the Michaelis constant, the velocity should be linear and maximal. Calculation is therefore straightforward. In any untried situation, however, tissue blanks should be obtained with reagent containing DPNH but no dihydroxyacetone-P, and corrections applied if necessary. In addition it is advisable to see if $DPN^+$ destruction by the tissue is completely prevented. This is accomplished by incubating standard $DPN^+$ with tissue in the blank reagent. If destruction occurs, but is of moderate degree, it may be possible to correct for it. For example, if there is a 10% loss of added $DPN^+$, it is justifiable to apply a 5% correction since $DPN^+$ formed during incubation would, on the average, be exposed to tissue destruction for only half the incubation period.

D. Indirect Measurement of Glycero-P dehydrogenase (0.3 to 3 × $10^{-12}$ Mole/Min)

This is carried out with the same reagents and (except for volume changes) in almost the same way as in the previous protocol. The first step is made with 10 $\mu$l of reagent in 5 × 60 mm tubes (3 mm i.d.). Tissue equivalent to 0.1–1 $\mu$g of brain (wet weight) is added (e.g., 1 $\mu$l of 1:2000 brain, or 0.1 $\mu$g of frozen dried brain tissue). Blanks and standards are included. If the samples are added as a solution or suspension it is convenient to add the standards with the same pipette as the sample. This simplifies calculation and can increase precision. For example, if the sample is 1 $\mu$l of 1:2000 brain (0.5 $\mu$g), the standards might consist of 10 $\mu$l volumes of reagent plus 1 $\mu$l volumes of 100 $\mu M$ $DPN^+$. If, instead, frozen dried samples are to be analyzed, it is often more convenient to add standard $DPN^+$ to a large portion of the reagent (1 ml or more) so that 10 $\mu$l volumes contain the required amount of $DPN^+$. Thus, for frozen dried brain samples averaging 0.1 $\mu$g dry weight (0.5 $\mu$g wet weight) the standards might consist of 10 $\mu$l of reagent made 10 $\mu M$ in DPNH.

To increase sensitivity the incubation time is extended to 60 min at 38°. The excess DPNH is destroyed with 1 $\mu$l of 1 *N* HCl. The whole sample, or a 10 $\mu$l aliquot, is then transferred to a fluorometer tube containing 100 $\mu$l of 6 *N* NaOH: 0.03% $H_2O_2$. After heating 10 min at 60°, 1 ml of $H_2O$ is added and readings are made.

The reason for the change in the last step, as mentioned before, is that, due to the low final concentration of fluorescent $DPN^+$ product, the exciting light may need to be very intense. This could cause significant fading unless the NaOH is diluted before reading.

## A MULTISTEP ENZYME ASSAY (P-GLUCOMUTASE)

An enzyme which is not DPN- or TPN-linked can usually be measured with as much precision as one that is so linked, even though one or more auxiliary enzyme steps are required. It is, however, necessary to understand the kinetics of the auxiliary enzymes and to eliminate or evaluate side reactions, which can be more troublesome than with one-step assays.

*Kinetic Considerations*

$$\text{Glucose-1-P} \xrightarrow{v_1} \text{glucose-6-P} \tag{5-7}$$

$$\text{Glucose-6-P} + TPN^+ \xrightarrow{v_2} \text{6-P-gluconolactone} + TPNH + H^+ \tag{5-8}$$

Ideally, in this and similar assays, Step 1 will be zero order or at least linear over the time of the assay. On the other hand, the auxiliary step (or steps) will, almost of necessity, be approximately first order. This is because the auxiliary enzyme will be added in sufficient quantity to give a $V_{max}$ which is much larger than the velocity of Step 1.

There will obviously be a lag between the formation of glucose-6-P and the appearance of TPNH. The problem is to evaluate this lag and to keep it small.

At the moment that reaction 1 is started, $v_2 = 0$, but accelerates, as the glucose-6-P concentration builds up, until a steady state is reached, i.e., $v_2 = v_1$. As discussed more fully in Chapter 2, the time taken to reach the steady state is for practical purposes 5 half-times ($5t_{1/2}$) for Step 2, whereas the maximum lag time between Step 1 and Step 2 is $1.44t_{1/2}$, or $K_m/V_{max}$ [Eq. (2-39)]. For yeast glucose-6-P dehydrogenase, $K_m$ is about 20 $\mu M$ under the assay conditions given below. Therefore, with 20 U/liter (20 $\mu M$/min), the maximum lag would be 1 min, and the rate of TPNH formation should be linear after 3.5 min ($5t_{1/2}$).

The Michaelis constant for glucose-1-P in mammalian tissues is quite low (10 $\mu M$ or less under usual assay conditions). An excessive concentration of this substrate is avoided since it may be contaminated with a trace of glucose-6-P. (If necessary this contaminant can be removed by making the stock glucose-1-P solution 0.1 $N$ in NaOH and heating 15 min at 100°.)

*Side Reactions*

In the assay of P-glucomutase in tissues there are several potentially disturbing side reactions, but only one of these is likely to be serious. This is the further oxidation of 6-P-gluconate by 6-P-gluconate dehydrogenase present in the tissue. This is most likely to be significant for tissues with a high ratio of 6-P-gluconate dehydrogenase to P-glucomutase. There are several ways to avoid this problem. The simplest is to carry out the analysis at a high tissue dilution, which is easily possible with the fluorometric procedures. This can keep the 6-P-gluconate concentrations so low that further oxidation is insignificant (see below). A second possibility is to deliberately add sufficient 6-P-gluconate dehydrogenase to oxidize all the 6-P-gluconate, giving 2 moles of TPNH per mole of glucose-6-P formed. This maneuver will fail if the tissue does not contain enough 6-P-gluconate lactonase, since the spontaneous hydrolysis of the lactone is quite slow. A third possibility is to carry out the assay in two steps. The P-glucomutase is allowed to react for a measured period, without added $TPN^+$ or glucose-6-P dehydrogenase, after which tissue enzymes are destroyed with heat, and the glucose-6-P is then measured without danger of complications.

Two other possible side reactions are the diversion of glucose-6-P to fructose-6-P, by P-glucoisomerase (which is an extremely active enzyme in most tissues), and the oxidation of some of the TPNH formed by one of several enzymes that might be present. TPNH oxidation is usually not a problem for the measurement of an enzyme as active as P-glucomutase.

(If this should be a problem it can also be minimized by high dilution or avoided by conducting the assay in two steps.) P-Glucoisomerase should not be a problem in a direct assay. The equilibrium ratio is about 3:1 favoring glucose-6-P, therefore tissue isomerase would merely increase the lag period by a third or less. However, in the case of an assay conducted in two steps, 25% of the product could be present as fructose-6-P at the end of the first step. It would therefore be necessary to add isomerase as well as glucose-6-P dehydrogenase to measure the total activity, or to apply a 33% correction if it was established that equilibration of the hexose phosphate would occur.

## Spectrophotometric Procedure

### A. Spectrophotometric Measurement of P-Glucomutase ($1\text{–}5 \times 10^{-9}$ Mole/Min)

*Reagent.* Imidazole-HCl buffer, pH 7.2 (35 m$M$ imidazole base, 15 m$M$ imidazole HCl), containing 5 $\mu M$ glucose-1,6-$P_2$, 2 m$M$ $MgCl_2$, 100 $\mu M$ EDTA, 500 $\mu M$ $TPN^+$, and 100 U/liter of yeast glucose-6-P dehydrogenase (i.e., 100 $\mu$moles/liter/min). Tissue homogenates are prepared in 20 m$M$ imidazole-HCl buffer (pH 7), containing 1 m$M$ $MgCl_2$ and 100 $\mu M$ EDTA (to maintain full activity, probably by eliminating free $Zn^{2+}$).

To 500 $\mu$l of reagent in the spectrophotometer cell is added sufficient homogenate to give a velocity of 2 to 10 $\mu M$/min. This would, for example, correspond to 0.08 to 0.4 mg of mouse brain (1.5 to 8 $\mu$l of a 1:20 homogenate). The reaction is started by adding 5 $\mu$l of 50 m$M$ glucose-1-P (giving a 500 $\mu M$ concentration). Readings are taken starting at 1 min (i.e., after the lag period) and continued for 5 min, or until an optical density change of at least 0.1 is obtained.

With the amount of auxiliary enzyme added it can be calculated from what was said above that for Step 2, $t_{1/2} = 0.7 \times 20\ \mu M/100\ \mu M/\text{min} = 0.14$ min. Therefore the steady state should be reached (i.e., the rate of O.D. change should become linear) in about 0.7 min. Actually it will require a little longer than this at higher levels of P-glucomutase since Step 2 will no longer be strictly first order. For example, if the velocity of Step 1 is 20 $\mu M$/min, the steady-state glucose-6-P level will be about 5 $\mu M$ or 25% of the $K_m$ for Step 2 [calculated from $v = V_{max}[S]/([S] + K_m)$]. In practice, the effectiveness of the auxiliary enzyme is confirmed by measuring the rate of TPNH formation upon addition of glucose-6-P.

Checks on the possibility of disturbing side reactions are easily made. Reagent is prepared with glucose-1-P omitted. Tissue homogenate is added in the amount required for the assay. The hazard from 6-P-gluconate dehydrogenase is assessed by measuring the rate of TPNH formation when 6-P-gluconate is added to give the approximate concentration expected in the middle of the assay (probably 20–40 $\mu M$). The hazard from possible TPNH oxidation is similarly assessed by adding TPNH instead of 6-P-gluconate and observing whether there is a significant rate of decrease in reading.

## Three Fluorometric Procedures

### B. Direct Fluorometric Measurement of P-Glucomutase ($2–100 \times 10^{-11}$ Mole/Min)

*Reagent.* The reagent is the same as that for the spectrophotometric method except for the reduction of $TPN^+$ to 50 $\mu M$ and of glucose-6-P dehydrogenase to 50 U/liter. If the glucose-1,6-$P_2$ should contain significant glucose-6-P the coenzyme can be decreased to 1 $\mu M$ with minimal decrease in P-glucomutase activity (the Michaelis constant is about 0.05 $\mu M$).

To 1 ml of reagent in a fluorometer tube is added sufficient homogenate to give a P-glucomutase velocity of 0.02 to 1 $\mu M$/min. The fluorometer sensitivity is adjusted according to the velocity anticipated, as described for the isocitric dehydrogenase assay above. With lowest activities it may be desirable to reduce the buffer concentration to 20 m$M$ to reduce the fluorescence blank.

The reaction is started by addition of 10 $\mu$l of 10 m$M$ glucose-1-P. Readings are taken starting at 2 min and continuing at intervals for 5 or 10 min, or until an increase of at least 20 divisions has occurred. The fluorometer is standardized by the increase in fluorescence upon the addition of standard glucose-6-P to give a concentration in the assay range. The rate of increase in fluorescence during this standardization will also check on the adequacy of the glucose-6-P dehydrogenase. With 50 U/liter the half-time should be in the order of 20 sec.

#### *Side Reactions and Effect of Tissue Dilution*

The possibility of disturbing side reactions may be checked in the manner described for the spectrophotometric method. It may be useful, however, to compare for a hypothetical case the expected error from one source, 6-P-gluconate dehydrogenase, at different tissue dilutions. Suppose a tissue contains amounts of P-glucomutase and 6-P-gluconate dehydrogenase which would be about equal if both enzymes were saturated with substrate. Furthermore, assume the Michaelis constant for 6-P-gluconate to be 20 $\mu M$. If the assay is conducted at a tissue dilution and time interval such that 40 $\mu M$ product is formed (average 6-P-gluconate concentration roughly 20 $\mu M$ or half-saturating) the error would be nearly 50%. If the tissue dilution is increased 10-fold so that only 4 $\mu M$ product is formed during the assay, the error would be reduced to about 10%. If the dilution is increased 100-fold only 0.4 $\mu M$ product would be formed, reducing the error to 1%.

## C. Indirect Fluorometric Assays for P-Glucomutase (Auxiliary Enzyme Present in First Step)

*Reagent.* The reagent is the same as for direct assays except that glucose-1-P is incorporated in the reagent rather than added separately. The concentrations of $TPN^+$ and glucose-1-P are chosen to provide a safe excess for the range of activity anticipated. However, to achieve nearly maximum velocity the glucose-1-P level could not be reduced much below 100 $\mu M$.

Indirect assays for P-glucomutase can follow almost exactly the protocols given above for indirect isocitric dehydrogenase assays. Thus with enzyme activities of 30 pmoles/min, or more, the fluorescence of the TPNH generated can be read directly after stopping the reaction with EDTA in carbonate buffer. For smaller amounts of enzyme, the reaction volume is reduced as necessary (to provide for at least 5% reduction of the $TPN^+$ present), and the TPNH generated is measured by the indirect strong alkali method after destroying excess $TPN^+$ at pH 12. In these indirect assays the lag due to the auxiliary reaction has to be either reduced to insignificance or evaluated, if absolute rates are to be determined. If glucose-6-P dehydrogenase is added at the level of 50 U/liter, the maximum lag time in TPNH generation would be about 0.4 min (see above). Thus after a 30 min incubation, the TPNH generated would equal the glucose-6-P formed by the mutase in 29.6 min, i.e., there would be 1.5% negative error. A safe procedure is to check the half-time for glucose-6-P directly in the fluorometer with 1 ml of the complete reagent.

## D. Indirect Fluorometric Assays for P-Glucomutase ($0.2\text{–}10 \times 10^{-11}$ Mole/Min) (Auxiliary Enzyme Absent in First Step)

If side reactions are difficult to control when the auxiliary enzyme system is present, the mutase assay can be divided into two steps. The following is a protocol designed for measuring the amount of P-glucomutase present in 0.2–5 $\mu$g of mammalian brain (0.04–1 $\mu$g dry weight).

### *Step 1*

The reagent is 50 m$M$ imidazole-HCl buffer (pH 7.2) containing 5 $\mu M$ glucose-1,6-$P_2$, 2 m$M$ $MgCl_2$, 100 $\mu M$ EDTA, and 2 m$M$ glucose-1-P.

Ten microliters of reagent are placed in a 5 × 60 mm tube (3 mm i.d.). The sample is added in a volume of 1 $\mu$l or less (or a frozen dried sample is added directly). The tube is immediately placed in a rack in ice water. Included are reagent blanks and standards (reagent plus 1 $\mu$l of 0.5–5 m$M$

glucose-6-P). The rack of tubes is incubated for 60 min in a $H_2O$ bath at 38°, then heated for 1 min in a water bath at 100°.

*Step 2*

The reagent is Tris-HCl buffer, pH 8.1 (10 m*M* Tris base, 10 m*M* Tris-HCl), containing 0.05 m*M* $TPN^+$, 10 U/liter of glucose-6-P dehydrogenase, and 1 μg/ml of P-glucoisomerase. (Isomerase is added because some of the glucose-6-P will have been converted to fructose-6-P, see above).

The samples from the first step are transferred into fluorometer tubes containing 1 ml of this reagent. The whole sample or an 8 or 9 μl aliquot may be transferred. The fluorescence is measured when the dehydrogenase reaction is complete, calculated to be about 10 min.

CHAPTER 6

# IMPROVEMENT, MODIFICATION, ADAPTATION, TROUBLE SHOOTING, AND DEVELOPMENT OF NEW METHODS

This chapter is intended to offer suggestions for simplifying or improving methods, for adapting methods to a different order of sensitivity, and for developing new methods. It also discusses some of the difficulties that may arise and ways to circumvent them.

## Simplification and Improvement

It is a rare method that cannot be improved or simplified for the purposes at hand. Once methods have been published they tend to crystallize and most users hesitate to tamper with the procedure even though it might be easy to make changes that would save time and increase precision.

### ELIMINATING STEPS

Often one of the easiest things to do is eliminate a step, or combine several steps in a procedure. Merely asking if this can be done will usually suggest a way to do so. Elimination of even one step is worthwhile as a time saver when there are many assays to run, and will often decrease the chances for error. If the reaction requires many components these can frequently be combined ahead of time into a single reagent. It may even be possible to prepare a complete or partially complete reagent and store it frozen for repeated use. The advent of commercial freezers which maintain temperatures of $-50°$ or below has made this even more practicable. Reagent components which do not store well at $-20°$ usually are quite stable at $-50°$.

### Neutralization

Often a method calls for acidification or alkalinization followed by neutralization to a particular pH. To neutralize each sample by titration would be a tedious process, whereas it is almost always possible to accomplish the same result by adding a single calculated volume of buffer or of acid or base containing buffer. For example, 1 ml of unbuffered solution is acidified with 100 $\mu$l of 0.5 *N* HCl (50 $\mu$equiv). This can subsequently be brought to pH 8.1 by adding 100 $\mu$l of 1 *M* Tris base (100 $\mu$equiv, $pK_a = 8.1$) or if less buffer is wanted, 100 $\mu$l of 0.4 *N* NaOH (40 $\mu$equiv) containing 0.2 *M* Tris base (20 $\mu$equiv).

A variation of this would be a situation calling for neutralization to pH 8.1 of a sample, such as the above, followed by addition of two other components unstable in alkali, for example, an enzyme and $DPN^+$. Instead of making three separate additions, all three components could be added together by neutralizing with a buffer composed of 1.1 *M* Tris base: 0.1 *M* Tris-HCl (pH 9.2).

The enzyme and $DPN^+$ would be added to the buffer shortly before use. Unless the enzyme were unusually sensitive to alkali it should be able to withstand brief exposure to pH 9.2, and as a precaution the buffer at this time could be chilled to 0°.

### Simplification by Reducing the Scale

Decreasing the scale and increasing sensitivity for its own sake will be discussed in another section. The point to be made here is that many methods can be carried out faster on a reduced scale. Smaller volumes are quicker to pipette and mix. More small samples can go into a centrifuge than large ones. It is quicker to clean 1 ml tubes than 100 ml flasks.

### Substituting Less Time Consuming Steps

There are several analytical manipulations which can be accomplished in alternative ways: one slow, one fast. To make a quantitative transfer from one vessel to another in the classical manner involves pouring or pipetting out of the first vessel, rinsing the first vessel, and diluting to volume. It is much faster to transfer a large measured aliquot from vessel 1 to vessel 2. The fraction left behind can usually be calculated with precision. (Alternatively, as described in Chapter 3, with volumes of 100 $\mu$l or less, essentially total transfer can be accomplished rapidly without rinsing, using a constriction pipette).

Another classic operation, filtration, can usually be replaced by centrifugation with a great saving in time and glassware, and with less risk of contamination.

If a sample is to be diluted, it is much more time consuming to "dilute to volume," than it is to dilute by addition of a measured volume. For example, a 1 ml sample is to be diluted 10-fold. The classic procedure would be to pipette 1 ml into a 10 ml volumetric flask, dilute to volume, and mix (hard to do in a small volumetric flask). Far quicker is to pipette 1 ml into a 25 ml vessel and add 10 ml with a pipette (with easy subsequent mixing). The dilution in the second case is 1:11 instead of 1:10, but the slide rule is quicker than the hand.

## Improvement of an Enzyme Method

In addition to eliminating steps a method may often be improved in other ways. It frequently pays to observe the effects of modest changes in the concentrations of the reactants or in the pH to see if any of the conditions are critical or suboptimal. This is important in any assay, but is usually of particular importance in an enzyme assay.

### *Reagent pH*

Ordinarily it is desirable to measure an enzyme at its optimal pH. This is not just because it gives maximal sensitivity, but also because small deviations in pH will usually cause the least change in velocity in the neighborhood of the pH optimum. However, there is no fundamental reason for measuring an enzyme under optimal conditions; few if any enzymes operate under optimal conditions *in vivo*! The actual velocity observed in the assay is arbitrary in any event. The important thing is reproducibility to permit accurate comparisons, or possibly calculation of the absolute amount of enzyme present. There may be good reasons to measure the enzyme at a pH different from the optimum. At some other pH, the enzyme or one of the other reactants may be more stable, or the equilibrium may be more favorable, thereby making it easier to achieve linearity, or a disturbing side reaction may cause less interference.

### *Incubation Temperature*

Similarly, there is no inherent reason why an enzyme has to be measured at some particular temperature. There is a tradition for the use of 38°, but other temperatures may be preferable. Occasionally a rugged enzyme of low activity can be measured at much higher temperature. More frequently there is an advantage in a temperature lower than 38°. Usually enzyme *stability* has a much larger temperature coefficient than enzyme *activity*. There are cases where it is actually advantageous to carry out the incubation at 0°. Some enzymes at 0° have 20% as much activity as at 38°, or even more.

*Substrate Level*

Usually in an enzyme assay an effort is made to "saturate" the enzyme with substrate. The fact is often overlooked that to achieve 99% saturation for an enzyme with normal kinetics requires a substrate level 100 times the $K_m$. It is very often undesirable to use so much substrate. There may be trace impurities in the substrate that can cause high blanks or even erroneous results, the substrate may itself inhibit, etc. Again, there is nothing sacred about measuring the maximum velocity. It may actually be better in some cases to use much less than saturating levels of substrate, and to calculate the $V_{max}$, if that is the wish. An example of a case in which low substrate is an advantage is the measurement of malic dehydrogenase with oxalacetate and DPNH. Not only is oxalacetate inhibitory at high levels, but it is unstable and breaks down partially to pyruvate. Since most tissues contain rather high levels of lactic dehydrogenase, there is clearly a danger of a substantial positive error. The more oxalacetate used the greater the amount of pyruvate that will be present and the greater this danger. Compare the case of two reagents which contain oxalacetate, in one instance at a concentration equal to the $K_m$ and in the other at a concentration 20 times the $K_m$. The second reagent will give 90% greater velocities for malic dehydrogenase but 2000% greater errors due to lactic dehydrogenase.

*Improvement by Dilution*

As pointed out for the P-glucomutase assay in Chapter 5, a method for measuring an enzyme in a crude tissue preparation can often be improved by increasing the tissue dilution. The activity of the enzyme being measured should not suffer by the dilution if the concentration of its substrates and cofactors remain the same. Other enzymes, however, which might interfere are often rendered less effective. An example of this is the measurement of hexokinase via glucose-6-P dehydrogenase. There are several tissue enzymes which could cause trouble: ATPase could deplete one of the substrates; P-glucoisomerase and P-fructokinase could sidetrack some of the glucose-6-P, converting it to fructose diphosphate; 6-P-gluconate dehydrogenase could partially convert 6-P-gluconate to ribulose-5-P, producing an uncertain extra amount of TPNH; and other enzymes could reoxidize part of the generated TPNH. Glucose-6-P (a potent inhibitor of hexokinase) if generated too fast might accumulate in sufficient concentration to slow the reaction. At high dilution all these difficulties would be reduced without any necessary decrease in hexokinase activity.

*Improvement by Changing the Number of Steps*

In Chapter 5 it was pointed out that a potential side reaction in the P-glucomutase assay could be eliminated by splitting the assay into two steps. Sometimes the opposite is true. A good example is the assay just cited for hexokinase. Here it is highly desirable to combine the two enzyme steps. Otherwise there is almost certain to be conversion of much of the glucose-6-P to fructose diphosphate and triose phosphates. Glucose-6-P dehydrogenase added in generous amount can successfully compete with P-glucoisomerase (and indirectly with P-fructokinase) for the glucose-6-P as it is generated.

There are in most tissues enzymes capable of cleaving $DPN^+$ and $TPN^+$ (DPNase and organic pyrophosphatases). These become important for enzyme assays if the final product is $DPN^+$ or $TPN^+$, if the enzyme is of moderate to low activity, and if the assay is indirect. (Obviously neither enzyme would be disturbing in a direct assay in which the decrease in absorption or fluorescence is to be measured.) DPNase can usually be inhibited sufficiently by nicotinamide. The product of organic pyrophosphatase is nicotinamide monophosphate, which gives the same strong fluorescent product in 6 *N* NaOH as $DPN^+$ or $TPN^+$. Therefore, this enzyme would not interfere if the nucleotide product were to be measured in strong alkali. It would, however, interfere if the product had to be amplified by enzymatic cycling, which, of course, requires the intact pyridine nucleotide. If, in an assay with two or more enzyme steps, tests show that $DPN^+$ ($TPN^+$) cannot be recovered satisfactorily, the assay can usually be divided into two parts, with destruction of tissue enzymes by acid, alkali, or heat after the first part.

There is in all tissues another group of enzymes which is potentially disturbing. This consists of enzymes which catalyze direct oxidation of DPNH or TPNH by $O_2$. Direct oxidation of DPNH is the most troublesome. The greater part of this can be blocked with amytal as well as with several other electron transport inhibitors. Most of the activity has also been found to disappear in frozen dried tissue. There is, however, always a sufficient residue of DPNH oxidase activity to be highly disturbing for measurement of an enzyme with low activity. Here, as with DPNase, the solution in the case of a multistep assay may be to divide the assay into two steps, with destruction of tissue enzymes after the first step before addition of the pyridine nucleotide.

## Improvement of an Enzymatic Metabolite Assay

Conditions are usually less critical for a metabolite assay than for an enzyme assay, but it may be harder to achieve adequate sensitivity and specificity. Because enzymes are never completely pure nor absolutely specific, it is best to use no more enzyme than necessary. The more favorable the conditions for the enzyme action, the less enzyme will be required. Conditions which give maximum enzyme velocity with saturating levels of substrate may be very different from conditions most favorable with the low levels of substrate encountered in an assay. A striking example is the case of α-ketoglutarate measurement given in Chapter 4. For metabolite assays the important kinetic parameter is not $V_{max}$ or $K_m$ but the apparent first-order rate constant $k$ which is equal to $V_{max}/K_m$ (see Chapter 2). To measure this constant an elaborate kinetic study is not required. It is not even essential to know the $V_{max}$ or the $K_m$ (although it is well to know these under final assay conditions.) This constant, or simply the half-time, can be determined directly with a low level of the substrate. Optimal conditions can be found by making this determination with a suitable variety of pH's, buffers, etc.

In a multistep metabolite assay, as in a multistep enzyme assay, there may be an advantage in breaking the assay into two parts. This is frequently true when one of the enzymes has a disturbing contaminant or side action. For example, a method for measuring glycerol consists of converting it to glycero-P with glycerokinase and ATP, and then measuring the glycero-P with glycero-P dehydrogenase and $DPN^+$. If the glycerokinase were contaminated with lactic dehydrogenase, malic dehydrogenase or DPNH oxidase, there could be a serious error in a one-step assay, whereas the error could be eliminated if the assay were split into two steps.

## Adaptation of Pyridine Nucleotide Methods to Increase Sensitivity

Chapters 4 and 5 have given examples of how a method can be adapted to a wide range of sensitivity needs. These can best serve to illustrate the principles. Only a few general remarks will be offered here. The spectrophotometer is well suited to measuring reduced pyridine nucleotides in the 10–150 $\mu M$ range, or a total in 0.5 ml of 5–75 × $10^{-9}$ moles. The fluorometer is suited to measuring reduced pyridine nucleotides directly or oxidized nucleotides indirectly in the 0.05–10 $\mu M$ range, or a total in 1 ml of 0.05–10 × $10^{-9}$ moles. Through the use of enzymatic cycling, sensitivity can be increased another 10,000-fold or more.

Conversion of a spectrophotometric method to a fluorometric one of moderate sensitivity is usually simple and straightforward, as has been seen. Often it is possible and desirable to reduce the concentration of cofactors and auxiliary enzymes. Fewest changes are needed if the reaction is measured in the direction of DPNH or TPNH formation, because it is then usually possible to retain nearly saturation levels of the coenzyme ($DPN^+$ or $TPN^+$). With reactions proceeding in the opposite direction it is necessary, in the case of direct assays in the fluorometer, to reduce the DPNH or TPNH levels, possibly to concentrations far below enzyme saturation. This means that the amount of auxiliary enzyme may have to be increased to compensate, or in the case of a metabolite assay that the reaction time must be prolonged.

Prolonging the time may present no problem in a fluorometric assay. In the spectrophotometer, with expensive cuvettes, it is usually desirable to complete the reaction promptly. In the fluorometer, with an unlimited number of tubes, a great many samples can be followed if necessary for an hour or more without disadvantage.

In adapting a method to the fluorometer, the alternative to reducing DPNH or TPNH levels is to measure the increase in $DPN^+$ or $TPN^+$ (see α-ketoglutarate method, Chapter 4). This introduces an extra step, but, because it is unnecessary to follow the reaction, it actually decreases analytical time with large numbers of samples.

## Troubleshooting

Surely the most important thing in quantitative analysis is to recognize trouble, and the next most important is to know how to cure it. It is too much to expect that a method can be applied to all biological materials without ultimately encountering difficulty. This is especially true of methods for substances present in small quantity, accompanied as they are in biological specimens, by innumerable other substances many of them unknown.

### Detecting Trouble in Metabolite Assays

Trouble can consist of nonspecificity (high results), incomplete reaction (low results), loss of product (low results), high or/and erratic blank values (erratic results). In a metabolite assay one of the simplest and most effective tests is to carry out simultaneous time curves with a standard, a blank, and a biological sample. In a valid assay the sample will follow the same time curve as the standard and there will be no further change in either direction after the reaction is finished. The reagents used for making the extract may affect the enzyme activities, therefore the best comparison is made with a standard which includes simulated extract equal to the sample. If the assay is indirect, it is still possible to obtain several points on a time curve with replicate samples.

The following examples illustrate the value of this kind of test: (1) When $HClO_4$ extracts of brain were analyzed for pyruvate (with lactic dehydrogenase and DPNH) it was noted that after a rapid drop in DPNH at the same rate as for a pyruvate standard, there occurred a long slow fall which finally stopped when DPNH consumption had nearly doubled. This second step could be accelerated by increasing the amount of enzyme. (This phenomenon was not observed with extracts of liver or kidney.) The unknown substance, presumably another keto acid, was not identified. In consequence of this test, brain pyruvate assays were made with carefully controlled amounts of lactic dehydrogenase just sufficient to allow complete reaction with pyruvate. (2) When liver extracts were analyzed for $P_i$ with glycogen phosphorylase,

P-glucomutase, and glucose-6-P dehydrogenase (Chapter 9) it was observed that TPNH rose at the expected rate but fell thereafter. This fall did not occur with standards and was less noticeable in extracts of several other tissues. The difficulty was attributed to oxidized glutathione in the liver extract reacting with a glutathione reductase contaminant of the glucose-6-P dehydrogenase to oxidize the TPNH. It was cured by addition of dithiothreitol to reduce the glutathione. (3) In an enzymatic method for P-creatine (Chapter 9) the tissue extract, made with $HClO_4$, is assayed with a reagent containing hexokinase, glucose, glucose-6-P dehydrogenase, and $TPN^+$. After TPNH has been produced in stoichiometric yield, creatine kinase and extra ADP are added, resulting in extra TPNH equivalent to the P-creatine present. It was observed that with certain tissue samples the reaction did not stop at the same time as the standards. The difficulty was traced to failure to remove the last traces of protein (inadequate centrifugation). This protein included some adenylokinase, known to be very acid-resistant, which generated extra ATP from the ADP present.

One might anticipate increased opportunities for trouble in a complex assay system with one or more sequential enzyme steps. Although it is true that more things could go wrong, such an assay system actually offers greater specificity and more opportunities for testing that specificity. Omission of an auxiliary enzyme provides an excellent null test. Intermediate standards can be introduced to test each step individually. For example, if there should be difficulty in measuring $P_i$ in the example given above, the system could be quickly checked with standard solutions of glucose-1-P, glucose-6-P, and TPNH.

### Erratic Results

There are many possible causes of erratic results. One of the most common is an erratic blank ("cherchez le blanc"). If the blanks are the trouble they will vary as much as the standards in absolute terms. The most common cause is contamination, but its source may not be obvious (glassware, dust, fingers, one of the reagent components, even the distilled $H_2O$). The higher the blank the greater the error it is likely to introduce; it is therefore well to know where the blank comes from and if possible to reduce it.

Experience has shown that one of the most likely causes of error is failure to mix properly. It is difficult to describe how much mixing is enough for all volumes and all sizes and shapes of containers, but until proven otherwise faulty mixing should be suspected as the cause of erratic results. In some enzyme assays mixing that is too violent can be almost as bad as too little mixing, because of surface denaturation. It cannot be overemphasized that samples or reagents stored frozen become unmixed (the $H_2O$ freezes out) and must be thoroughly remixed after complete thawing.

In chasing down the source of erratic results it is helpful to know whether the faulty values are high or low or both. If they are all high it is probably contamination, if all are low it is probably something interfering with one of the enzymes or causing destruction of the product.

With reagents containing many components it may sometimes be difficult to locate the source of a high reagent blank, or the cause of a slow reaction rate. The trouble cannot always be located by omitting one component at a time, since the faulty component may be essential for the reaction to occur. Usually, however, the offender can be identified by increasing or decreasing each component one at a time by a factor of 2 or 3.

## Detecting Trouble in Enzyme Assays

Troubles in an enzyme assay can include any of those listed for metabolites but there is a special hazard of underestimation due to inhibition or lag in an auxiliary enzyme step or inactivation during the reaction.

### *Specificity*

Specificity may be easier to check than in a metabolite assay. Often a good test is to run a "tissue" blank, that is to incubate the tissue (or other) sample with reagent from which the specific substrate has been omitted. This may occasionally fail. For example, in assays for glucose-6-P dehydrogenase in liver and muscle, the addition of sample to a reagent without glucose-6-P may not provide a valid tissue blank. The reason is that breakdown of the rich amounts of glycogen in these tissues can provide enough glucose-6-P to give substantial dehydrogenase rates. (In this case a test can be made by running the reaction at very high tissue dilution which, except in rare cases, will lower the endogenous substrate level to insignificance.)

### *Proportionality with Time and Amount*

In any new situation or where trouble is suspected, a time curve should be made as well as a test of proportionality between product formation and amount of sample. Fall off with time could be due enzyme destruction, product inhibition, depletion or destruction of substrate, or destruction of the product that is being measured (e.g., reoxidation of DPNH produced). Enzyme destruction can usually be detected by preincubating the sample with reagent from which the substrate(s) has been omitted and then determining whether this affects the subsequent rate when substrate is finally added. Product inhibition and product destruction are easily tested directly. Depletion or destruction of substrate can be detected by adding more substrate when the rate begins to fall off; however, in this test, if acceleration should occur it could also be due to counteracting product inhibition.

*Acceleration*

Acceleration with time, rather than fall off, can occur. In a multienzyme reaction acceleration is expected to at least a small degree, the amount depending on the activity of the auxiliary enzyme(s). Acceleration can also be due to activation of the enzyme in the assay, or to further reaction of the product if this yields more of the substance being measured. (For example, in a glucose-6-P dehydrogenase assay the product 6-P-gluconate can be further oxidized, yielding TPNH at an accelerating rate.)

It is usually very informative to measure velocity (and fall off or acceleration, if any) at several enzyme dilutions. (Big differences in dilution tell more than small differences.) At increased dilution the effects of possible stimulatory, inhibitory, or destructive factors are all reduced, and velocity will be affected accordingly. If, however, velocity is proportional to concentration, this would be taken as evidence that none of these factors are operative.

*Cure of Trouble in an Enzyme Assay*

The location of the difficulty will often suggest the cure. As already said, perhaps *ad nauseum*, running the reaction at greater dilution may be the solution. If the trouble is enzyme stability there is no guaranteed cure, but it is worth trying the addition of plasma albumin (total protein should not fall below 0.02% in any event), mercaptoethanol or dithiothreitol (0.2–5 m*M*), or/and EDTA (0.1–1 m*M*). If $Mg^{2+}$ is required by the enzyme, EDTA can nevertheless be used at a lower level). Stability might also be better at another pH. It is worth sacrificing sensitivity for an increase in stability. The same is true of lowering the temperature, which, as said elsewhere, may greatly increase stability with comparatively small decrease in enzyme velocity.

## Development of a New Method

The increasing commerical availability of crystalline enzymes makes it relatively easy to develop new specific sensitive methods. Most "new" methods are really adaptations of old methods, or the application of well-established reactions to analytical use. Once the basic reaction is selected and the necessary enzymes are available, a tentative procedure is set up and tested. The method is then worked out on the most *convenient* sensitivity scale, which is often not the same as the scale ultimately required. If preliminary trials make the method appear feasible, the rest is simply a matter of improvement, troubleshooting, and finally adapting to the scale of sensitivity required see above).

A typical example is the method for $P_i$ in Chapter 4. A method was required for measuring $P_i$ in the $10^{-13}$ mole range (for analyses of 0.1 μg histochemical samples). Existing colorimetric methods fell far short of the necessary sensitivity. A number of enzymatic possibilities were considered. The basic reaction selected (glycogen phosphorolysis) was discovered by the Cori's in the 1930s. The availability of crystalline phosphorylase and P-glucomutase, and sufficiently pure glucose-6-P dehydrogenase made it appear feasible to devise a practical enzymatic pyridine nucleotide method.

Most of the experimental work was carried out by direct observation of fluorescence changes using 1 ml volumes of reagent and $5 \times 10^{-9}$ mole of $P_i$, i.e., 10,000 times as much $P_i$ as would have to be finally measured in a much smaller volume by an indirect procedure. In this case the kinetics of all three enzymes were well known, otherwise a cursory kinetic study would have been conducted. Rates were merely confirmed under assay conditions and minor adjustments made. As described in Chapter 4, the pH and buffer were a compromise based on the different behavior of the three enzymes concerned. One of the major troubles encountered was contamination with $P_i$, which seemed to be everywhere, including the enzyme preparations. The presence in the phosphorylase of a phosphatase for 5′-AMP was discovered during work with standard solutions. The presence of an ATPase in the phosphorylase and glutathione reductase in the glucose-6-P dehydrogenase were not discovered until trials were made with tissues. Cures for these troubles were not difficult once they were spotted. The final operation was to increase the sensitivity 10,000-fold. This was straightforward and consisted essentially of decreasing all volumes and introducing an enzymatic cycling step. The chief problems had all been solved on the macro level.

## CHAPTER 7

# PREPARATION OF TISSUES FOR ANALYSIS

Often the preparative steps in a tissue analysis are the most critical. In the case of a metabolite assay the most hazardous period is usually between the moment the blood supply is cut off and the moment enzyme action is finally stopped. In the case of an enzyme, with a few important exceptions, there is not likely to be much change for many minutes or even hours. Instead, the biggest problem may be to render the enzyme fully accessible for assay or to prevent loss during or after homogenization. In a few cases, if the objective is to determine the state of activity as it was *in vivo*, the problem may be to prevent a specific activity change of the enzyme (e.g., the conversion of phosphorylase *b* to *a*, etc.).

### Preparation for Enzyme Assays

It is easier to give specific directions for preparing tissues for metabolite assays than for enzyme assays. Each enzyme is a separate problem and only trial can determine the conditions necessary for releasing and preserving full activity. In consequence, this section will be limited to some generalities. In specific cases the literature should be consulted if possible.

The majority of enzymes are not affected by freezing when the tissue is intact. This is even true of many enzymes which are unstable to freezing after isolation Therefore suitable homogenates can usually be prepared from either fresh or frozen tissues. If the frozen tissue is to be stored before homogenization, the temperature should be $-50°$ or below, since losses are much more likely to occur at higher temperature (e.g., $-20°$).

Many soluble enzymes can be prepared in fully active form by merely homogenizing tissues at 0° to 4° in hypotonic buffer. Water alone is sometimes used, but in general a buffer at the pH of maximum enzyme stability is much safer. This is particularly true of a tissue with an active glycolytic system, which may continue to form acid in the homogenate. The use of hypotonic rather than isotonic buffer favors disruption of cells and organelles. A glass homogenizer, hand or motor driven, is preferable to a mechanical blender, which is likely to denature some of the enzyme. Even though an enzyme may be soluble it is strongly urged that whenever possible the assay made be on the whole homogenate rather than an extract because a portion, and possibly a variable portion, may be lost due to incomplete tissue disintegration.

Many insoluble enzymes are also fully active in simple homogenates. If not, they can frequently be solubilized or made accessible to the substrate by sonication, freezing, and thawing, or by the use of a detergent. Unfortunately each of these could also cause some enzyme destruction.

A number of enzymes require stabilization by additives. The most common additives are chelators such as EDTA and sulfhydryl compounds such as mercaptoethanol or the more stable dithiothreitol (Cleland's reagent). Stability is sometimes increased by addition of the substrate or the coenzyme.

Stability is, or course, favored almost always by keeping the temperature close to 0°. If the enzyme withstands freezing and thawing it can usually be stored in the homogenate without loss for long periods at $-50°$ or below.

Stability can also be affected by dilution. The original homogenate is usually prepared for convenience at a tissue dilution of not more than 10- or 20-fold. Often further dilution is required before assay. Many enzymes become unstable due to surface denaturation if the total protein concentration falls much below 0.2 mg/ml (corresponding to 500–1000-fold tissue dilution). Protection can usually be afforded by adding bovine plasma albumin to give this concentration.

## Preparation for Metabolite Analyses

Most metabolite analyses are performed on extracts prepared from rapidly frozen tissues. There are two objectives: to freeze the tissue fast enough to prevent significant change in the levels of the metabolites concerned, and to extract the frozen tissue without allowing opportunity for enzyme action.

The first objective is difficult to achieve except with very small animals or with anesthetized animals. The reason, as emphasized by Dawson (1948), is the slow rate of heat flow through water-filled material. If the surface is brought to the temperature of liquid $N_2$ it requires 2–5 sec to freeze the tissue at a depth of 1 mm, and 10–20 sec at a depth of 10 mm. Wollenberger *et al.* (1960) introduced the use of metal tongs, chilled to liquid $N_2$ temperature, with which a small organ, or portion of a larger one, can be crushed flat with nearly instantaneous freezing.

This method is not suitable for all organs and satisfactory results may often be obtained by quickly removing a small piece of tissue from an anesthetized animal and immersing it in a very cold liquid. It may even be satisfactory in some cases to freeze in this manner an entire small animal such as a mouse. Alternatively the tissue can be exposed under anesthesia and flooded with the cold liquid. Less of the liquid will be needed if it is possible to place a wall of foil around the organ to retain the freezing liquid.

Possibly the earliest valid analyses in brain for such labile metabolites as P-creatine and ATP were made by Kerr and Ghantus (1937), who exposed the brain of the dog under anesthesia and poured over it liquid $N_2$ while maintaining artificial respiration. In this way the circulation to deeper parts of brain was maintained until the freezing front arrived.

Liquid $N_2$ is not as satisfactory for rapid freezing of small samples as an organic liquid chilled with liquid $N_2$, because the gas evolved from the boiling $N_2$ forms an insulating layer around the tissue. Isopentane and even liquid propane have been used, but constitute a fire hazard since $O_2$ will eventually condense in the liquid $N_2$. For many purposes Freon-12 ($CCl_2F_2$), which freezes at $-150°$, is a satisfactory and readily available substitute. The Freon in a metal container is suspended in liquid $N_2$ with stirring until it begins to freeze. To provide sufficient heat capacity it is recommended that at least 25–50 ml of Freon be used per gram of tissue. When the tissue is immersed the Freon should be stirred vigorously and, unless the sample is very small, the container lowered into the liquid $N_2$ (see also Chapter 10).

Ferrendelli (1972) has shown that with larger structures, liquid $N_2$ is superior to Freon for freezing at depths greater than 3 or 4 mm. Apparently the colder temperature of liquid $N_2$ ($-190°$) ultimately more than compensates for the brief early delay at the surface due to the gas evolution.

Once frozen the tissue can be safely stored for several weeks (and probably several months) at $-50°$ or below. At higher temperatures, changes due to enzyme action can occur surprisingly rapidly. Tests made with mouse brain showed a 20% loss in ATP in 24 hr at $-20°$, 1–4 hr at $-15°$, and 20–30 min at $-10°$.

Glucose-6-P was lost even more rapidly. It was at least 20% gone in an hour at $-20°$, 30 min at $-15°$, and 10 min at $-10°$. Changes were also observed in AMP, fructose-1,6-$P_2$, and 3-P-glycerate in a few hours at $-20°$. P-Creatine fell to half in 8 hr at $-15°$. Even glucose had fallen by a third after 6 hr at $-10°$. On the other hand, none of these metabolites changed during a week's storage at $-35°$.

Clearly, in the case of tissues to be analyzed for labile constituents such as these, the temperature should not be raised above −35° except as necessary for short periods during preparation of extracts.

## Preparation of Tissue Extracts

Most metabolites can be satisfactorily assayed in protein-free extracts prepared with $HClO_4$ or trichloroacetic acid. $HClO_4$ is usually preferred because most of it can be easily removed by precipitation as the potassium salt. Trichloroacetic acid is also a potent inhibitor of several enzymes used in metabolite assays. (An important example is glucose-6-P dehydrogenase, which is used in many different analytical systems and which is exceedingly sensitive to trichloroacetate inhibition.)

Once a tissue has been frozen it must not be thawed even for a few seconds if drastic changes in many metabolite levels are to be avoided. If frozen tissue is added to an acid precipitant at 0°, the tissue must in principle thaw, however briefly, before the acid penetrates and stops enzyme action. The following procedure, using $HClO_4$ as the precipitant, avoids this.

### Perchloric Acid Extracts

If the tissue needs to be dissected before extraction this is conveniently done in a room or cryostat maintained at −15° to −20° since at much lower temperatures tissues are exceedingly hard and brittle. As the data shown above indicate, the tissue should be held no longer than necessary in this warmer temperature range. Samples are either powdered at liquid $N_2$ temperatures or cut into pieces no larger than 10 mg. A weighed portion of the powdered or fragmented tissue is placed on top of approximately 3 volumes of 3 *M* $HClO_4$ previously frozen and kept on dry ice in a tube of appropriate size. The tubes are transferred to an alcohol bath maintained at −8° to −10°, and the samples stirred or agitated until the acid completely penetrates the powder or tissue fragments (5–15 min). The temperature must not be below −10° for complete extraction of the ice (2 *M* $HClO_4$ freezes at about −12°). When it is certain that no frozen tissue remains, 1 ml of $H_2O$ is added for each 0.3 ml of $HClO_4$ and the sample is repeatedly mixed at 4° for 5 or 10 min. If there is a possibility of bone contaminating the sample 1 m*M* EDTA is included in the water. When even a small amount of bone is present, calcium phosphate will precipitate later upon neutralization and remove part of the ATP, fructose diphosphate, and probably other phosphorylated compounds.

The samples are centrifuged at 5000*g* for 10 min and the supernatant fluid removed from the protein precipitate. The removal of all protein is critical because a number of enzymes survive the brief acid treatment at this low temperature. The presence of enzymes in the neutralized extract may lead to disturbances in the subsequent analyses for metabolites. For example, during the analysis of P-creatine, which requires added ADP, the presence of myokinase, which is very stable toward acid, would lead to erroneously high results.

Similarly, at least some of the ATPase can survive the acid treatment and give low ATP results if not completely removed.

As a rule the $HClO_4$ extracts are partially or completely neutralized before proceeding with analyses. If acid-labile metabolites are to be assayed this should be done promptly. In the case of P-creatine, for example, the tolerance limit is about an hour in either 2 *M* $HClO_4$ at $-10°$ or in 0.6 *M* $HClO_4$ at 0°.

Neutralization can be made in several ways, depending on the properties of the particular metabolite. One standard procedure is to neutralize with a 5 or 10% excess of 2 *M* $KHCO_3$. (This is about 0.35 volume per volume of acid extract, but the exact amount should be checked with the particular $HClO_4$ and $KHCO_3$ solutions used.) After most of the evolved $CO_2$ has been dissipated the $KClO_4$ precipitate is removed at 0° to 4° by centrifugation. Approximately 0.05 *M* $KClO_4$ remains in solution at 0°; the solubility is considerably greater at room temperature. The pH at this point is about 6 but later may fall to 7 or 7.5 as more $CO_2$ comes off. The majority of metabolites, including ATP, P-creatine, and most members of the citric acid cycle, are stable in such an extract and may be stored indefinitely at $-50°$. There is always the danger, however, that traces of certain enzymes may remain. Consequently it is recommended that extracts not be allowed to stand for long periods unfrozen even at 0°. A case in point is ATP, which has on occasion been found erroneously low in certain samples as the result of the action of ATPase that had not been completely removed.

An alternative method of neutralization which avoids $CO_2$ evolution is to use a mixture containing 2 *N* KOH, 0.4 *M* imidazole base, and 0.4 *M* KCl. This is added in such proportion (0.29 ml per milliliter of extract) as to neutralize 90% of the $HClO_4$ with KOH and the rest with imidazole, leaving the solution buffered at pH 7. The KCl is added to favor precipitation of the perchlorate.

Glyceraldehyde-P and dihydroxyacetone-P are more stable at a slightly acidic pH and should be stored at pH 3.5. For these triose phosphates, to a portion of the acid extract is added slightly less than the equivalent amount of 2 *M* $KHCO_3$, followed, after $CO_2$ has been evolved, by 10 or 15% as much 2 *M* formate buffer (1 *M* formic acid: 1 *M* sodium formate). Still other buffers can be added to maintain the extract at a particular pH. The use of Tris should be avoided if pyruvate analyses are to be made. If stored at $-18°$ in the presence of Tris there is 30–100% loss of pyruvate in 3 days.

## The Use of Methanol-HCl in Preparing Extracts

At the risk of being repetitious, one more variant on the preparation of acid extracts is offered. This permits somewhat quicker extraction of frozen tissue at even lower temperature than that above and offers the possibility of measuring glycogen on the same tissue sample used for other metabolite analyses (Nelson *et al.*, 1966). (Glycogen in many tissues appears partly in the acid extract and partly in the acid precipitate.) The weighed frozen tissue (fragment or powder) is homogenized at $-20°$ to $-30°$ with 2 volumes of 0.1 *N* HCl in methanol in a loose fitting glass homogenizer of suitable size. (Or with a little more trouble, the tissue can be dispersed with a stirring rod in an ordinary test tube.) After the solvent has penetrated the tissue completely, the temperature is raised to 0°, 1 ml of 0.02 *N* HCl in $H_2O$ is added for each 0.1 ml of methanol, and the homogenization or stirring is continued briefly to ensure thorough mixing.

A portion of the sample can now be set aside for glycogen assay. The enzymatic assay is made on this whole homogenate after heating 10 min at 100° (Chapter 9). Protein is precipitated from the rest of the solution by adding per milliliter of homogenate 0.125 ml of 10 m*M* EDTA in 3 *M* $HClO_4$. After centrifuging the extract is neutralized by one of the procedures given above, except that because the total acid is only 60% as great as above the amount of base is reduced in proportion.

If the methanol concentration exceeds 15% after $HClO_4$ addition, protein precipitation will not be complete; methanol, like a number of polar organic solvents, increases solubility of proteins in acid.

## Fluorescence of Tissue Extracts

Extracts prepared as described are somewhat fluorescent and contribute to the total fluorescence blank. Brain extracts, for example, are equivalent to at least 50 μmoles of DPNH per kilogram. Extracts of other tissues, such as liver and kidney, are even more fluorescent (the equivalent of 300–500 μmoles of DPNH). This presents a serious problem in direct measurement of metabolites present at levels of 20 μmoles/kg or less, of which there are many. The problem can be reduced by using the strong alkali method for measuring the final pyridine nucleotide, and with enzymatic cycling it can be practically eliminated. However, there are, as has been discussed, certain advantages of following the reactions directly in the fluorometer. A substantial fraction of the tissue blank is contributed by riboflavin nucleotides FMN and FAD. FAD is only about 10% as fluorescent as FMN at pH 7. It is easily hydrolized in acid to FMN. Therefore acid extracts that are not neutralized promptly will show increased fluorescence.

### Imidazole

In the case of analyses conducted near neutrality the riboflavin component of the blank can be specifically reduced by the use of 50–200 m*M* imidazole buffer. Concentrations of 50, 100, and 200 m*M* imidazole reduce the fluorescence of FMN at pH 7 approximately 40, 60, and 80%. Imidazole preparations usually contain fluorescent impurities. Sigma Chemical Co. offers a preparation with greatly reduced fluorescence, but even this contributes substantially to the blank reading when the concentration is 200 m*M*. The fluorescence of imidazole solutions can be reduced by charcoal treatment. [This capacity of imidazole to quench riboflavin fluorescence is apparently related to the fact that flavinadenine dinucleotide is much less fluorescent than flavin monophosphate. The imidazole portion of the adenine ring must associate with the isoalloxazine ring to form a nonfluorescent adduct (Bessey *at al.*, 1949).]

### Adsorption of Blank Materials

An alternative and more effective method of reducing the tissue fluorescence blank is to treat the extract with Florisil or charcoal. In either case, certain metabolites will be adsorbed and the treated extracts are useful only for special analyses.

#### *Florisil*

To prepare Florisil for use it is boiled in 10 volumes of 1 *N* HCl for 10 min, rinsed acid-free with water, and dried at 50°. The "fines" are removed in the rinsing process. The acid extract before neutralization is passed through a 2 mm internal diameter column containing 50–60 mg of Florisil per milliliter of extract (approximately 7 mg of tissue). The diameter of the column should be such as to provide a column at least 20 mm long. Alternatively, the extraction can be carried out in a test tube with similar proportions of extract and Florisil. The sample is repeatedly mixed for a total of 5 min, and the Florisil allowed to settle or spun down in a centrifuge. Florisil treatment reduces the tissue fluorescence as much as 90% or more, and permits quantitative recovery of dicarboxylic and tricarboxylic acids. Phosphorylated sugars and nucleotides are partially adsorbed.

#### *Charcoal*

The $HClO_4$ extract can be treated with 10 mg of dried acid-washed charcoal (Norit) per milliliter of extract. After adding the charcoal, the sample is brought to pH 6.8–7.0 with the KOH-imidazole-KCl mixture and filtered through acid-washed glass fiber paper. Treatment with charcoal also removes up to 90% of the tissue fluorescence and allows complete recovery of the phosphorylated intermediates, but removes nucleotides.

## Measurement of Pyridine Nucleotides in Tissues

The measurement of pyridine nucleotides in tissues (Burch *et al.*, 1967) has presented special difficulties in preparing the material for analysis without loss due to enzymatic action, and without oxidation of DPNH and TPNH.

It seems worthwhile to present the preparation in detail in order to illustrate the kinds of unexpected problems that may arise and to emphasize the point made earlier that the preliminary preparation may be more crucial than the assay itself.

From the stability properties of oxidized and reduced pyridine nucleotides (Chapter 1) it would be supposed that an acid extract would be suitable for measuring $DPN^+$ and $TPN^+$ and that a heated alkaline homogenate or extract would be suitable for measuring DPNH and TPNH. Unfortunately, if any blood (hemoglobin) is present during heating in alkali there may be some oxidation of reduced nucleotides leading to low results. Conversely, during the preparation of acid homogenates there is a special danger of oxidation which leads to erroneously high values for the oxidized forms. This second phenomenon is again attributable to hemoglobin, which if present can stoichiometrically oxidize the reduced nucleotides before they are completely destroyed.

These problems are solved by making a single homogenate in cold dilute alkali containing cysteine. One portion is acidified after adding ascorbic acid, which protects DPNH and TPNH from oxidation by acid decomposition products of hemoglobin. This acid portion is then heated at a pH and temperature chosen to give maximal destruction of the very acid stable enzyme DPNase and minimal destruction of $DPN^+$ and $TPN^+$. Acidification of the alkaline homogenate must be done promptly, before there is time for significant destruction of $DPN^+$ or $TPN^+$.

Another portion of the alkaline homogenate is heated to destroy $DPN^+$ and $TPN^+$. The cysteine present serves to protect DPNH and TPNH from oxidation in the process.

The preparation of the tissue is facilitated by the fact that the actual assays are made by enzymatic cycling and therefore the tissue can be highly diluted.

Since the cycling assay does not distinguish between oxidized and reduced nucleotides, total DPN and total TPN can be measured directly on a third portion of the homogenate which is not heated and which is analyzed promptly. Because of the extreme dilution during cycling, DPNase need not interfere even though it is not destroyed. Liver is used as the specific example, but the same procedure can be applied to other tissues providing the final tissue dilution in the cycling reagent is adjusted to give the proper concentration.

Rats are lightly anesthetized with ether, the abdominal cavity opened, and a piece of liver removed and quick-frozen in Freon 12 ($CCl_2F_2$) chilled to its freezing point ($-150°$) in liquid nitrogen. A 50 mg sample (weighed in a room maintained at $-20°$) is homogenized rapidly at 0° in 5 ml of 0.04 *N* NaOH containing 0.5 m*M* cysteine (NaOH-cysteine). The homogenate is treated subsequently as outlined in Table 7-1. When the levels are much lower than in

**TABLE 7-1**

PREPARATION OF LIVER FOR ANALYSIS OF TRIPHOSPHOPYRIDINE AND DIPHOSPHOPYRIDINE NUCLEOTIDES[a]

| Nucleotide measured | Volume of homogenate or standard ($\mu$l) | NaOH–[b] cysteine (ml) | 1.2 $M$[c] ascorbic acid ($\mu$l) | 0.02 $N$ $H_2SO_4$–[c] 0.1 $M$ $Na_2SO_4$ (ml) | Minutes at 60° | Aliquot for cycling ($\mu$l) |
|---|---|---|---|---|---|---|
| Total TPN | 200 | 4 | — | — | — | 2 |
| TPNH | 200 | 4 | — | — | 10 | 2 |
| $TPN^+$ | 200 | — | 5 | 2 | 30 | 4 |
| Total DPN | 200 | 4 | — | — | — | 2 |
| DPNH | 200 | 1 | — | — | 10 | 4 |
| $DPN^+$ | 200 | — | 5 | 2 | 30 | 1 |

[a]The homogenates are prepared as described in the text. For total TPN (DPN) the samples are kept at 0° and added to the appropriate cycling reagent within 30 min. The $TPN^+$ ($DPN^+$) samples are kept on ice and acidified within 30 min. Standard cycling procedures in 100 $\mu$l of the appropriate cycling reagent are used as described in Chapter 8. Standards are provided by adding stock pyridine nucleotide solutions to NaOH-cysteine at 0° to give concentrations of 5 and 10 $\mu M$ for $DPN^+$, 0.5 and 1 $\mu M$ for DPNH, 0.5 and 1 $\mu M$ for $TPN^+$ and 3 and 6 $\mu M$ for TPNH. The standards are prepared at the same time as the homogenates and treated as much alike as possible.

[b]Added to the 200 $\mu$l sample before heating or taking the aliquot for cycling.

[c]Added to the 200 $\mu$l aliquot before heating.

liver the original homogenates can be made at dilutions of 1 : 50 or 1 : 20 instead of 1 : 100 and in the case of DPNH and TPNH the subsequent dilution in NaOH–cysteine can be reduced as well. However, when hemoglobin concentrations are very high the dilution at this step should not be reduced too far, otherwise some oxidation of DPNH and TPNH may occur on heating. For example, with whole blood, the dilution at this step must be at least 1 : 200. For $DPN^+$ and $TPN^+$ assays the dilution in $H_2SO_4$-$Na_2SO_4$ should probably not be reduced below that shown (11-fold) unless the ascorbic acid concentration is increased or hemoglobin is very low. The aliquot used for cycling can be increased to 4 or 5 $\mu$l. If it is increased more than this it is important this be done for standards as well, because the cycling rate may be affected. This is particularly true for $TPN^+$ assays since sulfate decreases the cycling rate.

Obviously much smaller tissue samples could be used, since only a minute fraction of the sample is taken for cycling.

CHAPTER 8

# ENZYMATIC CYCLING

## PRINCIPLE

A cycling system contains two enzymes which catalyze two interrelated reactions such as the following:

$$A + S_1 \xrightarrow{\text{Enzyme 1}} B + P_1 \qquad (8\text{-}1)$$

$$B + S_2 \xrightarrow{\text{Enzyme 2}} A + P_2 \qquad (8\text{-}2)$$

$$\text{Net reaction:} \quad S_1 + S_2 \longrightarrow P_1 + P_2 \qquad (8\text{-}3)$$

A can be a coenzyme which is oxidized or reduced to B or a second substrate for enzyme 1 which is converted to a second product B. With relatively large amounts of the two enzymes and of the substrates $S_1$ and $S_2$, a small amount of A (or B) can "catalyze" the formation of large amounts of the products $P_1$ and $P_2$. By measuring either $P_1$ or $P_2$ in a suitable second analytical step the system becomes a chemical amplifier for the measurement of A (or B). The amplification possible is very great. In some cases rates of 20,000 per hour can be obtained. With double cycling (see below) total amplification of 400,000,000 has been achieved.

There are many enzyme pairs which can be coupled in the manner shown. The usefulness of an enzymatic cycle will be determined by the turnover rate of each enzyme for the catalyst added, the degree to which product accumulation affects the reactions, and the ease of measurement of the product or products. Four enzymatic cycles will be described for measuring either TPN, DPN, ADP plus ATP, or glutamate plus α-ketoglutarate.

### Kinetics

During cycling, in each case considered, the substance to be measured (A or B) will be used at concentrations well below its Michaelis constant, so that reaction rates will be proportional to the amount of substance added. The important kinetic factor for each enzyme is, therefore, its apparent first-order rate constant $k = V_m/K_m$, where $V_m$ is the velocity with saturating levels of A or B. With the establishment of a steady state the rate of formation of B [Eq. (8-1)] equals the rate of regeneration of A [Eq. (8-2)], i.e., $k_1[A] = k_2[B]$. The overall cycling velocity is $v = k([A] + [B])$. The overall cycling constant can be calculated from $k_1$ and $k_2$.*

$$k([A]+[B]) = k_1[A] \qquad \text{or} \qquad k = \frac{k_1[A]}{[A]+[B]}$$

Substituting $[B] = (k_1/k_2)[A]$,

$$k = \frac{k_1[A]}{[A]+(k_1/k_2)[A]} = \frac{k_1 k_2}{k_1 + k_2}$$

$$k = \frac{k_1 k_2}{k_1 + k_2} \tag{8-4}$$

For example, if $k_1$ is held constant at 100/min and $k_2$ is made to be 50, 100, and 200/min, the respective cycling rates will be 33, 50, and 67/min or 2000, 3000, and 4000/hr.

It might be expected that the cycling rate would increase proportionately with concomitant increases in both enzymes. Although this is true at moderate enzyme levels, it ceases to be true at higher levels. This results from the fact that in a cycling system the enzymes are frequently used in amounts which, on a molar basis, greatly exceed the substance to be measured. In this case the rates are limited, no matter how much enzyme is used, by the turnover numbers of the respective enzymes.

### TPN Cycle

The cycling system for TPN measurement utilizes glucose-6-P dehydrogenase and glutamate dehydrogenase.

$$\text{TPN}^+ + \text{glucose-6-P} \longrightarrow \text{TPNH} + \text{6-P-gluconolactone} + \text{H}^+ \tag{8-5}$$

$$\text{TPNH} + \alpha\text{-ketoglutarate} + \text{NH}_4^+ \longrightarrow \text{TPN}^+ + \text{glutamate} \tag{8-6}$$

(The 6-P-gluconolactone breaks down to 6-P-gluconate, in part during the cycling steps, and *in toto* during the heating step.) Under the conditions given,

* When the steady state is established.

for each molecule of TPN present 15,000–20,000 molecules of 6-P-gluconate (and glutamate) are formed in 1 hr. The 6-P-gluconate is then measured in a second step with 6-P-gluconate hydrogenase and extra $TPN^+$.

SAMPLE PROCEDURE (1 to 10 × $10^{-13}$ mole)

*TPN Cycling Reagent.* Tris-acetate buffer, pH 8 (50 m*M* Tris base, 50 m*M*Tris acetate); α-ketoglutarate, 5 m*M*; glucose-6-P, 1 m*M*; ammonium acetate, 10 m*M* (may be added with enzymes, see below); 5′-ADP, 100 μ*M*.

The reagent can be stored indefinitely at −50°, or lower. At −20° there is some loss of α-ketoglutarate within 2 weeks. Just before use, 200 μg/ml (9 U/ml) of beef liver glutamate dehydrogenase and 6 U/ml of yeast glucose-6-P dehydrogenase (equivalent to 15 μg/ml of crystalline enzyme) are added to the cycling reagent kept at 0°.

If the enzymes are suspended in $(NH_4)_2SO_4$ solution, the $(NH_4)_2SO_4$ is removed by centrifugation and the enzymes are dissolved in 2 *M* ammonium acetate which can provide the necessary ammonium ion for the reagent (the ammonium ion concentration is not critical). This reduces the sulfate concentration to 5 m*M* or less, which is desirable because sulfate inhibits glucose-6-P dehydrogenase (with low $TPN^+$ concentrations) in proportion to the square of its concentration.

*6-P-Gluconate Reagent.* Tris-HCl, pH 8.1 (20 m*M* Tris base, 20 m*M* Tris-HCl); EDTA, 100 μ*M*; ammonium acetate, 30 m*M*; $MgCl_2$, 5 m*M*; $TPN^+$, 30 μ*M*; 6-P-gluconate dehydrogenase (yeast), 0.5 μg/ml (0.006 U/ml).

The $TPN^+$ and 6-P-gluconate dehydrogenase are added within an hour or two of use. A 10-fold concentrated stock reagent, complete except for the $TPN^+$ and enzyme, can be prepared and stored at −50°.

*Step 1*

Volumes of 100 μl of the complete cycling reagent are pipetted into 3 ml fluorometer tubes (10 × 75 mm) kept in a rack in an ice bath. $TPN^+$ or TPNH standards and samples are added in volumes of 1–20 μl to give concentrations of 1–10 × $10^{-9}$ *M*. (The standards and blanks are carried through the entire analysis including any significant procedure necessary before cycling.) Water, or a medium identical to that containing the TPN, is added to bring all samples to the same volume ±2%. The rack is transferred to a 38° water bath for an hour, and then to a 100° bath for 2 min.

*Step 2*

To each tube is added 1 ml of 6-P-gluconate reagent. 6-P-Gluconate standards, in the range anticipated, are included as a control test for this step. Readings are made of the fluorescence after the reaction is complete (30 min or less at room temperature).

### Precautions

The cycling rate is of course affected by the temperature (in the neighborhood of 38° the overall reaction has a coefficient of about 8% per degree). Nevertheless, to achieve precision it is not necessary to keep the temperature exactly constant, but it is necessary to make sure that all samples and standards are incubated at the same temperature. The safest procedure is to use a well-stirred water bath and to have samples and standards in the same racks.

By the procedure outlined for Step 1, the first tubes in the rack are exposed to the cycling reagent at 0° for a longer period than the last samples, whereas the reaction is terminated simultaneously with heat for all samples. Therefore, since the reaction rate at 0° is appreciable (about 8% of that at 38°), the time required to add the reagent should be taken into account in planning the assay. For example, if the difference in cycling between the first and last sample is not to exceed 2%, the time to add reagent to all samples should not be more than a quarter of the time of incubation at 38°.

The TPN concentrations during incubation are exceedingly low, and therefore TPN contamination could be a problem. Suitable precautionary measures are described below under "Purity of Reagents and Sources of Blanks."

### Optimal Conditions

The amounts of glucose-6-P dehydrogenase and glutamate dehydrogenase specified in the example procedure give nearly maximal cycling rates. For the reason given in the section on Kinetics above, further increases in enzyme do not substantially increase the yield of 6-P-gluconate.

The α-ketoglutarate concentration cannot be altered much without a decrease in cycling rate. Lower levels decrease the turnover of TPNH by glutamate dehydrogenase; higher concentrations inhibit glucose-6-P dehydrogenase. The ADP is included to stabilize the glutamate dehydrogenase. The $NH_4^+$ level is not critical, but at a concentration of 100 m$M$ cycling is decreased, owing to an increase in the apparent $K_m$ of $TPN^+$ for glucose-6-P dehydrogenase.

The glucose-6-P level is also not critical since the Michaelis constant is about 20 $\mu M$ and levels up to at least 5 m$M$ are not inhibitory. The glucose-6-P in the reagent, 1 m$M$, would be completely used up at the full cycling rate with TPN concentrations of $50\text{–}70 \times 10^{-9}$ $M$. Therefore, if higher levels are to be measured, the glucose-6-P concentration may be increased to 5 m$M$ to permit cycling of nucleotide concentrations as high as $300 \times 10^{-9}$ $M$.

### Purity of Reagents and Sources of Blanks

Although commercial preparations of the enzymes used in cycling have been practically free of TPN, certain lots of glucose-6-P dehydrogenase have been found to have a contaminant which appears to bind $TPN^+$ or TPNH even more tightly than does glucose-6-P dehydrogenase. This contaminant has the effect of decreasing the cycling rate at lowest levels and the yield of product is proportionately greater at higher TPN levels. If a departure from linearity is observed between nucleotide concentration added and product measured, other lots of enzyme are tested until a suitable one is found.

Certain preparations of ADP have been found to contain enough $TPN^+$ to interfere with analyses. The $TPN^+$ can be removed by bringing the ADP solution to pH 12 with NaOH, heating 10 min at 60°, then neutralizing to pH 7 with HCl. The brief alkaline treatment does not destroy significant amounts of ADP. A solution of 100 m*M* ADP may be treated in this manner and stored at −20° or below for months without deterioration.

Adventitious contamination with TPN is another possible cause of high or variable blanks. With concentrations in the $10^{-9}$ *M* range, contamination with 1 part in 10 million from a 1 m*M* TPN solution could represent a 10% error. It is therefore desirable to set aside pipettes for use with solutions at the cycling step which are not used for stronger TPN solutions. These may be cleaned initially, if contamination is suspected, by soaking them for an hour, inside and out, in 0.1 *N* NaOH and then rinsing them with 0.1 *N* HCl.

The overall blank in the TPN cycle, by the sample procedure, need not exceed the equivalent of $10^{-9}$ *M* TPN during cycling or roughly 1.5 $\mu M$ TPNH in the final reagent. Part of this blank is due to the 6-P-gluconate reagent; the contribution from this source can be kept down to the equivalent of 0.3 $\mu M$ TPNH or less. Part of the blank is due to 6-P-gluconate formed during incubation of the cycling reagent (presumably due to a trace of TPN); this should not exceed the equivalent of $0.5 \times 10^{-9}$ *M* TPN during cycling. Assessment of contributions to the blank can be easily made. The fluorescence of the final reagent (cycling reagent omitted) can be checked directly with and without addition of 6-P-gluconate dehydrogenase. The cycling reagent can be tested with and without incubation to determine whether the blank it adds is due to the cycling procedure itself. If the incubated cycling reagent shows an increment more than the equivalent of $0.5 \times 10^{-9}$ *M* TPN, the source of contamination should be determined. This is easily done by comparing non-incubated and incubated blanks and standards in which the components have been varied individually over a two- or three-fold range.

The absolute blank can be reduced by decreasing the cycling volume as discussed below.

### Variations in Procedure

In principle the amplification achieved by a single cycling step can increase sensitivity 20,000-fold or more beyond that provided by simple measurement of $TPN^+$ or TPNH. This means that as little as $10^{-15}$ mole of TPN can be satisfactorily measured. However, much lower sensitivity will often suffice. Consequently, no one procedure will be appropriate for all situations.

Table 8-1 may be helpful in adapting the cycling step to a particular situation. Ordinarily it is well to keep TPN levels during cycling between $10^{-9}$ and $10^{-7}$ $M$. At lower levels the blank may be troublesome; at higher levels linearity may suffer, and with full cycling rates the glucose-6-P will be exhausted.

The time of incubation can be varied from 15 to 60 min with proportionate yields of 6-P-gluconate. However, if large numbers of samples are to be analyzed the shorter incubation presents a problem because of the substantial cycling rate at 0° (see Precautions section above). With longer cycling time than 1 hr the rate may fall off slightly but this need not affect precision and greater amplification can be obtained.

The yield of 6-P-gluconate can be reduced by lowering the cycling rate as shown in Table 8-1.

The volume may be varied from 50 to 200 μl in a fluorometer tube, or decreased to as little as 1 μl in appropriately smaller tubes. Decreasing the volume of the cycling reagent decreases the contribution of the cycling reagent to the total blank, thereby increasing the useful sensitivity. Reduction of volume is particularly advantageous when the TPNH generated at the final step is to be measured by the strong alkali method (see below).

If the 6-P-gluconate formed in the cycling step exceeds $10^{-8}$ mole, an

**TABLE 8-1**

TYPICAL VARIATIONS IN CYCLING CONDITIONS FOR TPN

| TPN amount (moles) | Cycling Volume (μl) | TPN concentration ($10^{-9}$ $M$) | Enzyme levels (% of max)[b] | Cycling[a] rate (per hr) | Concentration in fluorometer (μ$M$) |
|---|---|---|---|---|---|
| $10^{-11}$–$10^{-12}$ | 100[d] | 10–100 | 5 | 1200 | 1–10 |
| $10^{-12}$–$10^{-13}$ | 100[d] | 1–10 | 25 | 5000 | 0.5–5 |
| $10^{-12}$–$10^{-13}$[c] | 10 | 10–100 | 100 | 15,000 | 1.5–15 |
| $10^{-13}$–$10^{-14}$ | 10 | 1–10 | 100 | 15,000 | 0.15–1.5 |
| $10^{-14}$–$10^{-15}$ | 2 | 0.5–5 | 100 | 15,000 | 0.015–0.15[e] |

[a] Approximate.

[b] Both enzymes used at these percentages of the levels recommended in the sample procedure in the text.

[c] Glucose-6-P level in reagent increased to 5 m$M$.

[d] Cycling in fluorometer tube.

[e] Final TPNH measured by indirect strong alkali method (see text).

aliquot may be used for the final step so that the concentration of TPNH formed does not exceed the linearity range of the fluorometer. (Note this is an alternative procedure to reducing the cycling rate as shown in examples of Table 8-1.)

When necessary, sensitivity can be increased 10-fold by converting the TPNH in the terminal step to the 10-fold more fluorescent product in strong alkali (see Chapter 1). This permits measurement of as little as $2 \times 10^{-11}$ mole of 6-P-gluconate or $10^{-15}$ mole of TPN. In the example of this indirect procedure given in Table 8-1, the cycling step is made in 2 $\mu$l in a test tube of about 2.5 mm bore and 50 mm length. After stopping the cycling by heating at 100°, each sample is incubated for 30 min at room temperature (25°) with 20 $\mu$l of 6-P-gluconate reagent. To this is next added 20 $\mu$l of solution containing 0.3 $M$ $Na_3PO_4$ and 0.3 $M$ $K_2HPO_4$ and the tubes are heated 10 min at 60° to destroy excess $TPN^+$. An aliquot, or the entire sample, is then added with immediate mixing to 1 ml of 6 $N$ NaOH containing 0.03% $H_2O_2$ in a fluorometer tube. After heating 10 min at 60°, the tubes are cooled and the fluorescence measured.

### Assessment of the Reagent

If difficulty is encountered in achieving the desired cycling rate, the cycling process can be assessed directly in the fluorometer. This permits determination of the two rate constants under actual cycling conditions (Fig. 8-1). A low but measurable concentration of $TPN^+$, e.g., 1 $\mu M$ is added to 1 ml of the cycling reagent (no enzymes added). The blank fluorescence of the tube is read, and glucose-6-P dehydrogenase alone is added in the desired concentration. The increment in fluorescence of TPNH is a measure of the total nucleotide present, since all of the $TPN^+$ present becomes reduced. The desired amount of glutamate dehydrogenase is now added, which starts the cyclic process. There is an immediate partial drop in fluorescence due to partial TPNH oxidation. From this decrease in fluorescence the steady-state $TPN^+$ : TPNH ratio can be calculated. The steady state will persist for a few minutes, followed by a further sharp decrease in TPNH when the glucose-6-P is nearly used up. The time at which this occurs determines the overall cycling rate. This rate, in combination with the steady-state ratio, gives the individual rate constants (see legend of Fig. 8-1). This procedure will test the adequacy of the cycling reagent in regard to each enzyme separately, and permits choice of optimal conditions and enzyme concentrations. It also can detect a source of trouble such as a faulty enzyme preparation. (This test at room temperature gives a lower cycling rate than the same reagent would give at 38° under analytical conditions. The rate at 25° is only 40% of that at 38°, and the rate suffers in the test because of the excessively high TPN level. The fact that the rate is lower is taken into account in evaluating the test.)

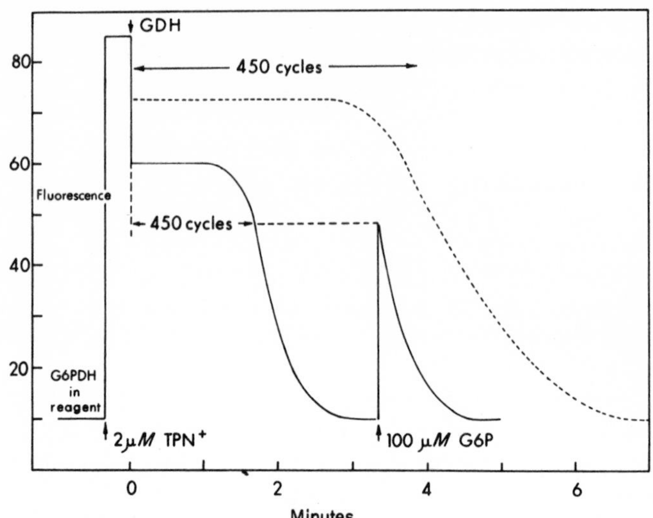


***Fig. 8-1.*** Direct test of TPN cycling reagent (simulated example). To 1 ml of reagent containing 1 m$M$ glucose-6-P were added 50 $\mu$g of glucose-6-P dehydrogenase. When 2 $\mu M$ $TPN^+$ was added the fluorescence increased 75 divisions. Upon subsequent addition of 200 $\mu$g of glutamic dehydrogenase (GDH) the fluorescence decreased 25 divisions. At this time, therefore, the steady-state $TPN^+$:TPNH ratio was 25:50 or 0.67 $\mu M$:1.33 $\mu M$. Thus the rate constant for glutamic dehydrogenase was only half as great as that for glucose-6-P. After the fluorescence had fallen to the base line, glucose-6-P was added to give a 100 $\mu M$ concentration (one-tenth of the original amount). The magnitude of the sudden jump in fluorescence indicates the point on the original curve at which 100 $\mu M$ glucose-6-P was left. This point was reached originally in 1.67 min. Therefore, 1.67 min were required to oxidize 900 $\mu M$ glucose-6-P. Since the total TPN concentration was 2 $\mu M$, the cycling rate was 450 per 1.67 min, or 16,000 per hour. Let the respective rate constants for glucose-6-P dehydrogenase and glutamic dehydrogenase be $k_1$ and $k_2$. From the steady-state $TPN^+$:TPNH ratio $k_2 = 0.5k_1$. Substituting in Eq. (8-4), $16{,}000 = 0.5\ k_1^2/1.5\ k_1 = k_1/3$. Therefore $k_1 = 48{,}000$/hr and $k_2 = 24{,}000$/hr. The dotted curve in the Figure indicates what would happen if a third as much glutamic dehydrogenase had been used.

## DPN Cycle

DPN is cycled with lactate dehydrogenase and glutamate dehydrogenase.

$$DPN^+ + \text{lactate} \longrightarrow DPNH + \text{pyruvate} + H^+ \quad (8\text{-}7)$$

$$DPNH + \alpha\text{-ketoglutarate} + NH_4^+ \longrightarrow DPN^+ + \text{glutamate} \quad (8\text{-}8)$$

Pyruvate and glutamate are produced in yields up to 6000- to 8000-fold per hour. After stopping the reaction, one of the products is measured. In the case of glutamate this is done by reversing reaction (8-8), which requires addition of $DPN^+$, glutamic dehydrogenase and $H_2O_2$. The last is to remove α-ketoglutarate. Pyruvate is measured by reversing reaction (8-7) by adding DPNH and lactate dehydrogenase. The resulting $DPN^+$ is measured fluorometrically in strong alkali.

### Sample Procedure ($2–30 \times 10^{-13}$ mole)

*Cycling Reagent*. Tris-HCl buffer, pH 8.4 (70 m*M* Tris base, 30 m*M* Tris-HCl); sodium lactate, 100 m*M*; ADP, 0.3 m*M*, α-ketoglutarate, 5 m*M*; ammonium acetate, 10 m*M* [or 5 m*M* $(NH_4)_2SO_4$, but see below].

The reagent can be stored at −50° indefinitely. As with the TPN reagent, storage at −20° results in loss of α-ketoglutarate.

To the reagent at 0° are added 400 μg/ml of glutamate dehydrogenase and 50 μg/ml of crystalline beef heart lactate dehydrogenase. (Charcoal treatment to remove $DPN^+$ is required for lactic dehydrogenase and may be required for both enzymes; see below.) If the enzymes are suspended in $(NH_4)_2SO_4$ this can furnish the necessary $NH_4^+$. However, if glutamate is to be measured in the final step the $NH_4^+$ concentration is reduced to 2 m*M* and it will be necessary to remove most of the $(NH_4)_2SO_4$ from the enzymes by centrifugation.

#### *Step 1*

Volumes of 100 μl of the complete cycling mixture are placed in 3 ml fluorometer tubes in a rack of ice. $DPN^+$, DPNH, or samples are added in volumes of 1–20 μl to give a concentration in the range of 2 to $30 \times 10^{-9}$ *M*. Since pyruvate formation is not strictly linear with DPN concentration, standards should be included which increase in steps of two to cover the range of the assay. Water, or the medium in which the standards and samples are prepared, is added to bring the contents of each tube to the same volume ±2%. The rate of cycling at 0° is 10–15% of that at 38°; therefore, for the reason given in the case of the TPN cycle, the difference in time between the addition of the first and last sample should not exceed 10 min. The rack is transferred to a bath at 38° for 1 hr, then heated at 100° for 2 min.

After cycling, either product may be measured. It is simpler to measure glutamate, but for greatest sensitivity and lowest blank there is advantage in measuring pyruvate. Therefore, procedures for both products will be presented.

### Glutamate Measurement

*Reagent*. Tris-HCl buffer, pH 8.4 (35 m*M* Tris base, 15 m*M* Tris-HCl); $DPN^+$, 1 m*M*; ADP, 100 μ*M*; $H_2O_2$, 9 m*M* (0.03%).

*Step 2*

After the cycling step and after the tubes have been cooled to room temperature, 1 ml of glutamate reagent is added. Twenty minutes later 5 $\mu$l of 10 mg/ml solution of glutamate dehydrogenase are added. [Since it is undesirable to add $NH_4^+$, a glycerol preparation of the enzyme is preferred; otherwise most of the $(NH_4)_2SO_4$ is removed by prior centrifugation.]

The fluorescence generated is read after 20 or 30 min. A second reading 20 or 30 min later may be desirable to make sure the reaction is complete. After long standing (an hour or more) some loss of fluorescence is expected from oxidation of DPNH by the $H_2O_2$. For lower levels of glutamate (less than 2 $\mu M$ final concentration) there is advantage in making a reading before the dehydrogenase addition to correct for possible differences in blank fluorescence between individual tubes.

*Comment on Glutamate Measurement*

The analytical reaction is

$$\text{Glutamate} + \text{DPN}^+ \rightleftharpoons \alpha\text{-ketoglutarate} + \text{NH}_4^+ + \text{DPNH} + \text{H}^+ \quad (8\text{-}9)$$

$$\alpha\text{-Ketoglutarate} + \text{H}_2\text{O}_2 \longrightarrow \text{succinate} + \text{CO}_2 + \text{H}_2\text{O} \quad (8\text{-}10)$$

For the reaction to proceed to completion it is necessary to remove $\alpha$-ketoglutarate [reaction (8-10)], both the large amount contributed by the cycling reagent and the small amount formed from glutamate. Since glutamic dehydrogenase, and to a lesser degree DPNH, are slowly destroyed by $H_2O_2$, the enzyme is not added until the bulk of the $\alpha$-ketoglutarate has been removed.

Oxidation of $\alpha$-ketoglutarate by $H_2O_2$ proceeds rapidly at room temperature and is affected very little by $H^+$ concentration between pH 7 and 11. At pH 8 the half-time of oxidation with 9 m$M$ $H_2O_2$ is 2 or 3 min.

For more details concerning glutamic dehydrogenase kinetics and the equilibrium constant for the reaction, see the glutamate method in Chapter 9.

## Pyruvate Measurement

*Pyruvate Reagent.* Imidazole-HCl buffer, pH 6.2 (70 m$M$ imidazole base, 330 imidazole-HCl); DPNH, 20–250 $\mu M$ (3–10 times expected pyruvate), see below); lactic dehydrogenase (beef heart or skeletal muscle), 1.5 $\mu$g/ml.

The reagent is chilled in ice and both the DPNH and lactic dehydrogenase are added within 30 min of use, because of the acid pH.

*Step 2*

To each tube in ice from the cycling step is added an equal volume (100 $\mu$l) of the pyruvate reagent. (The pH is planned to be 6.5 after addition.) The samples are incubated at 20°–30° for 15 min and returned to the ice bath. To each sample are added 25 $\mu$l of 5 *N* HCl with thorough mixing, and the tubes brought to room temperature. The destruction of the DPNH with the prescribed amount of acid is very rapid.

*Step 3*

Finally, 1 ml of 6 *N* NaOH is added rapidly to each tube with immediate mixing. A 1 ml syringe pipette is useful at this step. After heating 10 min at 60°, the tubes are cooled exactly to room temperature in a water bath, dried, and the fluorescence read. (See Chapter 1 for precautions in using this strong alkali procedure.)

*Comment on Pyruvate Measurement*

Pyruvate is measured by reversing the lactic dehydrogenase reaction:

$$\text{Pyruvate} + \text{DPNH} \longrightarrow \text{lactate} + \text{DPN}^+ \qquad (8\text{-}11)$$

Although the high lactate helps to balance the unfavorable equilibrium during cycling, it becomes a disadvantage during the subsequent pyruvate measurement. Any pyruvate remaining represents a negative error. The low pH (6.5) chosen for pyruvate measurement is to help compensate for the high lactate level. At this pH,

$$\frac{[\text{pyruvate}]\,[\text{DPNH}]}{[\text{lactate}]\,[\text{DPN}^+]} = 10^{-5} \qquad (8\text{-}12)$$

With a lactate concentration of 50 m*M*, which is the level at Step 2 in the sample procedure,

$$\frac{[\text{pyruvate}]\,[\text{DPNH}]}{[\text{DPN}^+]} = 0.5\ \mu M \qquad (8\text{-}13)$$

To provide sufficient DPNH becomes a problem only after cycling at the lowest DPN levels. For example, a DPN concentration of $10^{-9}$ after 2000-fold cycling results in 1 $\mu M$ pyruvate at the pH 6.5 step and thus would require, as Table 8-2 shows, a 10-fold excess of DPNH to keep the error down to 5%. Aside from this consideration it is desirable to avoid an unnecessary excess of DPNH due to possible contamination with $DPN^+$ and to the usual residual blank fluorescence of DPNH preparations.

**TABLE 8-2**

EFFECT OF DPNH CONCENTRATION ON PYRUVATE REDUCTION IN STEP 2[a]

| Initial pyruvate ($\mu M$) | Initial DPNH ($\mu M$) | Final pyruvate ($\mu M$) | Error (%) |
|---|---|---|---|
| 50 | 500 | 0.05 | 0.1 |
| 50 | 150 | 0.25 | 0.5 |
| 10 | 100 | 0.05 | 0.5 |
| 10 | 30 | 0.24 | 2.4 |
| 10 | 15 | 0.8 | 8.0 |
| 2 | 20 | 0.05 | 2.5 |
| 2 | 6 | 0.21 | 10.5 |
| 1 | 10 | 0.05 | 5 |
| 1 | 3 | 0.19 | 19 |

[a] The concentrations shown are those after mixing equal volumes of the DPN cycling reagent and the Step 2 "pyruvate reagent." The table applies to the use of 100 m$M$ lactate in the cycling reagent (50 m$M$ during Step 2).

### COMMENT ON CYCLING PROCEDURE AND PERMISSIBLE VARIATIONS

Although the optimal pH of the DPN cycle is 8.4, the pH is not critical; the rate is 2% less at pH 8.8 and 7% less at pH 8.0. The temperature coefficient is 7% per degree between 0° and 25°, but only 2.2% per degree between 32° and 38°. Therefore, temperature control is less critical than in the TPN system.

For the reason discussed in the Kinetics section above, enzyme concentrations higher than those given above do not materially increase the yield of pyruvate.

Because of greater danger of contamination than in the case of TPN, if maximum sensitivity is not required it is often desirable to decrease the cycling rate and increase the DPN concentrations. For example, a cycling rate of about 1000/hr can be attained by reducing glutamate dehydrogenase to 100 $\mu$g/ml and the lactate dehydrogenase to 5 $\mu$g/ml. In general, in decreasing the cycling rate it is desirable, as in this example, to lower the lactate dehydrogenase more than the glutamate dehydrogenase to keep the $DPN^+$ : DPNH ratio high and minimize fall off in pyruvate formation.

Lactate dehydrogenase operates in the unfavorable direction as far as equilibrium is concerned. Therefore, in spite of the high lactate level, there is a decrease in cycling rate as pyruvate accumulates. A pyruvate concentration of 100 $\mu M$ at the end of incubation has been set as a somewhat arbitrary upper limit.

Commercial lactate preparations contain small amounts of pyruvate. (The best samples tested contained 1 part of pyruvate in 40,000.) Therefore, when measuring very low DPN levels ($1–5 \times 10^{-9}$ *M*), with pyruvate determination as the final step there is advantage in removing the pyruvate contamination with $H_2O_2$ (see below). Alternatively, with low DPN levels the lactate concentration can be lowered to 50 m*M* without reducing the initial cycling rate. Both the α-ketoglutarate and $NH_4^+$ concentrations are less critical than in the TPN cycle. Substantial reduction in α-ketoglutarate will decrease glutamate dehydrogenase activity, but since lactate oxidation is the limiting step, there is little effect on the cycling rate. As in the TPN cycle, ADP is included to stabilize the glutamate dehydrogenase.

## Purity of the Reagents

As mentioned above, all preparations of lactate tested were found to contain pyruvate. Pyruvate can be easily removed with $H_2O_2$. A 1 *M* lactate solution is made 2 m*M* in $H_2O_2$ and allowed to incubate an hour at 38°. Any excess $H_2O_2$ will either be decomposed during storage or will be destroyed by α-ketoglutarate in the cycling reagent.

All lots of beef heart lactate dehydrogenase, as commercially supplied, have been found to be contaminated with $DPN^+$. The coenzyme can be removed as follows. The enzyme as supplied [usually a 2–4% suspension in 2.5 *M* $(NH_4)_2SO_4$] is diluted to 0.4% protein concentration with 0.02 *M* phosphate buffer at pH 7 containing 2% Norit. After incubating for 15 min at 37° with occasional mixing, the charcoal is removed by centrifugation, and the lactate dehydrogenase precipitated by adding solid $(NH_4)_2SO_4$ to give a 3.0 *M* concentration. The enzyme is collected by centrifugation and suspended in 2.5 *M* $(NH_4)_2SO_4$ at an approximate protein concentration of 20 mg/ml. (There may be a 20 or 30% loss during this treatment, therefore, the protein concentration or the enzyme activity should be determined before cycling.)

Glutamic dehydrogenase may also contain $DPN^+$. This may be removed as follows. A 2% suspension of Norit is prepared in 20 m*M* phosphate buffer, pH 7. A volume of this equal to the volume of 1% dehydrogenase to be treated is centrifuged and the supernatant fluid discarded. A 1% solution of the enzyme, prepared in 50% glycerol, is now added, mixed with the washed charcoal, incubated 30 min at 38°, and centrifuged. The supernatant fluid may be used without further treatment.

## Blank Values

The major source of blank in the DPN cycle is the pyruvate in the lactate. In the purest preparations this should not exceed an equivalent of $10^{-9}$ *M*

DPN at the cycling step. The beef heart lactate dehydrogenase treated with charcoal should not contribute more than $2 \times 10^{-10}$ *M* DPN. The DPNH used in pyruvate measurement may contribute a blank equal to 1–2% of the DPNH present. If the DPNH has become oxidized on storage the blank will be much greater. The DPNH prepared in 100 m*M* carbonate buffer (80 m*M* $NaCO_3$, 20 m*M* $NaHCO_3$) should be heated 15 min at 60° each time before use to destroy $DPN^+$ that may have accumulated.

The possibilities for DPN contamination appear to be much greater than for TPN. Scrupulous care should be taken to avoid chance contamination. A procedure is given in the TPN section for initial decontamination of pipettes to be used in relation to the cycling step.

## ATP–ADP Cycle

This is an example of an enzymatic cycle which does not use a pyridine nucleotide except in a final step. This was described by Breckenridge (1964). ADP is converted to ATP and back again:

$$\text{ADP} + \text{P-pyruvate} \xrightarrow[\text{kinase}]{\text{Pyruvate}} \text{ATP} + \text{pyruvate} \tag{8-14}$$

$$\text{ATP} + \text{glucose} \xrightarrow{\text{Hexokinase}} \text{glucose-6-P} + \text{ADP} \tag{8-15}$$

After an hour at 37°, under the conditions stipulated, the yield of glucose-6-P is at least 1500 moles per mole of ATP plus ADP. The glucose-6-P generated is measured in a second step with $TPN^+$ in the usual manner. In some applications the two steps can be combined into one.

Although this cycle measures ADP plus ATP it can be used for measuring total adenylate by first converting any AMP present to ATP with myokinase, pyruvate kinase, P-pyruvate and a trace of ATP. Finally, the high sensitivity provided has made possible the measurement of tissue levels of 3′, 5′-cyclic adenylate (Breckenridge, 1964; Goldberg *et al.*, 1969a). Other adenylates are first eliminated, after which the cyclic adenylate is converted to 5′-AMP with P-diesterase.

In the procedure given below the analysis takes place in two steps, $TPN^+$ preparations (needed for the final reaction) ordinarily contain substantial amounts of ADP. If the ADP is removed the reaction can be conducted in one step and followed directly in the fluorometer.

### Sample Procedure (3 to $50 \times 10^{-13}$ mole)

*Cycling Reagent*. Tris-HCl, pH 7.5 (20 m*M* Tris base, 80 m*M* Tris-HCl); KCl, 50 m*M*; $MgCl_2$, 2 m*M*; P-enolpyruvate, 250 μ*M*; glucose, 2 m*M*; bovine plasma albumin, 0.01%; pyruvate kinase (rabbit muscle), 80 μg/ml; hexokinase (yeast), 250 μg/ml.

The reagent can be made in bulk and stored without the enzymes at −50°.

*Glucose-6-P Reagent.* Tris-HCl, pH 8.1 (25 m*M* Tris base, 25 m*M* Tris-HCl); $TPN^+$, 100 $\mu M$; glucose-6-P dehydrogenase, 0.05 U/ml. Both $TPN^+$ and the enzyme should be added shortly before use and not frozen in the reagent.

Volumes of 100 $\mu$l of complete cycling reagent are pipetted into 3 ml fluorometer tubes in an ice bath. Samples, standards, and blanks in volumes up to 20 $\mu$l are added to give concentrations of $3–50 \times 10^{-9}$ *M*. Water, or medium identical to that containing the ADP or ATP, is added to bring all samples to the same volume $\pm 2$ $\mu$l. The samples are incubated 1 hr at 37°, after which the reaction is stopped by heating 2 min at 100°.

To each tube is added 1 ml of glucose-6-P reagent. After 15 min at room temperature, the fluorescence of the resulting TPNH is read. The cycling yield can be checked with standards containing glucose-6-P in the range anticipated.

### Comment on Procedure

The ratio of pyruvate kinase to hexokinase, empirically determined, was found to be optimal. Adding more of both enzymes in the same ratio will increase the cycling rate at least twofold, although not in direct proportion to the enzyme concentration. The formation of glucose-6-P was found to be linear over a 40-fold range of added ADP.

## Glutamate–α-Ketoglutarate Cycle

$$\text{Glutamate} + \text{oxalacetate} \xrightarrow{\text{Transaminase}} \alpha\text{-ketoglutarate} + \text{aspartate} \quad (8\text{-}16)$$

$$\alpha\text{-Ketoglutarate} + NH_4^+ + \text{DPNH} + H^+ \xrightarrow[\text{dehydrogenase}]{\text{Glutamic}} \text{glutamate} + \text{DPN}^+ \quad (8\text{-}17)$$

This is a cycle for measuring glutamate (or α-ketoglutarate). It is analogous to the above ATP–ADP cycle, but differs in that here a pyridine nucleotide is one of the substrates. The cycling rate is comparatively low, but it provides a simpler way to increase sensitivity than if glutamate is oxidized by $DPN^+$ and the resulting DPNH is cycled.

### Sample Procedure ($1–5 \times 10^{-11}$ mole)

*Cycling Reagent.* Imidazole-acetate, pH 6.7 (17 m*M* imidazole, 33 m*M* imidazole acetate); ammonium acetate, 10 m*M*; ADP, 200 $\mu M$; DPNH, 150 $\mu M$; oxalacetic acid, 200 $\mu M$; bovine plasma albumin, 0.02%; glutamic-oxalacetate transaminase (pig heart), 5 $\mu$g/ml (0.9 U/ml); glutamic dehydrogenase (beef liver), 5 $\mu$g/ml (0.22 U/ml).

The oxalacetic acid is added directly from a 200 m*M* stock solution in 1 *N* HCl, stored frozen (it does not keep except in strong acid). The transaminase, as currently supplied, contains added α-ketoglutarate which must be removed. This is accomplished by centrifuging the $(NH_4)_2SO_4$ suspension with subsequent washing three times by resuspension and centrifugation, using equal volumes of fresh 3 *M* $(NH_4)_2SO_4$. The enzyme is finally dissolved and stored in 50% glycerol containing 10 m*M* imidazole buffer, pH 7.

Fifty $\mu$l volumes of complete cycling reagent are transferred to fluorometer tubes in an ice bath. Samples and standards are added in volumes up to 10 $\mu$l to give concentrations of 0.2 to 1 $\mu M$ glutamate or $\alpha$-ketoglutarate. The total volumes should not differ by more than 1 $\mu$l. The rack of tubes is incubated 60 min at 38°, chilled in ice then acidified with 10 $\mu$l of 1 *N* HCl to destroy excess DPNH. Fluorescence is developed from the $DPN^+$ by adding 1 ml of 6 *N* NaOH (with rapid mixing) and heating 10 min at 60°.

The sample procedure provides only about 25-fold amplification. Greater sensitivity when needed can be achieved by increasing the concentration of the two enzymes and by longer incubation periods. Cycling rates of 150–200 per hour have been achieved. However, it has not been possible to obtain the amplification provided by the previous cycles described. This is in part because of the low first-order rate constant for glutamate with the transaminase. In addition, as the enzymes are increased there is a great deal of difficulty in avoiding major glutamate contamination.

#### Removing $\alpha$-Ketoglutarate

In most applications of this cycle, $\alpha$-ketoglutarate will be negligible compared to glutamate. If not, $\alpha$-ketoglutarate can be destroyed by treatment with $H_2O_2$ (e.g., 2 min at 100° with 1 m*M* $H_2O_2$).

### Other Cycles

Only four of the many possible cycling schemes have been presented. Matschinsky *et al.* (1968b) have described a cycle for DPN which uses yeast glyceraldehyde-P dehydrogenase in place of lactic dehydrogenase. Goldberg *et al.* (1969b) have devised a cycle for measuring GTP plus GDP which is similar to that for ATP plus ADP.

### Double Cycling

In the case of both the DPN and TPN cycles it is possible to repeat the cycling step and obtain exceedingly great amplification (up to 400,000,000-fold). It is beyond the scope of this book to present specific procedures for double cycling. However, the additional steps are straightforward. In the TPN cycle the last step is the oxidation of 6-P-gluconate with $TPN^+$. To double cycle, the excess $TPN^+$ at this step is destroyed with alkali and the TPNH is amplified in a second cycling step in the normal manner.

An amplification of 400,000,000-fold permits the measurements in the mole range of $10^{-17}$ to $10^{-18}$. To take advantage of this tremendous sensitivity, the first analytical steps must be carried out in very small volumes.

Suppose the problem is to measure $10^{-17}$ mole of ATP, the amount in one colon bacillus. The analytical steps could be as follows (abbreviated).

*Step 1.* The specific step; 0.1 nl volume (ATP concentration 0.1 $\mu M$).

*Step 2.* First cycling step; 10 nl (TPN concentration $10^{-9}$ $M$).

*Step 3.* First 6-P-gluconate step; 1 $\mu$l (TPNH formed at 0.2 $\mu M$ concentration).

*Step 4.* Second cycling step; 100 $\mu$l (TPN concentration $2 \times 10^{-9}$ $M$).

*Step 5.* Second 6-P-gluconate step; 1 ml (TPNH formed at 4 $\mu M$ concentration).

CHAPTER 9

# A COLLECTION OF METABOLITE ASSAYS

This chapter contains protocols for measuring 31 different metabolites. In each case a spectrophotometric assay and at least one fluorometric assay have been presented. In many cases additional fluorometric procedures are given for increasing sensitivity to permit measurement of as little as $10^{-14}$ moles of material. The reader should have no difficulty designing similar adaptations of any of the methods to achieve whatever sensitivity is required. Chapters 4 and 6 may be useful for this purpose.

The spectrophotometric assays, unless noted otherwise, are designed for measuring 20–150 $\mu M$ concentrations of the metabolite. The direct fluorometric methods are planned for 1 ml volumes of reagent. Chapter 4 gives fuller procedural details.

Unless otherwise noted, enzyme activities under reagent headings given as U/ml. (international units/ml) are those reported by Boehringer Corp.

Most of the methods are adaptations of spectrophotometric methods from many authors. Although we will not try to trace the history of each one, the following is an attempt to record the first edition of most of the methods.

ADP, Kornberg and Pricer (1951b); alanine, Pfleiderer *et al.* (1955a); aspartate, Pfleiderer *et al.* (1955b); ATP, Kornberg (1950); citrate, Moellering and Gruber (1966); dihydroxyacetone-P, Thorn *et al.* (1955); fructose and fructose-6-P, Slein (1950); fructose-1,6-$P_2$ via glycero-P-dehydrogenase, Slater (1953); fructose-1,6-$P_2$ via glyceraldehyde-P dehydrogenase, Vishniac and Ochoa (1952); glucose, Slein *et al.* (1950); glucose-6-P, Slein (1950); glucose-1,6-$P_2$, Paladini *et al.* (1949); glutamate, Albers *et al.* (1961); glycero-P, Bublitz and Kennedy (1954); glycogen, Bueding and Hawkins (1964), Gatfield and Lowry (1963); isocitrate, Ochoa (1948); lactate (method I), Noll (1966); P-pyruvate, Kornberg and Pricer (1951b); pyruvate, Kubowitz and Ott (1943); ATP via P-fructokinase as in total nucleotide triphosphate method, Slater (1953); UDP-glucose, Strominger *et al.* (1957).

## ADP and AMP

$$\text{AMP} + \text{ATP} \xrightarrow{\text{Myokinase}} 2\ \text{ADP} \qquad (9\text{-}1)$$

$$\text{ADP} + \text{P-pyruvate} \xrightarrow[\text{kinase}]{\text{Pyruvate}} \text{ATP} + \text{pyruvate} \qquad (9\text{-}2)$$

$$\text{Pyruvate} + \text{DPNH} + \text{H}^+ \xrightarrow[\text{dehydrogenase}]{\text{Lactic}} \text{lactate} + \text{DPN}^+ \qquad (9\text{-}3)$$

### A. Spectrophotometer

*Reagent.* Imidazole-HCl buffer, pH 7.0 (30 m*M* imidazole base, 20 m*M* imidazole-HCl); $MgCl_2$, 2 m*M*; KCl, 75 m*M*; ATP, 100 $\mu M$ (needed only for AMP assay); DPNH, 50–150 $\mu M$ (at least a 20% excess); P-pyruvate, 300 $\mu M$; myokinase (rabbit muscle), 1 $\mu$g/ml (0.36 U/ml); pyruvate kinase (rabbit muscle), 2 $\mu$g/ml (0.3 U/ml); lactic dehydrogenase (beef heart), 2 $\mu$g/ml (0.4 U/ml).

*Reaction times.* Pyruvate, 1 min or less; ADP, 2–4 min; AMP, 3–6 min.

*Conduct of the Assay*

Unless pyruvate is to be determined on the sample, lactic dehydrogenase can be incorporated in the reagent. After adding the sample, time is allowed for pyruvate to be reduced and a reading is made. ADP and AMP are then measured in sequence by the decrease in reading upon addition of first pyruvate kinase and then myokinase (2 moles of DPNH are produced per mole of AMP).

*Comment : ADP Method*

Because P-pyruvate may contain traces of pyruvate its concentration is kept low, particularly when measuring low levels of ADP. Samples which contain more than 1% pyruvate should not be used. If a satisfactory sample is not available, pyruvate can be removed at neutral pH by adding a slight molar excess of $H_2O_2$ over the pyruvate and heating for a few minutes at 100°.

The method for ADP is not completely specific since other nucleotide diphosphates react with P-pyruvate, although at slower rates. According to

Strominger (1955) the relative rates with ADP, GDP, IDP, UDP, and CDP are, respectively, 100, 19, 12, 3, and 2. In general, therefore, what is measured is the sum of ADP, GDP, and IDP, although Goldberg *et al.* (1966) distinguished ADP and GDP by careful control of time and enzyme concentration.

*Comment: AMP method*

Most DPNH preparations contain substantial amounts of 5′-AMP, sufficient in some cases to give a very high blank and use up a major fraction of the DPNH in the assay. A satisfactory lot can probably be selected. If not it is possible to remove the AMP with phosphatase in the following manner:

To a 5 m$M$ DPNH solution at pH 9.0 add 70 $\mu$g/ml (3 U/ml) of alkaline phosphatase (Sigma, type III, bacterial) and incubate 20 min at 38°. Add NaOH to give a 20 m$M$ excess (pH must be at least 12); heat 2 min at 100° to destroy the phosphatase. The DPNH can be stored at −50° at this pH. If the ATP contains disturbing amounts of either ADP or AMP, the amount of ATP can be lowered with compensatory increase in myokinase. ATP actually increases during the overall reaction sequence, therefore the initial level can be less than that of the AMP to be measured.

### B. Fluorometer Direct Assay, 0.1 to $8 \times 10^{-9}$ Mole of ADP, 0.1–$4 \times 10^{-9}$ Mole of AMP

*Reagent.* The reagent is the same as in A except as follows: ATP, 5 $\mu M$; DPNH, 0.2–10 $\mu M$ (20–75% excess); P-pyruvate, 20 $\mu M$ (10 $\mu M$ with less than 1 $\mu M$ substrate); pyruvate kinase, 5 $\mu$g/ml.

*Reaction times.* Pyruvate, 2 min or less; ADP, 3–6 min; AMP, 5–10 min.

*Conduct of the Assay*

See A above. Ordinarily in living systems ADP is present at substantially higher levels than is AMP. This tends to decrease the precision of AMP measurement. It is important to be sure that the reaction with ADP is complete before adding myokinase. If other dinucleotides are present in considerable amounts there may be a drift as they slowly react. In this case it may be wise to increase the amount of pyruvate kinase. If necessary two parallel samples can be run; both receive pyruvate kinase until ADP has reacted, after which one only receives myokinase. The difference should provide a more accurate measure of AMP.

*Kinetics*

Pyruvate kinase has a $V_{max}$ of about 150 $\mu$moles/mg/min. The $K_m$ for ADP is 200$\mu M$, that of P-pyruvate is 30 $\mu M$. From this the calculated half-time for conversion of ADP to ATP with saturating P-pyruvate, and 1 $\mu$g/ml of enzyme, would be about 1 min ($0.7 \times 200/150$). With 20 $\mu M$ P-pyruvate the half-time would increase to 2 or 3 min. This is why more pyruvate kinase is recommended in the fluorometric method. Because of the relatively large $K_m$ for ADP, the reaction is approximately first order with all ADP levels. Pyruvate kinase requires both $Mg^{2+}$ and $K^+$.

Myokinase is a very active enzyme. The $V_{max}$ of commercial preparations tested has been up to 600 μmoles/mg/min. Noda and Kuby (1963) report a $V_{max}$ of three times this. Under analytical conditions the $K_m$ for AMP is about 50 μ*M*, that of ATP close to 100 μ*M*. ATP samples almost always contain significant amounts of ADP. Therefore ATP levels are kept relatively low. With 5 μ*M* ATP and 1μg/ml of myokinase, observed half-times for AMP are approximately 1 min (the other two enzymes in excess). Myokinase dilutions keep better in the presence of 2 m*M* dithiothreitol.

## Alanine

$$\text{Alanine} + \alpha\text{-ketoglutarate} \xrightarrow[\text{transaminase}]{\text{glut-pyr}} \text{pyruvate} + \text{glutamate} \quad (9\text{-}4)$$

$$\text{Pyruvate} + \text{DPNH} + \text{H}^+ \xrightarrow[\text{dehydrogenase}]{\text{lactic}} \text{lactate} + \text{DPN}^+ \quad (9\text{-}5)$$

### A. Spectrophotometer

*Reagent.* Tris-HCl buffer, pH 8.1 (25 m*M* Tris base, 25 m*M* Tris-HCl); DPNH, 50–150 μ*M* (at least a 20 % excess); α-ketoglutarate, 200 μ*M*; glutamic–pyruvic transaminase (pig heart), 100 μg/ml (8 U/ml); lactic dehydrogenase (beef heart), 2 μg/ml (0.4 U/ml).

*Reaction time.* Twenty minutes.

*Conduct of the Assay*

For standardization purposes, both enzymes can be incorporated in the reagent. For samples which may contribute appreciably to the absorption at 340 nm, or which may contain pyruvate, it is preferable to add the transaminase last.

### B. Fluorometer Direct Analysis, 0.1–8 × $10^{-9}$ Mole

*Reagent.* Tris buffer as for the spectrophotometer; DPNH, 0.2–10 μ*M* (20–70 % excess over the expected alanine); α-ketoglutarate, 50 μ*M*; glutamic–pyruvic transaminase, 100 μg/ml; lactic dehydrogenase, 1 μg/ml.

*Reaction time.* Twenty minutes.

*Conduct of the Assay*

Many kinds of biological material contain pyruvate at levels that are significant relative to alanine. Therefore the sample is added to the reagent containing lactic dehydrogenase only. After allowing a few minutes for pyruvate to react, a reading is made and the reaction is started with transaminase. If it is wished to also evaluate pyruvate, lactic dehydrogenase can be omitted from the reagent and added to the fluorometer tube after the sample.

*Comment*

The large amount of transaminase used adds appreciable fluorescence to the samples. Therefore with highest sensitivities the blanks become especially important. Lactic dehydrogenase enhances the fluorescence of DPNH. Because the effect may not be linear, excessive amounts of this enzyme should be avoided and the linearity checked.

*Kinetics*

The Michaelis constant for alanine is very large, the order of 30 m*M*; that for α-ketoglutarate is very low, less than 20 μ*M* with alanine levels in the analytical range. The high Michaelis constant for alanine requires that an unusually large amount of enzyme be used. The $V_{max}$ is about 80 μmoles/mg/min. From the formula $k = V_{max}/K_m$, it can be calculated that $k = \mu g/ml \times 80/30{,}000$ per minute, or about 0.25/min with 100 μg/ml.

## Aspartate

$$\text{Aspartate} + \alpha\text{-ketoglutarate} \xrightarrow[\text{transaminase}]{\text{glut-oxal}} \text{oxalacetate} + \text{glutamate} \quad (9\text{-}6)$$

$$\text{Oxalacetate} + \text{DPNH} + \text{H}^+ \xrightarrow[\text{dehydrogenase}]{\text{malic}} \text{malate} + \text{DPN}^+ \quad (9\text{-}7)$$

Levels of aspartate in biological materials are ordinarily several orders of magnitude higher than those of oxalacetate. Consequently no appreciable blank from oxalacetate should be encountered. The instability of oxalacetate is no problem in the coupled assay.

### A. Spectrophotometer

*Reagent*. Imidazole-HCl buffer, pH 7 (30 m*M* imidazole, 20 m*M* imidazole-HCl); DPNH, 50–150 μ*M* (at least a 20% excess); α-ketoglutarate, 200 μ*M*; glutamic–oxalacetic transaminase (pig heart), 10 μg/ml (1.8 U/ml); malic dehydrogenase (pig heart), 2 μg/ml (1.4 U/ml).

*Reaction time*. Ten minutes.

*Conduct of the Assay*

For standardizing aspartate solutions, both enzymes can be incorporated in the reagent and the sample added last, after an initial reading. For samples which may contribute appreciable absorption at 340 nm, it is preferable to add the transaminase last.

B. Fluorometer Direct Analysis, 0.1–8 × $10^{-9}$ Mole

*Reagent.* Imidazole buffer as for the spectrophotometer. DPNH, 0.2–10 $\mu M$ (20–75% excess over expected aspartate); α-ketoglutarate, 50 $\mu M$; glutamic–oxalacetic transaminase, 10 μg/ml; malic dehydrogenase, 0.5 μg/ml (see comment).

*Reaction Time.* Ten minutes.

*Conduct of the Assay*

Malic dehydrogenase is incorporated in the reagent. The reaction is ordinarily started by addition of the transaminase after the sample has been added and an initial reading made.

*Comment*

Malic dehydrogenase enhances the fluorescence of DPNH. Because the effect is nonlinear under most circumstances, the amount of this enzyme used should be kept low, and the linearity of the assay checked.

*Kinetics*

The Michaelis constant for aspartate is large and increases with high levels of α-ketoglutarate. It ranges from about 1 m$M$ with 50 $\mu M$ α-ketoglutarate or less, to 4 m$M$ with 1 m$M$ α-ketoglutarate or more. Therefore the reaction is first order at all aspartate levels in the analytical range. On the other hand, the Michaelis constant for α-ketoglutarate is very small, less than 1 $\mu M$ at low levels of aspartate, and approaches 75 $\mu M$ with very high (20 m$M$) aspartate levels. Because of this situation the α-ketoglutarate requirement is no more than that which will provide a safe excess over the aspartate to be measured. High levels actually inhibit.

The $V_{max}$ at 25° is about 180 μmoles/mg/min. However, because of the peculiarity of the kinetics, in analytical situations with modest α-ketoglutarate concentrations, rates are those expected for a $V_{max}$ of about 60 μmoles/mg/min. Thus, the apparent first-order rate constant with 10 μg of enzyme per milliliter would be 10 × 60/1000 = 0.6/min (from $k = V_{max}/K_m$).

## ATP and P-Creatine

$$\text{P-Creatine} + \text{ADP} \xrightarrow{\text{creatine kinase}} \text{creatine} + \text{ATP} \tag{9-8}$$

$$\text{ATP} + \text{glucose} \xrightarrow{\text{hexokinase}} \text{ADP} + \text{glucose-6-P} \tag{9-9}$$

$$\text{Glucose-6-P} + \text{TPN}^+ \xrightarrow[\text{dehydrogenase}]{\text{glucose-6-P}} \text{6-P-gluconolactone} + \text{TPNH} + \text{H}^+ \tag{9-10}$$

A number of tissues contain ATP and P-creatine at comparable concentrations. In this case it is convenient to measure them in the same sample. With direct analyses this is done by making successive readings after each enzyme addition. With indirect analyses the sample is divided after ATP has reacted. TPNH is measured in one portion as a measure of ATP. In the other portion TPNH is destroyed, after which all three enzymes are added to give TPNH equal to P-creatine. Alternatively, creatine kinase can be added to the second sample without prior acidification to give TPNH equal to the sum of P-creatine and ATP, from which P-creatine can be calculated by difference. Glucose-6-P can also be measured on the same sample but usually the level is much lower than that of ATP, consequently a larger sample will be required.

A. SPECTROPHOTOMETER

*Reagent.* Tris-HCl, pH 8.1 (25 m*M* Tris base, 25 m*M* Tris-HCl); $MgCl_2$, 1 m*M*; dithiothreitol, 0.5 m*M*; $TPN^+$, 500 $\mu M$; glucose, 1 m*M*; ADP, 500 $\mu M$ (for P-creatine assay only); creatine kinase (rabbit muscle), 50$\mu$g/ml (0.9 U/ml)*; hexokinase (yeast), 2 $\mu$g/ml (0.28 U/ml); glucose-6-P dehydrogenase (yeast), 0.07 U/ml equivalent to 0.15 $\mu$g/ml of crystalline enzyme.

*Reaction Time.* Three to 5 minutes for glucose-6-P, 4–6 min for ATP, 10 min for P-creatine.

*Conduct of the Assay*

One of the greatest dangers in analyzing biological material is that myokinase, which is acid-stable, is not completely removed in preparing the samples for analysis. This would give enormously high results. To minimize this danger it is recommended that ADP be ommitted from the reagent until after the hexokinase reaction is complete. The ADP can conveniently be added with the creatine kinase. Careful checks should be made to see if there is any drift in sample readings after the ATP and P-creatine reactions are complete, as shown by standards.

*Comment*

In addition to the danger of myokinase contamination, two other dangers have been encountered in ATP analysis of tissues. The first is failure to remove ATPase completely. This can cause loss of ATP from tissue extracts prior to analysis rather than during the assay itself. The second danger, described in Chapter 7, is that ATP may be lost in preparing $HClO_4$ tissue extracts if there is a substantial amount of $Ca^{2+}$ present. Calcium phosphate

* Measured in the direction of P-creatine formation.

or carbonate precipitates, formed on neutralization, can absorb ATP. This can be prevented with EDTA.

ADP almost invariably contains some ATP. A sample should be chosen, if possible, that contains not more than 0.5% ATP. (For removal of ATP from ADP see comment after protocol B).

## B. Fluorometer Direct Analysis, 0.1–10 × $10^{-9}$ Mole

*Reagent.* This is the same as in A except as follows: $TPN^+$, 50 $\mu M$; glucose, 100 $\mu M$; ADP (for P-creatine assay only), 100 $\mu M$ (30 $\mu M$ for less than 1 $\mu M$ P-creatine); hexokinase, 1 $\mu$g/ml; glucose-6-P dehydrogenase, the equivalent of 0.08 $\mu$g/ml of crystalline enzyme (0.03U/ml).

*Reaction Time.* Two to 3 min for glucose-6-P, 4–8 min for ATP, 15–20 min for P-creatine with 100 $\mu M$ ADP, 25–40 min with 30 $\mu M$ ADP.

*Conduct of the Assay*

See the spectrophotometer protocol.

*Comment*

Glucose and glucose-6-P dehydrogenase are reduced to decrease the progressive blank from the glucose dehydrogenase side activity (Chapter 5).

The ADP is decreased for the lowest P-creatine levels to decrease the blank from ATP contamination. If necessary, the ATP can be removed from the ADP as follows: To 1 ml of reagent in a fluorometer tube containing 0.1 *M* Tris-HCl, pH 8.0, 100 $\mu M$ glucose, 1 m*M* $Mg^{2+}$, 30 $\mu M$ $TPN^+$, and 1 m*M* ADP, add 0.14 U/ml of glucose-6-P dehydrogenase and 2 $\mu$g/ml of hexokinase. The ATP present will be converted to ADP, and TPNH will be formed in equivalent amount. After TPNH formation ceases it is destroyed by adding 20 $\mu$l of 5 *N* HCl, and then neutralizing after a few minutes with an equivalent amount of NaOH. The enzymes will also be destroyed, and the remaining components will not interfere with subsequent analyses. Stronger solutions of ADP can be treated in a similar fashion with TPNH formation followed in a spectrophotometer. If the ADP is 10 m*M*, the glucose should be increased to 1 m*M* and the $TPN^+$ to 0.3 m*M*. If the reaction is too slow the enzymes can be increased.

## C. Cycling, 1–10 × $10^{-12}$ Mole

(This and the other cycling methods for ATP and P-creatine are described for the analysis of frozen dried sections. See Comment below in regard to other applications.)

*ATP Reagent.* Tris-HCl buffer, pH 7.5 (10 m*M* Tris base, 40 m*M* Tris-HCl); $MgCl_2$, 1 m*M*; dithiothreitol, 0.5 m*M*; glucose, 100 $\mu M$; $TPN^+$,

20 $\mu M$; bovine plasma albumin, 0.02%; hexokinase, 2 $\mu$g/ml; glucose-6-P dehydrogenase, 0.15 U/ml.

*P-creatine Reagent.* Tris-HCl buffer, pH 8.6 (150 m$M$ Tris base, 50 m$M$ Tris-HCl); ADP, 80 $\mu M$; bovine plasma albumin, 0.02%; hexokinase, 4 $\mu$g/ml; glucose-6-P dehydrogenase, 0.35 U/ml; creatine kinase, 100 $\mu$g/ml.

*Cycling Reagent.* Basic TPN cycling reagent (Chapter 8) with 4 $\mu$g/ml of glutamic dehydrogenase and 0.25 U/ml of glucose-6-P dehydrogenase.

*Reagent for Final 6-P-Gluconate Measurement.* See Chapter 8.

*Reaction Vessels.* Tubes 4 × 50 mm, 2–2.5 mm i.d.

*Procedure*

*Step 1.* NaOH, 1 $\mu$l of 0.1 $N$ for samples and blanks, 1 $\mu$l of 3–10 $\mu M$ ATP or P-creatine in 0.1 $N$ NaOH for standards. Heat 10 min at 60°.

*Step 2.* Add 5 $\mu$l of ATP reagent; 20 min at room temperature. Transfer 3 $\mu$l to another tube for P-creatine analysis.

*Step 3.* To the portion for P-creatine add 1 $\mu$l of 0.125 $N$ HCl (to destroy TPNH); 10 min at room temperature. Add 3 $\mu$l of P-creatine reagent; 30 min at room temperature.

*Step 4.* To both the ATP and P-creatine samples add 3 $\mu$l of 0.2 $N$ NaOH, 15 min at 60°.

*Step 5.* Transfer samples ("total transfer," Chapter 3) to 200 $\mu$l volumes of cycling reagent in fluorometer tubes. Incubate 1 hr at 38°, 2 min at 100°.

*Step 6.* Add 1 ml of 6-P-gluconate reagent and read when reaction has stopped.

*Comment*

The bovine plasma albumin is added to protect the enzymes in the small volumes. The buffers used for the ATP and P-creatine are designed so that after addition to the sample the resulting pH will be close to 8. If the original sample is not treated with alkali (or with less alkali as in D below) a buffer at pH 8 would be used for the ATP reagent as in protocols A and B. The reaction times should be checked directly in the fluorometer using all reagents in 200-fold larger volumes.

D. CYCLING, 1–10 × $10^{-13}$ MOLE

*ATP Reagent.* This is the same as in C, but the buffer is 50 m$M$ Tris-HCl, pH 8.1 as in A and B.

*P-Creatine Reagent.* Tris-HCl buffer, pH 8.1 (100 m$M$ Tris base, 100 m$M$ Tris-HCl); ADP, 25 $\mu M$; $TPN^+$, 20 $\mu M$; glucose, 100 $\mu M$; bovine plasma albumin, 0.02%; hexokinase, 2 $\mu$g/ml; glucose-6-P dehydrogenase, 0.15 U/ml; creatine kinase, 125 $\mu$g/ml.

*Cycling Reagent.* Basic TPN cycling reagent (Chapter 8) with 100 $\mu$g/ml of glutamic dehydrogenase and 2.5 U/ml of glucose-6-P dehydrogenase.

*Reagent for 6-P-Gluconate.* See Chapter 8.

*Reaction Vessels.* Oil wells.

*Procedure*

*Step 1.* NaOH, 0.06 $\mu$l of 0.1 $N$, for samples and standards, 0.06 $\mu$l of 5–20 $\mu M$ ATP or P-creatine in 0.1 $N$ NaOH for standards; 20 min at 80°.

*Step 2.* ATP reagent, 0.6 $\mu$l, 20 min at room temperature.

*Step 3.* Transfer 0.3 $\mu$l into 5 $\mu$l of 0.05 $N$ NaOH in another oil well rack; heat 20 min at 80°. (These are the ATP samples.)

*Step 4.* To the remainder of the sample in the original rack add 0.1 $\mu$l of 0.2 $N$ HCl; 10 min at room temperature.

*Step 5.* Add 0.6 $\mu$l of P-creatine reagent, 45 min at room temperature.

*Step 6.* Add 5 $\mu$l of 0.05 $N$ NaOH to P-creatine samples, 20 min at 80°.

*Step 7.* Transfer 4 $\mu$l aliquots to 50 $\mu$l of cycling reagent in fluorometer tubes. Incubate 1 hr at 38°; 2 min at 100°.

*Step 8.* Add 1 ml of 6-P-gluconate reagent and read when reaction has stopped.

*Comment*

Because of the larger relative volume of P-creatine reagent here than in C, $TPN^+$ and glucose are added to the reagent. This is preferable to increasing the concentrations in the ATP reagent.

E. Cycling, 1–5 × $10^{-14}$ Mole

*ATP Reagent.* This is the same as in C except as follows: Tris buffer, pH 7.4 (10 m$M$ Tris, 50 m$M$ Tris-HCl); bovine plasma albumin, 0.03%.

*P-Creatine Reagent.* This is the same as in C except as follows: ADP, 60 $\mu M$; bovine plasma albumin, 0.06%; glucose-6-P dehydrogenase, 0.15 U/ml.

*Cycling Reagent.* Basic TPN cycling reagent (Chapter 8) with 400 $\mu$g/ml glutamic dehydrogenase and 12 U/ml of glucose-6-P dehydrogenase.

*Reagent for 6-P-Gluconate.* See Chapter 8.

*Reaction Vessels.* Oil wells.

*Procedure*

*Step 1.* NaOH, 0.025 $\mu$l of 0.1 *N* for samples and blanks; 0.025 $\mu$l of 1–4 $\mu M$ ATP and P-creatine, for standards; 20 min at 80°.

*Step 2.* Add 0.08 $\mu$l of ATP reagent; 15 min at room temperature.

*Step 3.* Transfer 0.05 $\mu$l into 0.2 $\mu$l of 0.1 *N* NaOH in another oil well rack, 20 min at 80°. (This is the portion for ATP.)

*Step 4.* Add 0.005 $\mu$l of 0.5 *N* HCl to the remaining part of the sample for P-creatine; 10 min at room temperature.

*Step 5.* Add 0.025 $\mu$l of P-creatine reagent to this portion; 45 min at room temperature.

*Step 6.* Add 0.2 $\mu$l of 0.1 *N* NaOH; 20 min at 80°.

*Step 7.* To all samples add 5 $\mu$l of cycling reagent. Incubate 2 hr at 38°; add 1 $\mu$l of 1 *N* NaOH; heat 10 min at 90°.

*Step 8.* Transfer a 4.5 $\mu$l aliquot into 1 ml of 6-P-gluconate reagent. (Make reading of the tubes before the transfer to increase precision.)

*Kinetics*

The $V_{max}$ of hexokinase is about 140 $\mu$moles/mg/min. The $K_m$ for ATP with 1 m$M$ glucose is 30 $\mu M$ and that for glucose with 300 $\mu M$ ATP is 160 $\mu M$. From this it may be calculated that with 10 $\mu M$ ATP or less and saturating glucose levels the half-time with 1 $\mu$g/ml of enzyme should be $0.7 \times 30/140 = 0.15$ min. With 100 $\mu M$ glucose the half-time should be 0.4 min [0.7(30/140)(260/100)]. Under assay conditions the rates are about half this fast.

The kinetic situation is much less favorable for P-creatine. The $V_{max}$ for creatine kinase of a commercial preparation under assay conditions was about 40 $\mu$moles/mg/min with a very large $K_m$ for P-creatine, 2.2. m$M$. With P-creatine in the analytical range the rates are even lower than this would indicate. With a saturating level of ADP the half-time for P-creatine with 1 $\mu$g/ml of creatine kinase would be about 60 min. The $K_m$ for ADP is approximately 150 $\mu M$, which constitutes no problem in assays for higher P-creatine levels, but is a problem at the lowest levels because of possible contamination of ADP with ATP.

## Citrate

$$\text{Citrate} \xrightarrow[\text{lyase}]{\text{citrate}} \text{oxalacetate} + \text{acetate} \qquad (9\text{-}11)$$

$$\text{Oxalacetate} + \text{DPNH} + \text{H}^+ \xrightarrow[\text{dehydrogenase}]{\text{malic}} \text{malate} + \text{DPN}^+ \qquad (9\text{-}12)$$

A method based on this principle was originally described by Moellering and Gruber (1966). Citrate lyase has the peculiarity of being inactivated by its own substrate. In the absence of $Mg^{2+}$ or $Zn^{2+}$ 30 $\mu M$ citrate will inactivate the enzyme completely in 5 min. The divalent cations protect the enzyme, but at high levels will inhibit by chelating the substrate. The reagent composition is a compromise between these extremes. At the low levels of citrate in the fluorometric assays, the inactivation problem is much less serious than it is at levels required in the spectrophotometer.

### A. Spectrophotometer

*Reagent*. Tris-HCl buffer, pH 7.6 (25 m$M$ in the free base, 75 m$M$ in the hydrochloride); DPNH, 50–150 $\mu M$ (at least a 20% excess); $ZnCl_2$, 40 $\mu M$; citrate lyase, 15 $\mu$g/ml (0.12 U/ml) (see "Comment" below); malic dehydrogenase (pig heart), 0.4 $\mu$g/ml (0.3 U/ml).

*Reaction Time*. Ten minutes.

*Conduct of the Assay*

The malic dehydrogenase can be incorporated in the reagent, but because of the danger of inactivation of the citrate lyase it is preferable to add the lyase last after the sample has been added and a preliminary reading made. If the reaction is not complete within 20 min it may stop because of enzyme inactivation by unreacted citrate. This would only be expected to occur if the enzyme stock solution had deteriorated. The weaker the enzyme the longer it is exposed to citrate, and therefore the greater the inactivation in the test system.

*Comment*

Citrate lyase, as a stock 0.5% solution, keeps for several weeks at 4° if dissolved in imidazole or triethanolamine buffer, pH 7–7.4, containing 300 $\mu M$ $ZnCl_2$. When necessary it is diluted further in either of these solutions. The enzyme is usually supplied as a mixture with a protective salt (e.g., $MgCl_2$).

In addition to citrate, oxalacetate and malate can also inactivate citrate lyase. However, malate, the final product, is not noticeably inhibitory at levels encountered during analyses.

B. FLUOROMETER DIRECT ASSAY, $0.1–8 \times 10^{-9}$ MOLE

*Reagent.* The same buffer is used as in A except that at lowest citrate levels the concentration may be cut in half to reduce the blank. DPNH, 0.2–10 $\mu M$ (20–75% excess over expected oxalacetate); $ZnCl_2$, 40 $\mu M$; citrate lyase, 2 $\mu$g/ml (0.016 U/ml); malic dehydrogenase, 0.4 $\mu$g/ml.

*Reaction Time.* Five to 15 min.

*Conduct of the Assay*

The citrate lyase is added last after an initial reading. At citrate levels under 5 $\mu M$ there is little danger of inactivation of the enzyme unless it is very weak and therefore exposed to substrate for a prolonged period.

*Kinetics*

Study of the kinetics of citrate lyase is complicated by the required metal which protects the enzyme against its substrate and at the same time chelates the citrate, thereby reducing its free concentration. Thus, increasing $Zn^{2+}$ concentration from 40 to 400 $\mu M$ prolongs the activity in the presence of 100 $\mu M$ citrate, but reduces the initial velocity 50%. With 1 $\mu M$ citrate the same increase in $Zn^{2+}$ concentration reduces initial velocity 80%. Under prescribed assay conditions the apparent Michaelis constant for citrate is about 15 $\mu M$. The metal concentration requirement is very low, at least at low citrate levels. With 1 $\mu M$ citrate the rate was essentially the same with 6 and 40 $\mu M$ $ZnCl_2$.

## Creatine

$$\text{Creatine} + \text{ATP} \xrightarrow[\text{kinase}]{\text{creatine}} \text{P-creatine} + \text{ADP} \qquad (9\text{-}13)$$

$$\text{ADP} + \text{P-pyruvate} \xrightarrow[\text{kinase}]{\text{pyruvate}} \text{ATP} + \text{pyruvate} \qquad (9\text{-}14)$$

$$\text{Pyruvate} + \text{DPNH} + \text{H}^+ \xrightarrow[\text{dehydrogenase}]{\text{lactic}} \text{lactate} + \text{DPN}^+ \qquad (9\text{-}15)$$

Because of the high Michaelis constants for both substrates, and low inherent activity in the assay direction, a large amount of creatine kinase and relatively high levels of ATP are required. This introduces problems which will be discussed below.

A. SPECTROPHOTOMETER

*Reagent.* Imidazole-HCl, pH 7.5 (40 m$M$ imidazole, 10 m$M$ imidazole-HCl); $MgCl_2$, 5 m$M$; KCl, 30 m$M$; DPNH, 50–150 $\mu M$ (at least a 20% excess); P-pyruvate, 200 $\mu M$; ATP, 1 m$M$; creatine kinase (rabbit muscle), 200 $\mu$g/ml (3.6 U/ml); pyruvate kinase (rabbit muscle), 5 $\mu$g/ml (0.75 U/ml); lactic dehydrogenase (beef heart), 1 $\mu$g/ml (0.2 U/ml).

*Reaction Time.* Ten to 20 min.

*Conduct of the Assay*

Lactic dehydrogenase and pyruvate kinase are ordinarily incorporated in the reagent. It is usually preferable to add the creatine kinase after the sample has been introduced and an initial reading made. In any event, the kinase should not be added to the reagent long before use, because of the slow splitting of ATP by present creatine kinase preparations.

*Comment*

ATP preparations inevitably contain some ADP. This will use up some of the DPNH in the reagent. If the contamination is not excessive, it can be ignored, or more DPNH added to compensate. However, if desirable most of the ADP can be removed (see below). Contamination of P-pyruvate with free pyruvate is ordinarily no problem because only a modest excess is required.

### B. Fluorometer Direct Assay, 1–8 × $10^{-9}$ Mole

*Reagent.* The reagent has the same composition as in A, except for the following: DPNH, 1–10 $\mu M$ (20–75% excess); P-pyruvate, 25 $\mu M$; ATP, 200 $\mu M$.

*Reaction Time.* Fifteen to 30 min.

*Conduct of the Assay*

The creatine kinase is added last. Because of the large amount used and the liberation of ADP from ATP, reagent blanks become very important especially at lowest levels. The timing of samples and blanks must be made the same.

*Comment*

The most serious problem is the slow ATP splitting by ATPase in current creatine kinase preparations. This may be a side reaction of creatine kinase itself. It appears to have about the same Michaelis constant for ATP. In any event, several lots have all given about the same rate of ATP cleavage, i.e., 1 part in 100,000 parts per minute with the amount of enzyme in the protocol (200 $\mu$g/ml). With 200 $\mu M$ ATP this yields 0.02 $\mu M$ ADP/min or 0.3 $\mu M$ in a 15 min incubation (30% of the reading for a 1 $\mu M$ sample, but only 4% of an 8 $\mu M$ sample). This limits the useful sensitivity and means that assays should be carefully timed, with final readings taken as soon as the reaction is complete. There is no gain in reducing the ATP concentration since this prolongs the incubation time without decreasing the proportionate blank.

A larger relative ATP excess is prescribed for the fluorometer in order to keep the assay time and enzyme requirement in bounds. This makes ADP contamination more serious. The best ATP preparations stored at neutral pH may be satisfactory except for the lowest concentrations. It is, however, possible to reduce the ADP almost to zero as follows. A 100 m*M* ATP solution is diluted with 10 volumes of reagent having the composition of that given in A above, except for omission of creatine kinase. Additional DPNH is added if necessary until a check in the spectrophotometer shows that DPNH consumption has ceased. The excess DPNH is then destroyed by adding 0.02 volume of 5 *N* HCl and allowing this to stand for 10 min (not more) at room temperature.

If the presence of $DPN^+$ is not objectionable the solution is simply neutralized with 0.02 volume of 5 *N* NaOH. Otherwise 0.04 volume of 5 *N* NaOH is added and the mixture is heated 10 min at 60° then finally neutralized with 0.02 volume of 5 *N* HCl.

*Kinetics*

The $V_{max}$ of creatine kinase in this direction is only 18 μmoles/mg/min. The Michaelis constant is at least 15 m*M* for creatine and 0.5–1 m*M* for ATP. Increase in one substrate decreases the apparent Michaelis constant for the other. With 200 μ*M* ATP the half-time at 25° is 2–3 min with 200 μg of creatine kinase per milliliter.

## Dihydroxyacetone Phosphate Method I (with glycero-P-dehydrogenase)

$$\text{Dihydroxyacetone-P} + \text{DPNH} + \text{H}^+ \xrightarrow[\text{dehydrogenase}]{\text{glycero-P-}} \text{glycero-P} + \text{DPN}^+$$

This is the simpler of the two methods presented, but the alternate, glyceraldehyde-P dehydrogenase method, has advantages for *direct* measurement of very low levels of dihydroxyacetone-P, particularly in the presence of fluorescent tissue extracts. For *indirect* measurements of very small amounts of the metabolite the glycero-P-dehydrogenase method has some advantages. (See Comment after the indirect procedures.)

A. Spectrophotometer

*Reagent.* Imidazole buffer pH 7.0 (30 m*M* imidazole, 20 m*M* imidazole-HCl); DPNH, 50–150 μ*M* (20–50% excess); glycero-P-dehydrogenase (rabbit muscle), 2 μg/ml (0.08 U/ml).

*Reaction Time.* Two to 4 min.

*Conduct of the Assay*

The enzyme is added separately after the sample has been added and an initial reading has been made.

### B. Fluorometer Direct Assay, 1–8 × $10^{-9}$ Mole

*Reagent.* Same as for spectrophotometer except: DPNH, 0.2–10 $\mu M$ (20–75% excess); glycero-P-dehydrogenase, 0.5 $\mu$g/ml (1 $\mu$g/ml with DPNH below 1 $\mu M$).

*Reaction Time.* Two to 8 min depending on final DPNH concentration (see "Kinetics" below).

*Conduct of the Assay*

Ordinarily the reaction is started by addition of the enzyme. Glycero-P-dehydrogenase strongly enhances the fluorescence of DPNH. Therefore, in the fluorometric assays the amount of enzyme is kept minimal. Correction for enhancement, if necessary, is made on the basis of the change in reading of blanks (containing the same amount of DPNH) upon addition of the enzyme. (With the maximal amount of enzyme indicated in protocol B, 1 $\mu$g/ml, enhancement is not more than 3% with 1 $\mu M$ DPNH.)

*Comment*

If the imidazole concentration is increased to 0.2 $M$ to quench flavin fluorescence of biological extracts, the rate of reaction will be decreased by a factor of 3 or 4. To compensate, the enzyme concentration is increased accordingly, or the reaction time is prolonged.

Because of the very favorable kinetic situation (see below), the reaction will follow nearly first-order kinetics except with very low DPNH concentrations. Nevertheless with lowest initial DPNH levels (0.2–0.5 $\mu M$) the rate will fall off badly if more than half of the DPNH is used up and care must be taken to increase the enzyme or the time of reaction.

### C. Indirect Assay, 1–10 × $10^{-11}$ Mole

*Reagent.* The buffer is the same as for spectrophotometer. DPNH, 2$\mu M$; bovine plasma albumin, 0.01%; glycero-P-dehydrogenase, 1 $\mu$g/ml.

The reagent should be prepared within an hour of use and kept in ice. The DPNH is as free as possible of $DPN^+$ (see Chapter 1).

*Procedure*

*Step 1.* Reagent volumes of 100 $\mu$l in fluorometer tubes at 0°. Samples and standards are added in volumes of 20 $\mu$l or less. Incubate 15 min at 20°–25° (water bath); return to ice bath.

*Step 2.* DPNH destroyed with 20 $\mu$l of 2 $N$ HCl (room temperature 5–10 min).

*Step 3.* NaOH, 1 ml of 6 $N$ (rapid mixing); 10 min at 60°.

*Step 4.* Fluorescence reading with tubes exactly at room temperature.

(In regard to tissue analyses, see Comment in next section.)

D. INDIRECT ASSAY, 0.5–10 × $10^{-12}$ MOLE (OIL WELLS)

*Specific Reagent.* The buffer is the same as for the spectrophotometer. DPNH, 1 $\mu M$ for 0.5–2 × $10^{-12}$ mole, 5 $\mu M$ for 10 × $10^{-12}$ mole; bovine plasma albumin, 0.02%; ascorbic acid, 2 m*M*; glycero-P-dehydrogenase, 1 $\mu$g/ml.

The ascorbic acid is added before the DPNH from a stock 9% (500 m*M*) solution which is stored at −20°. (See also directions for protocol C.)

*Cycling Reagent.* DPN cycling reagent (Chapter 8) designed for final glutamate measurement, i.e., $(NH_4)_2SO_4$ concentration reduced to 1 m*M*. Glutamic dehydrogenase and lactic dehydrogenase are 400 $\mu$g/ml and 50 $\mu$g/ml, respectively. [The latter is centrifuged before use to remove most of the $(NH_4)_2SO_4$]

*Glutamate Reagent.* See Chapter 8.

*Reaction Vessels.* Oil wells.

*Procedure*

*Step 1.* The sample in 1 $\mu$l or less is added to 5 $\mu$l of reagent in an oil well and incubated 20 min at room temperature. Standards consist of 1 $\mu$l of 1–10 $\mu M$ dihydroxyacetone-P, depending on range of samples.

*Step 2.* Add 2 $\mu$l of 0.5 *N* HCl and allow to stand at least 30 min.

*Step 3.* For samples less than 2 × $10^{-12}$ mole transfer 5 $\mu$l into 100 $\mu$l of cycling reagent in a fluorometer tube. In the case of larger samples use 1 $\mu$l aliquots of samples, standards, and blanks. Incubate 1 hr at 38°; heat 3 min at 100°.

*Step 4.* Add 1 ml of glutamate reagent. After 20 min add 5 $\mu$l of 1% glutamic dehydrogenase in 50% glycerol. Read after 20–30 min.

*Comment*

If the samples are not all 1 $\mu$l, a correction is made in the calculation for the differences in volume at the end of Step 2.

E. INDIRECT ASSAY, 2–5 × $10^{-13}$ MOLE

*Specific Reagent.* The same as in D, except that all components are double strength; DPNH is 4 $\mu M$.

*Cycling Reagent.* As in D.

*Glutamate Reagent.* As in D.

*Reaction vessels.* Oil Wells

*Procedure*

*Step 1.* To 0.25 μl volumes of samples, blanks, and standards (1–2 μ*M*) in oil wells are added 0.25 μl volumes of reagent. Incubate 20 min at room temperature.

*Step 2.* Add 0.2 μl of 0.5 *N* HCl and allow to stand at least 20 min.

*Step 3.* Add 10 μl of cycling reagent to each oil well and incubate 2 hr at 38°.

*Step 4.* Add 1 μl of 1 *N* NaOH to each sample; heat the rack 5 min at 95°.

*Step 5.* Transfer 10 μl aliquots to fluorometer tubes containing 100 μl of glutamate reagent. Incubate 20 min then add 1 μl of 0.5% glutamic dehydrogenase. After 30 min add 1 ml of $H_2O$, mix, and read.

*Comment*

In the indirect assays of C and D, the final analysis is for $DPN^+$. In general, dihydroxyacetone-P levels in tissues and microorganisms are much lower than $DPN^+$ levels. Although $DPN^+$ can be destroyed with alkali, dihydroxyacetone-P will be destroyed even more rapidly. Although one can imagine enzymatic ways of destroying $DPN^+$ alone (reduction to DPNH, cleavage by DPNase or pyrophosphatase) for analysis of very small tissue samples a more satisfactory solution would appear to use the alternative assay with glyceraldehyde-P dehydrogenase. In this case, the final analysis is for DPNH, which is easily destroyed in the tissue samples with acid without harming dihydroxyacetone-P.

*Kinetics*

The kinetic values are as follows: $V_{max}$, 40 U/mg; $K_{DHAP}$, 6 μ*M*; $K_{DPNH}$, 0.3 μ*M*. [In 20 m*M* imidazole buffer, pH 7. The $V_{max}$ is that found with current commercial preparations. The pure enzyme may be more active. Baranowski (1963) reported a value of 260 U/ml.]

At equilibrium, at pH 7,

$$\frac{[GOP]\,[DPN^+]}{[DHAP][DPNH]} = 12{,}500$$

Thus, the low Michaelis constants for both substrates and the equilibrium constant are all very favorable for reduction of dihydroxyacetone-P. Even a slight excess of DPNH is sufficient to drive the reaction to completion.

From the kinetic constants it may be calculated that for the reaction to be

98% complete in 10 min would require only 0.1 μg/ml with 10 μ*M* substrate or less, and only 0.3 μg/ml with 100 μ*M* substrate [Chapter 2, Fig. (2–4)]. Both sulfate and phosphate are inhibitory.

## Fructose

$$\text{Fructose} + \text{ATP} \xrightarrow{\text{hexokinase}} \text{fructose-6-P} + \text{ADP} \quad (9\text{-}16)$$

$$\text{Fructose-6-P} \xrightarrow[\text{isomerase}]{\text{P-gluco-}} \text{glucose-6-P} \quad (9\text{-}17)$$

$$\text{Glucose-6-P} + \text{TPN}^+ \xrightarrow[\text{dehydrogenase}]{\text{glucose-6-P}} \text{6-P-gluconolactone} + \text{TPNH} + \text{H}^+ \quad (9\text{-}18)$$

Both fructose and glucose are substrates for yeast hexokinase, but for fructose analysis more enzyme is required because the Michaelis constant is higher than for glucose.

### A. Spectrophotometer

*Reagent*. Tris-HCl buffer, pH 8.1 (25 m*M* Tris base, 25 m*M* Tris-HCl); $MgCl_2$, 2 m*M*; ATP, 500 μ*M*; $TPN^+$, 500 μ*M*; hexokinase (yeast), 5 μg/ml (0.7 U/ml); P-glucoisomerase (yeast), 1 μg/ml (0.35 U/ml); glucose-6-P dehydrogenase (yeast) equivalent to 0.3 μg/ml of crystalline enzyme (0.12 U/ml).

*Reaction Time*. Ten minutes.

*Conduct of the Assay*

If glucose is known to be absent and the sample does not contribute appreciably to the light absorption, all enzymes can be incorporated in the reagent and the sample added last after an initial reading. If glucose is also present, and its value required, only glucose-6-P dehydrogenase is added to the reagent. After adding the sample and making a reading, hexokinase is added and a second reading made after all the glucose has reacted (5 min or less). P-Glucoisomerase is then added to complete the reaction (5 min or less since fructose has been largely converted to fructose-6-P in the interim).

*Comment*

There is some danger of contamination of hexokinase or glucose-6-P dehydrogenase with P-glucoisomerase. Even the buffer can become contaminated with this enzyme through the growth of microorganisms. This can be detected by prolonged incubation of glucose with the isomerase omitted. Glucose contamination can also readily occur, but will not distort the fructose value if the assay is made in two steps.

B. FLUOROMETER, 0.1–10 × $10^{-9}$ MOLE

*Reagent.* The buffer, $MgCl_2$, hexokinase, and P-glucoisomerase are the same as in A. ATP, 200 $\mu M$; $TPN^+$, 50 $\mu M$; dithiothreitol, 200 $\mu M$; glucose-6-P dehydrogenase, 0.06 U/ml.

*Reaction Time.* Ten minutes.

*Conduct of the Assay*

Because of the danger of at least trace glucose contamination it is almost always preferable to incubate the sample for 5 or 10 min with glucose-6-P dehydrogenase and hexokinase and take a reading before adding P-gluco-isomerase (see part A above).

*Comment*

Dithiothreitol is added as a precaution in case of contamination with glutathione and glutathione reductase (see the glucose-6-P method). The danger of contamination with P-glucoisomerase has been mentioned above.

*Kinetics*

The $V_{max}$ for fructose (220 $\mu$moles/mg/min) is almost twice as great as that for glucose, but the $K_m$ is four to six times greater (700 $\mu M$). Therefore, three times as much enzyme (or time) is required for fructose as for glucose to react completely. Because the $K_m$ is large, the same amount of hexokinase is appropriate in the spectrophotometer and in the fluorometer.

## Fructose-6-Phosphate

$$\text{Fructose-6-P} \xrightarrow[\text{isomerase}]{\text{P-gluco}} \text{glucose-6-P} \qquad (9\text{-}19)$$

$$\text{Glucose-6-P} + TPN^+ \xrightarrow[\text{dehydrogenase}]{\text{glucose-6-P}} \text{6-P-gluconolactone} + TPNH + H^+ \qquad (9\text{-}20)$$

In direct assays fructose-6-P can be measured by first allowing reaction (9-20) to proceed to completion and then adding isomerase to give extra TPNH equivalent to the fructose-6-P present. However, the equilibrium position of reaction (9-19) favors glucose-6-P by a ratio of about 3 : 1. Consequently fructose-6-P levels in living systems are almost invariably much lower than those of glucose-6-P. This increases the potential error in fructose-6-P measurement. An alternate procedure is to destroy with acid the TPNH produced from glucose-6-P and then measure fructose-6-P alone. To illustrate both procedures, the simpler method is given for the spectrophotometer (since it will usually be used only for standardization purposes), and the two-step procedure is given for the fluorometer.

### A. Spectrophotometer

*Reagent.* Tris buffer, pH 8.1 (25 m$M$ Tris base, 25 m$M$ Tris-HCl); $TPN^+$, 500 $\mu M$; P-glucoisomerase (yeast), 1 $\mu$g/ml (0.35 U/ml); glucose-6-P dehydrogenase (yeast) equivalent to 0.3 $\mu$g/ml of crystalline enzyme (0.12 U/ml).

*Reaction Time.* Ten minutes.

*Conduct of the Assay*

The sample is added first and a reading made. Glucose-6-P dehydrogenase is added and a second reading is made after glucose-6-P has reacted completely (5 min or less). Isomerase is added last. Because, as stated above, the second increment is less than the first, any drift in reading will cause substantial error. It is therefore desirable to carry through an extra sample which receives only the glucose-6-P dehydrogenase. Any change in reading of this sample during the period of isomerase action can be used as a correction.

### B. Fluorometer, 0.1–8 × $10^{-9}$ Mole

*Reagent.* The buffer is the same as in A. $TPN^+$, 50 $\mu M$; dithiothreitol, 200 $\mu M$; glucose-6-P dehydrogenase, 0.06 U/ml.

*Reaction Time.* Five to 10 min.

*Conduct of the Assay*

*Step 1.* If a value for glucose-6-P is required, readings are made before and after addition of glucose-6-P dehydrogenase.

*Step 2.* TPNH is destroyed by adding 10 $\mu$l of 5 $N$ HCl (0.05 $N$ final concentration). After standing 10 min, 10 $\mu$l of 5 $N$ NaOH is added.

*Step 3.* Glucose-6-P dehydrogenase is added to give the same concentration as in the original reagent. Ordinarily, the fluorometer sensitivity would by increased two- or threefold at this point. After reading, P-glucoisomerase is added to give a concentration of 0.5 $\mu$g/ml. A final reading is made when the reaction is over (6–10 min).

*Comment*

As discussed more fully for glucose-6-P measurement, dithiothreitol is added to guard against the danger of TPNH reoxidation by oxidized glutathione reductase. This danger is lessened by the three-step procedure described. Another danger which must be guarded against is that of P-glucoisomerase contamination during the first step. P-Glucoisomerase is of high activity in most biological materials. It is essential to test a fructose-6-P standard with first step reagent.

Fructose-6-P can be measured successfully without the acidification step if special care is taken in making the reading and in checking for drift after glucose-6-P has reacted, as described for the spectrophotometric procedure.

*Kinetics*

The $V_{max}$ of yeast P-glucoisomerase is high, about 350 $\mu$moles/mg/min, and the Michaelis constant is of the order of 50 $\mu$M. Therefore, the enzyme requirement is low and only slightly more enzyme is required for spectrophotometer levels than for those in the fluorometer. If glucose-6-P is removed as fast as it is formed the calculated half-time for 1 $\mu$g/ml of isomerase ($V_{max}$ 350 $\mu$moles/$L$/min) would be 0.1 min. (0.7 × 50/350). 6-P Gluconate is a potent inhibitor, which increases the enzyme requirement.

## Fructose-1,6-Diphosphate: Method I

$$\text{Fructose-P}_2 \xrightarrow{\text{aldolase}} \text{dihydroxyacetone-P} + \text{glyceraldehyde-P} \quad (9\text{-}21)$$

$$\text{Glyceraldehyde-P} \xrightarrow[\text{isomerase}]{\text{triose}} \text{dihydroxyacetone-P} \quad (9\text{-}22)$$

$$2 \text{ Dihydroxyacetone-P} + 2 \text{ DPNH} \xrightarrow[\text{dehydrogenase}]{\text{glycero-P-}} 2 \text{ glycero-P} + 2 \text{ DPN}^+ + 2\text{H}^+ \quad (9\text{-}23)$$

Fructose diphosphate can be measured with the aid of either glycero-P-dehydrogenase or glyceraldehyde-P dehydrogenase. Each has advantages in certain situations.

### A. Spectrophotometer

*Reagent.* Imidazole, pH 7 (30 m$M$ imidazole, 20 m$M$ imidazole-HCl); DPNH, 50–150 $\mu$M (20–50% excess); aldolase (rabbit muscle), 2 $\mu$g (0.02 U/ml); triose-P isomerase (rabbit muscle), 0.5 $\mu$g/ml (1.2 U/ml); glycero-P-dehydrogenase (rabbit muscle), 4 $\mu$g/ml (0.16 U/ml).

*Reaction Time.* Two to 4 min.

*Conduct of the Assay*

The enzymes can be added stepwise to measure successively dihydroxyacetone-P, glyceraldehyde-P, and fructose diphosphate. If fructose diphosphate alone is to be measured, glycero-P-dehydrogenase and triose-P isomerase can be incorporated into the reagent.

### B1. Fluorometer, Direct Assay, 0.05–4 × $10^{-9}$ Mole

*Reagent.* The buffer and triose-P isomerase are the same as for the spectrophotometer. DPNH, 0.2–10 $\mu M$ (20–70% excess); aldolase, 0.4 $\mu$g/ml; glycero-P-dehydrogenase, 1 $\mu$g/ml.

*Reaction Time.* Two to 5 min.

*Conduct of the Assay*

As for spectrophotometer.

*Comment*

Because glycero-P-dehydrogenase enhances the fluorescence of DPNH, the concentration is kept relatively low. At the recommended concentration of 1 $\mu$g/ml enhancement is about 3% with 1 $\mu M$ DPNH, and less than this at higher DPNH concentrations.

B2. Fluorometer, Indirect Assays

It is simple enough to design indirect assays for fructose diphosphate with almost any desired sensitivity. However, with the use of glycero-P-dehydrogenase for the final step, $DPN^+$ is the product ultimately measured. Most living cells contain much more $DPN^+$ than fructose diphosphate. Therefore, to measure smaller amounts of fructose diphosphate, the alternative procedure with glyceraldehyde-P dehydrogenase as the auxiliary reaction is recommended.

When the presence of $DPN^+$ is not a problem, sensitive indirect procedures can follow the plan given for dihydroxyacetone-P (Method I), using for the first steps a reagent of the composition given in B1, but with added ascorbic acid (2 m$M$) and bovine plasma albumin (0.02%) as needed.

*Kinetics*

In imidazole buffers at pH 7 the $V_{max}$ for aldolase is relatively low, about 10 $\mu$moles/mg/min (10 U/mg), but this is compensated for by an unusually small $K_m$, 0.6 $\mu M$. Therefore very little enzyme is needed for assay of low levels of fructose diphosphate, but the amount must be increased at higher levels.

The $V_{max}$ for triose isomerase is very large, 2400 $\mu$moles/mg/min, but this is offset somewhat by the large $K_m$, about 50 $\mu M$. The kinetics of glycero-P dehydrogenase are given in the first method for dihydroxyacetone-P.

## Fructose-1,6-Diphosphate (Method II) Dihydroxyacetonephosphate (Method II), and Glyceraldehydephosphate

$$\text{Fructose-1,6-P}_2 \xrightarrow{\text{aldolase}} \text{dihydroxyacetone-P} + \text{glyceraldehyde-P} \quad (9\text{-}24)$$

$$\text{Dihydroxyacetone-P} \xrightarrow[\text{isomerase}]{\text{triose-P}} \text{glyceraldehyde-P} \quad (9\text{-}25)$$

$$\text{Glyceraldehyde-P} + \text{DPN}^+ \xrightarrow[\text{GAP dehydrogenase}]{\text{arsenate}} \text{3-P-glycerate} + \text{DPNH} + \text{H}^+ \quad (9\text{-}26)$$

Compared to the alternative methods for these compounds (with glycero-P-dehydrogenase) these methods have the advantage that readings increase with increasing amounts of substance. Each substrate can be measured in succession, if desired, although this may be impractical when the concentrations are widely different. Of the three, glyceraldehyde-P is the most difficult to measure in most biological materials because the levels are exceedingly low. (The equilibrium ratio between the two triose phosphates is about 25:1 in favor of dihydroxyacetone-P.) For standardization purposes, the glycerol-P dehydrogenase methods are usually preferable, since the reaction is much faster.

A. SPECTROPHOTOMETER

*Reagent.* Imidazole buffer, pH 7.5 (40 m$M$ imidazole, 10 m$M$ imidazole-HCl); $DPN^+$, 1 m$M$; $Na_2HAsO_4$, 1 m$M$; EDTA, 1 m$M$; mercaptoethanol, 2 m$M$; aldolase (rabbit muscle), 10 $\mu$g/ml (0.09 U/ml); triose-P-isomerase (rabbit muscle), 1 $\mu$g/ml (2.4 U/ml); glyceraldehyde-P dehydrogenase (rabbit muscle), 50 $\mu$g/ml (1.8 U/ml measured in the direction of glyceraldehyde-P formation).

The dehydrogenase stock suspension is centrifuged to remove most of the $(NH_4)_2SO_4$ and dissolved in buffer.

*Reaction Time.* Five to 10 min for glyceraldehyde-P, 20 to 25 min for dihydroxyacetone-P, and only a little longer for fructose diphosphate.

*Conduct of the Assay*

For standardization purposes the substrate concerned can be added last to the reagent containing the necessary enzymes. Otherwise, the sample is added first and the enzymes are added in succession with sufficient time after each addition to complete that step. One mole of fructose diphosphate, of course, yields 2 moles of DPNH.

*Comment*

Glyceraldehyde-P readily forms Schiff bases which may dissociate slowly. Consequently standard solutions may react in a biphasic manner with a rapid initial phase followed by a slower phase. In analyzing glyceraldehyde-P solutions it is necessary to be aware of the slightest contamination with triose isomerase. The presence of a trace of this enzyme would cause part of the substrate to be converted to dihydroxyacetone-P. Once formed, dihydroxyacetone-P would not be converted back to glyceraldehyde-P fast enough to be measured unless a very much larger amount of triose-P isomerase were to be added. The safe procedure is, in fact, to add triose-P isomerase, at the level shown above, after the dehydrogenase reaction has stopped, to see if there is a second increment.

The $(NH_4)_2SO_4$ is removed from the dehydrogenase because sulfate is inhibitory to both triose-P isomerase and glyceraldehyde-P dehydrogenase. It should not exceed 2 m*M* during the reaction.

B. Fluorometer Direct Analysis, 0.1–10 × $10^{-9}$ Mole (As Triose-P)

*Reagent.* The reagent is that given in A with the following changes: $DPN^+$, 400 $\mu M$ (100 $\mu M$ with 1 $\mu M$ substrate or less), aldolase, 2 μg/ml; glyceraldehyde-P dehydrogenase, 20 μg/ml.

*Reaction Times.* Two minutes for glyceraldehyde-P, 10–15 min for dihydroxyacetone-P, and 15–20 min for fructose diphosphate.

*Conduct of the Assay*

The sample is added first, followed by glyceraldehyde-P dehydrogenase, triose isomerase, and aldolase in succession, with delay between additions as indicated by the reaction times of appropriate standards. If fructose diphosphate is present in substantially higher amount than the triose phosphates, it may be desirable to use a more sensitive setting for the first two steps and to then reduce the sensitivity for the final aldolase step.

C. Fluorometer Indirect Analysis, 2–20 × $10^{-12}$ Mole (As Triose-P)

The method given is for the sum of fructose diphosphate and triose phosphates. For measuring these metabolites individually see Comment below.

*Specific Reagent.* The reagent is the same as in A except as follows: $DPN^+$, 30 μM; bovine plasma albumin, 0.02%.

*Cycling Reagent.* Basic cycling reagent (Chapter 8) designed for final glutamate measurement, i.e., $(NH_4)_2SO_4$ concentration reduced to 1 m*M*. Glutamic dehydrogenase, 100 μg/ml; lactic dehydrogenase, 4 μg/ml.

For last step use glutamate reagent (see Chapter 8).

*Reaction Vessels.* Oil wells or tubes 4 × 50 mm, 2–2.5 mm i.d.

*Procedure for Frozen Dried Tissue Samples*

*Step 1.* For samples and blanks, 1 μl of 0.02 *N* HCl; 1 μl of 1–10 $\mu M$ fructose diphosphate for standards, heat 10 min at 60° (20 min at 80° for oil wells).

*Step 2.* Specific reagent, 5 μl; 40 min at room temperature.

*Step 3.* NaOH, 5 μl of 0.1 *N*; 15 min at 60° (20 min at 80° in the case of oil wells).

*Step 4.* A 5 μl aliquot added to 100 μl of cycling reagent in fluorometer. Incubate 1 hr at 38°, heat 2 min at 100° (10 min at 95° for oil wells).

*Step 5.* Add 1 ml of glutamate reagent. After 20 min add 5 μl of 1% glutamic dehydrogenase in 50% glycerol. Read after 20–30 min.

*Comment*

Glyceraldehyde-P, or the sum of both triose phosphates, can be obtained by using glyceraldehyde-P dehydrogenase alone or in combination with triose-P isomerase. Fructose diphosphate can be measured alone if the samples are made alkaline after the acid treatment to destroy the triose phosphates. For example, after the acid treatment in Step 1, an equal volume (1 μl) of 0.04 *N* NaOH is added and the samples are heated 5 min at 80°.

Because glyceraldehyde-P levels are exceedingly low in living cells, the indirect procedure with glyceraldehyde-P dehydrogenase and amplification by cycling is the method of choice. However, great care must be taken if true values are to be obtained. The dehydrogenase is not very specific, and the possibility that other compounds are contributing to the result should be examined critically.

D. FLUOROMETER INDIRECT ASSAY, 1–10 × $10^{-13}$ MOLE (AS TRIOSE-P)

*Specific Reagent.* The same as in protocol C but all components double strength.

*Cycling Reagent.* As in protocol C but 400 μg/ml glutamic dehydrogenase and 50 μg/ml lactic dehydrogenase.

For final glutamate reagent see Chapter 8.

*Procedure for Frozen Dried Tissue Samples*

*Step 1.* HCl, 0.05 μl of 0.02*N*, for samples and blanks, and 0.05 μl of 2–10 μ*M* FDP for standards; heat 20 min at 80°.

*Step 2.* Specific reagent, 0.05 μl; 40 min at room temperature.

*Step 3.* NaOH, 0.5 μl of 0.1 *N*; 20 min at 80°.

*Step 4.* Add 10 μl of cycling reagent to each oil well and incubate 1 hr at 38°, heat 10 min at 95°.

*Step 5.* Transfer 8 μl to 1 ml of glutamate reagent in a fluorometer tube and complete reaction as in C.

*Kinetics*

The kinetic constants for glyceraldehyde-P dehydrogenase are favorable. The $V_{max}$ is 10 μmoles $mg^{-1}$ $min^{-1}$ and the apparent $K_m$ is about 10 μ*M* under analytical conditions. Thus with a concentration of 1 μg ml (10 μmoles/liter/min) the half-time for low levels of substrate is about 0.7 min. Probably because of partial dissociation, some loss of activity occurs on dilution. This is lessened but not overcome by the presence of $DPN^+$ (Velick and Furfine, 1963). The $K_m$ for $DPN^+$ is about 30 $\mu M$ under analytical conditions. Sulfate is inhibitory; a 5 m$M$ concentration inhibits the rate 65% (tested with 2 $\mu M$ glyceraldehyde-P).

With dihydroxyacetone-P analysis the problem is not triose-P isomerase. This is a very active enzyme, 2400 U/mg measured in the fast direction (dihydroxyacetone-P formation). It is perhaps a third as fast in the reverse direction and the $K_m$ is of the order of 250 $\mu M$. The problem is that, because of the unfavorable equilibrium constant, no matter how much isomerase is added, there can never be more than the 4% of the triose-P present as glyceraldehyde-P. Therefore, the conversion of dihydroxyacetone-P to 3-P-glycerate must be at least 25 times slower than for glyceraldehyde-P, and a high level of dehydrogenase is needed. Triose-P isomerase is inhibited 50% by 5 m$M$ $P_i$ and 65% by 5 m$M$ sulfate (tested with 2 $\mu M$ dihydroxyacetone-P).

The kinetics of aldolase have been given in Method I for fructose-1,6-diphosphate. The equilibrium of the aldolase step is concentration-dependent and at analytical levels is in favor of the triose phosphates. Hence with moderate amounts of aldolase the reaction is practically as fast as with dihydroxyacetone-P. Thus the rate-limiting step for all three substrates turns out to be glyceraldehyde-P dehydrogenase.

## Fumarate

$$\text{Fumarate} \xrightarrow{\text{fumarase}} \text{malate} \tag{9-27}$$

$$\text{Malate} + DPN^+ \xrightarrow[\text{dehydrogenase}]{\text{malic}} \text{oxalacetate} + DPNH + H^+ \tag{9-28}$$

$$\text{Oxalacetate} + \text{glutamate} \xrightarrow[\text{transaminase}]{\text{glut-oxal}} \text{aspartate} + \alpha\text{-ketoglutarate} \tag{9-29}$$

### A. Spectrophotometer

*Reagent.* Except for the addition of fumarase, the reagent is the same as that given for the spectrophotometer in Method I for malate.

Fumarase (pig heart), 50 $\mu$g/ml (18 U/ml, measured in the opposite direction). If the enzyme is supplied as suspension in $(NH_4)_2SO_4$, it is centrifuged and dissolved in neutral buffer to remove most of the $NH_4^+$.

*Reaction Time.* Eight to 15 min.

*Conduct of the Assay*

For standardization purposes the fumarate can be added last after a preliminary reading. For other purposes the reagent is prepared with the enzymes omitted. The sample is added; after a preliminary reading malic dehydrogenase and the transaminase are added. After allowing time for any malate present to react (5 or 10 min), a second reading is made and fumarase is then added to measure fumarate itself. Because the enzymes may contribute small increments to the reading, blanks are needed. All of the enzymes slowly lose activity at the alkaline pH, therefore they should not be left in the reagent for more than 20–30 min before starting the reaction.

*Comment*

Fumarate levels in biological materials are usually much lower than those of malate. At equilibrium in animal tissues the ratio would be about 1 to 4. Therefore to measure accurately the increment in absorption due to fumarate it is necessary to be sure the malate reaction is complete before adding fumarate. The best procedure may be to prepare extra samples which are treated identically except that fumarase is omitted. In order not to decrease the pH, the $(NH_4)_2SO_4$ contributed by the enzyme preparations should not exceed 10 m*M* (see Malate; Method I). (See also Comment for the fluorometric procedure.)

B. Fluorometer Direct Analysis, 0.1–10 × $10^{-9}$ Mole

*Reagent.* Except for fumarase the reagent is the same as that given for the fluorometer in Method I for malate.

Fumarase, 20 μg/ml (most of the $NH_4^+$ is removed as for the spectrophotometer).

*Reaction Time.* Ten to 20 min.

*Conduct of the Assay*

See spectrophotometric method. If fumarate is very low compared to malate, precision can be increased by destroying the DPNH from the malate as in the next protocol.

C. Fluorometer, Two-Step Direct Assay

*Reagent.* The first-step reagent is the same as in B except that fumarase is omitted and no more than 3 m*M* glutamate is used.

*Conduct of the Assay*

*Step 1.* The malate reaction is carried out as in the one-step procedure and readings are made if the value of malate is required.

*Step 2.* First 15 μl of 5 *N* HCl are added. After 10 min 15 μl of 5 *N* NaOH are added.

*Step 3.* Malic dehydrogenase and glutamic–oxalacetic transaminase are added first to make sure that no malate remains. A reading is made after 5 or 10 min, after which fumarase is added to complete the reaction.

*Comment*

The HCl and NaOH should be exactly equivalent. To be sure of this they are titrated against each other and the concentrations adjusted if necessary. The same micropipette can be used for acid and base additions. To destroy quickly the DPNH from the malate, sufficient HCl must be added to reach pH 2 or less. Glutamate and sulfate increase the HCl required. The adequacy can be checked by following the rate of decrease in fluorescence of one sample upon acidification.

*Kinetics*

In order to favor the malic dehydrogenase step the analytical reaction is conducted at pH 9.9, which is far from the pH optimum for fumarase (about 6.5; Massey and Alberty, 1954). Under analytical conditions with a commercial preparation we observed the $V_{max}$ to be 12 $\mu$moles/mg/min and the $K_m$ 1250 $\mu M$. Thus with 1 $\mu$g/ml of enzyme the half-time is about 70 min $(0.7 \times 1250/12)$. For these reasons a relatively high level of enzyme is required.

The kinetics for malic dehydrogenase and transaminase have been described in Malate: Method I.

## Glucose

$$\text{Glucose} + \text{ATP} \xrightarrow{\text{hexokinase}} \text{glucose-6-P} + \text{ADP} \tag{9-30}$$

$$\text{Glucose-6-P} + \text{TPN}^+ \xrightarrow[\text{dehydrogenase}]{\text{glucose-6-P}} \text{6-P-gluconolactone} + \text{TPNH} + \text{H}^+ \tag{9-31}$$

A. SPECTROPHOTOMETER

*Reagent.* Tris-HCl buffer, pH 8.1 (25 m$M$ Tris base, 25 m$M$ Tris-HCl); $MgCl_2$, 1 m$M$; ATP, 500 $\mu M$; $TPN^+$, 500 $\mu M$; hexokinase (yeast), 2 $\mu$g/ml (0.28 U/ml); glucose-6-P dehydrogenase (yeast), equivalent to 0.2 $\mu$g/ml of crystalline enzyme (0.08 U/ml).

*Reaction Time.* Four to 6 min.

*Conduct of the Assay*

For standardization purposes both enzymes can be incorporated in the reagent and the reaction started with glucose. Otherwise hexokinase is omitted from the reagent and added last after an initial reading. If glucose-6-P is present the initial reading is not made until time for it to react (2–4 min).

## B. Fluorometer Direct Assay, $0.1–10 \times 10^{-9}$ Mole

*Reagent*. The buffer and $MgCl_2$ are the same as in A. Dithiothreitol, 0.5 m$M$; ATP, 300 $\mu M$; $TPN^+$, 30 $\mu M$; hexokinase, 1 $\mu$g/ml; glucose-6-P dehydrogenase, 0.02 U/ml.

*Reaction Time*. Five to 10 min.

### *Conduct of the Assay*

Ordinarily glucose-6-P dehydrogenase is incorporated into the reagent. The sample is added; after allowing a few minutes for any glucose-6-P to react a reading is made and hexokinase is added. If glucose-6-P is to be measured in the same sample, both enzymes are omitted from the reagent. The sample is added first followed by the glucose-6-P dehydrogenase. Usually glucose-6-P levels are much lower than glucose levels. Therefore, the readings before and after the dehydrogenase addition are made at a relatively sensitive setting. Before adding hexokinase the sensitivity is reduced according to the glucose range expected and a new reading made, preferably with a stronger quinine standard. The glucose reaction is then initiated with hexokinase.

## C. Cycling Assay, $1–10 \times 10^{-12}$ Mole

*Reagent for Glucose Reaction*. Buffer, $MgCl_2$ and enzyme concentrations as in B. ATP, 100 $\mu M$; $TPN^+$, 5 $\mu M$; bovine plasma albumin, 0.02%.

*Cycling Reagent*. Basic TPN cycling reagent (Chapter 8) with 4 $\mu$g/ml glutamic dehydrogenase and 0.35 U/ml glucose-6-P dehydrogenase (cycling rate about 1000 hr).

For reagent for 6-P-gluconate measurement see Chapter 8.

*Reaction Vessels*. Tubes $4 \times 50$ mm, 2–2.5 mm i.d.

### *Procedure for Frozen Dried Tissue Samples*

*Step 1*. HCl, 1 $\mu$l of 0.02 $N$, for samples and blanks; 1 $\mu$l of 10 $\mu M$ glucose in 0.02 $N$ HCl for standards; 10 min at 60°.

*Step 2*. Glucose reagent, 5 $\mu$l; 20 min at room temperature.

*Step 3*. NaOH, 5 $\mu$l of 0.1 $N$; 15 min at 60°.

*Step 4*. Whole sample transferred into 200 $\mu$l of cycling reagent in a fluorometer tube. Incubate 1 hr at 38°, heat 2 min at 100°.

*Step 5*. Add 1 ml of 6-P-gluconate reagent and read when reaction is complete.

*Comment*

The acid treatment is sufficient to destroy preformed TPNH as well as tissue enzymes, but too mild to hydrolyze an appreciable amount of glycogen. It may, however, partially break down UDP-glucose to glucose and UDP. For analysis of glucose samples in which enzyme destruction is unnecessary, the acid treatment step can of course be omitted.

D. CYCLING ASSAY, 1–10 × $10^{-13}$ MOLE

*Reagent for Glucose Reaction.* The same as in C except: ATP, 200 $\mu M$; $TPN^+$, 20 $\mu M$.

*Cycling Reagents.* Basic cycling reagent of Chapter 8 with 100 $\mu$g/ml glutamic dehydrogenase and 2.3 U/ml of glucose-6-P dehydrogenase.

For reagent for 6-P-gluconate measurement see Chapter 8.

*Reaction Vessels.* Oil wells.

*Procedure for Frozen Dried Tissue Samples*

*Step 1.* HCl, 0.05 $\mu$l of 0.02 *N*, for samples and blanks; 0.05 $\mu$l of 2 to 20 $\mu M$ glucose in 0.02 N HCl for standards; 20 min in 80° oven.

*Step 2.* Glucose reagent, 0.5 $\mu$l, 20 min at room temperature.

*Step 3.* NaOH, 5 $\mu$l of 0.05 *N*; 20 min in 80° oven.

*Step 4.* A 4 $\mu$l aliquot transferred into 50 $\mu$l of cycling reagent in a fluorometer tube. Incubate 1 hr at 38°, heat 2 min at 100°.

*Step 5.* Add 1 ml of 6-P-gluconate reagent (Chapter 8) and read when reaction is complete.

E. CYCLING ASSAY, 1–5 × $10^{-14}$ MOLE

*Reagent for Glucose Reaction.* Buffer and $MgCl_2$ as in B. ATP, 200 $\mu M$; $TPN^+$, 20 $\mu M$; bovine plasma albumin, 0.04%; hexokinase, 5 $\mu$g/ml, glucose-6-P dehydrogenase, 0.14 U/ml.

*Cycling Reagent.* Basic TPN cycling reagent (Chapter 8) with 400 $\mu$g/ml glutamic dehydrogenase and 10 U/ml glucose-6-P dehydrogenase.

For reagent for 6-P-gluconate measurement see Chapter 8.

*Reaction Vessels.* Oil wells.

*Procedure*

*Step 1.* As in D except the volume is 0.015 $\mu$l and standard concentrations are 1–4 $\mu M$.

*Steps 2 and 3.* The same as in D except volumes added are reduced to 0.03 $\mu$l and 0.2 $\mu$l, respectively.

*Step 4.* Cycling reagent, 5 $\mu$l, is added to the sample *in the oil well.* Incubate 1 hr at 38°; add 1 $\mu$l of 1 *N* NaOH; heat 10 min at 90°.

*Step 5.* Transfer 4–5 $\mu$l aliquot to 1 ml of 6-P-gluconate reagent in fluorometer tube and read when reaction complete.

*Comment*

Alkali is added in Step 5 to dissolve the cycling reagent enzymes. Otherwise after heating the coagulum would clog the transfer pipette.

*Kinetics*

The $V_{max}$ for hexokinase is about 140 $\mu$moles/mg/min. The $K_m$ for ATP with 1 m*M* glucose is 30 $\mu M$; the $K_m$ for glucose with 300 $\mu M$ ATP is 160 $\mu M$. Thus with low substrate levels about five times more enzyme is required for measuring glucose than for ATP. With 1$\mu$g/ml of hexokinase, $t_{1/2}$ (for glucose) $= 0.7 \times 160/140 = 0.8$ min.

## Glucose-1-Phosphate

$$\text{Glucose-1-P} \xrightarrow{\text{P-glucomutase}} \text{glucose-6-P} \tag{9-32}$$

$$\text{Glucose-6-P} + \text{TPN}^+ \xrightarrow[\text{dehydrogenase}]{\text{glucose-6-P}} \text{6-P-gluconolactone} + \text{TPNH} + \text{H}^+ \tag{9-33}$$

Glucose-1-P levels in living material are almost invariably low compared to those of glucose-6-P because of the position of equilibrium between the two. Therefore, glucose-6-P must either be removed first or glucose-1-P measured as a small reading on top of a large reading from glucose-6-P (see Comment below).

### A. Spectrophotometer

*Reagent.* Tris-HCl buffer, pH 8.1 (25 m*M* Tris base, 25 m*M* Tris-HCl); 1 m*M* $MgCl_2$; 0.1 m*M* EDTA; TPN, 500 $\mu M$; glucose-1,6-$P_2$, 0.5 $\mu M$; P-glucomutase (rabbit muscle), 0.25 $\mu$g/ml (0.05 U/ml); glucose-6-P dehydrogenase (yeast), 0.04 U/ml, equivalent to 0.1 $\mu$g/ml of crystalline enzyme.

*Reaction Time.* Four to 6 min.

*Conduct of the Assay*

If glucose-6-P is present, the dehydrogenase is added first and a reading is made after the reaction is complete (3–5 min). The mutase is then added to measure glucose-1-P. If substantially more glucose-6-P than glucose-1-P is present, it should be made certain that all the glucose-6-P has reacted. As discussed elsewhere, because reaction (9–33) is reversible, the reaction is not complete until the 6-P-gluconolactone has been hydrolyzed. This can result in a slow drift after most of the glucose-6-P has reacted. This makes it desirable to use a generous excess of $TPN^+$, and pH 8 rather than 7, to favorably affect the equilibrium of reaction (9–33) and accelerate hydrolysis of the lactone.

*Comment*

P-Glucomutase is ordinarily partially phosphorylated, therefore if more enzyme is used (5 $\mu$g/ml) the coenzyme can be omitted. (See, however, comment on the fluorometric method below.) For biological samples it may be desirable to add 0.5 $\mu M$ dithiothreitol as in the fluorometric method. If large amounts of glucose-6-P are present in the samples, they can be destroyed by heating in 0.1 *N* NaOH for 15 min at 100° without any loss of glucose-1-P.

B. Fluorometer, Direct Assay, 0.1–10 × $10^{-9}$ Mole

*Reagent*. The same reagent as in A except as follows: $TPN^+$, 100 $\mu M$ (30 $\mu M$ with lowest levels); dithiothreitol, 500 $\mu M$; P-glucomutase, 0.1 $\mu$g/ml.

*Reaction Time*. Four to 8 min.

*Conduct of the Assay* see A

*Comment*

Sometimes glucose-1,6-$P_2$ preparations contain small amounts of glucose-6-P. If this is troublesome the level of the coenzyme can be reduced to 0.1 $\mu M$ with only a modest decrease in rate. Alternatively, the coenzyme can be heated in alkali without loss, to destroy glucose-6-P (see Comment under protocol A). Because glucose-1-P levels are usually very low compared to the other hexose phosphates, great care may have to be taken to obtain valid answers. A trace of P-glucoisomerase, for example, would lead to an erroneously high value. The best is to keep the amount of enzyme minimal and to control the reaction time.

If the sample contains a large amount of glucose-6-P, this can be removed with alkali, as described in A. Alternatively, glucose-6-P can be allowed to react and the resultant TPNH destroyed by acidification. After neutralization, fresh glucose-6-P dehydrogenase is added, a reading taken at a more sensitive fluorometer setting, and P-glucomutase added to complete the reaction.

*Kinetics*

The kinetics are very favorable. The $V_{max}$ of P-glucomutase is over 300 $\mu$moles/mg/min but may be a third of this in commercial preparations. The $K_m$ for glucose-1-P with saturating amounts of the coenzyme is 8 $\mu M$.

Assuming a $V_{max}$ of 100 $\mu$moles/mg/min, the half-time for very low glucose-1-P levels would be only 0.06 min with 1 $\mu$g/ml of enzyme (0.7 × 8/100). Because of the low Michaelis constant, more enzyme is required for glucose-1-P levels encountered in the spectrophotometer. The apparent dissociation constant for the coenzyme, glucose-1,6-$P_2$, is 0.06 $\mu M$. Its function is to phosphorylate the enzyme. When less than saturating levels of coenzyme are present, the Michaelis constant for glucose-1-P decreases. Consequently the effect of using suboptimal concentrations of coenzyme is less noticeable in the measurement of low levels of glucose-1-P than of high levels.

## Glucose-6-Phosphate

$$\text{Glucose-6-P} + \text{TPN} \xrightarrow[\text{dehydrogenase}]{\text{glucose-6-P}} \text{6-P-gluconolactone} + \text{TPNH} + \text{H}^+ \quad (6\text{-}34)$$

$$\text{6-P-gluconolactone} + \text{H}_2\text{O} \xrightarrow{\text{nonenzymatic}} \text{6-P-gluconate} \quad (6\text{-}35)$$

This is the indicator reaction for a family of analyses. The following protocols are designed to provide optimal conditions for measuring glucose-6-P itself. In other cases the conditions are altered according to the demands of the other substances to be measured. These methods were given in Chapter 4 *in extenso*, but are repeated here for convenience.

A. SPECTROPHOTOMETER

*Reagent.* Tris buffer, pH 8.1, (50 m$M$ Tris base, 50 m$M$ Tris-HCl); $TPN^+$, 500 $\mu M$; glucose-6-P dehydrogenase (baker's yeast), 0.06 U/ml (0.15 $\mu$g of crystalline enzyme or correspondingly more if only partially pure).

*Reaction Time.* Three to 5 min.

*Conduct of the Assay*

Ordinarily the enzyme is added separately to start the reaction after an initial reading has been made. For standardization purposes the reaction could be started equally well with the glucose-6-P solution.

B. FLUOROMETER DIRECT ASSAY, 0.1–10 × $10^{-9}$ MOLE

*Reagent.* Buffer, the same as in A except only half as strong (50 m$M$). $TPN^+$, 50 $\mu M$ for 2–10 $\mu M$ glucose-6-P, 10 $\mu M$ for 0.1–2 $\mu M$ glucose-6-P; dithiothreitol, 0.1 m$M$ (optional, see Comment below); glucose-6-P dehydrogenase, 0.02 U/ml.

*Reaction Time.* Two to 4 min.

*Comment*

The $TPN^+$ is kept as low as possible because of the likelihood of fluorescent trace impurities. However, an excess of at least threefold should be provided (see Kinetics below). Dithiothreitol is only required if the sample to be assayed contains oxidized glutathione, the enzyme is heavily contaminated with glutathione reductase, and the samples are not read promptly. In this case some of the TPNH initially formed will be reoxidized. The danger is much greater when glucose-6-P dehydrogenase is used as the auxiliary enzyme in an assay which takes longer to complete and if more glucose-6-P dehydrogenase is required.

In the assay of tissue or bacterial extracts, the blank fluorescence from flavins can be greatly reduced by changing the buffer to 100 or 200 m*M* imidazole, pH 7. The rate of reaction with low levels of glucose-6-P is practically unchanged, however at least a 10-fold excess of $TPN^+$ should be maintained to drive the reaction close to completion (see Kinetics below).

### C. Cycling, $1\text{–}10 \times 10^{-12}$ Mole

*Reagent for Glucose-6-P Reaction.* The same as in B (10 $\mu M$ $TPN^+$) but with 0.04% bovine plasma albumin.

*Cycling Reagent.* TPN cycling reagent of Chapter 8 with 0.25 U/ml of glucose-6-P dehydrogenase and 3 $\mu$g/ml (0.15 U/ml) of glutamic dehydrogenase (to give about 1000 cycles/hr).

*Reaction Vessels.* Tubes 4 × 50 mm, 2–2.5 mm i.d.

*Procedure for Dry Tissue Samples*

*Step 1.* HCl, 1 $\mu$l of 0.02 *N*, for samples and blanks; 1 $\mu$l of 2–10 $\mu M$ glucose-6-P in 0.02 *N* HCl for standards; 10 min at 60°.

*Step 2.* Reagent, 5 $\mu$l; 20 min at room temperature.

*Step 3.* NaOH, 5 $\mu$l of 0.2 *N*; 10 min at 60°.

*Step 4.* Whole sample to 100 $\mu$l of cycling reagent in fluorometer tubes. Incubate 1 hr at 38°, 2 min at 100°.

*Step 5.* Measurement of 6-P-gluconate in 1 ml (see Chapter 8).

*Comment*

This procedure and the two which follow are designed for measuring glucose-6-P in frozen dried fragments from tissue sections. The first step is for the purpose of destroying enzymes and preformed TPNH. There may be other occasions to measure glucose-6-P with cycling amplification when this is unnecessary. The samples may be acid extracts of tissue which either contain less glucose-6-P than can be measured directly in the fluorometer, or are so fluorescent as to make direct measurement unsatisfactory. In either case, Step 1 would be unnecessary. If the amount of glucose-6-P is greater than $10^{-12}$ moles, Steps 2 and 3 can be carried out with larger volumes of reagent, in larger vessels, and 10 $\mu$l aliquots taken for Step 4.

D. CYCLING, 1–10 × $10^{-13}$ MOLE

*Glucose-6-P Dehydrogenase Reagent.* The same as in C.

*Cycling Reagent.* TPN cycling reagent with 2.5 U/ml glucose-6-P dehydrogenase and 30 μg/ml glutamic dehydrogenase (7000–10,000 cycles/hr).

*Reaction Vessels.* Oil wells.

*Procedure*

*Step 1.* HCl, 0.1 μl of 0.02 *N*, for samples and blanks, and 0.1 μl of 1–10 μ*M* glucose-6-P in 0.02 *H* HCl for standards; 20 min at 80°.

*Step 2.* Reagent, 0.5 μl; 20 min at room temperature.

*Step 3.* NaOH, 5 μl of 0.05 *N*; 20 min at 80°.

*Step 4.* Add 4 μl to 50 μl of cycling reagent in fluorometer tubes. Incubate 1 hr at 38°; heat 2 min at 100°.

*Step 5.* Measurement of 6-P-gluconate in 1 ml (Chapter 8).

E. CYCLING, 1–5 × $10^{-14}$ MOLE

*Glucose-6-P Dehydrogenase Reagent.* The same as in C except all components at double concentration.

*Cycling Reagent.* TPN cycling reagent with 10 U/ml of glucose-6-P dehydrogenase and 200 μg/ml of glutamic dehydrogenase (15,000–20,000 cycles/hr).

*Reagent Vessels.* Oil wells.

*Procedure*

*Step 1.* Volumes of 0.015 μl for samples and blanks and 0.015 μl of 1 to 3 μ*M* glucose-6-P in 0.02 *N* HCl for standards; 20 min at 80°.

*Step 2.* Reagent, 0.03 μl; 20 min at room temperature.

*Step 3.* NaOH, 0.02 μl of 0.1 *N*; 20 min at 80°.

*Step 4.* Add 5 μl of cycling reagent to each oil well. Incubate 1 hr at 38°; add 1 μl of 1 *N* NaOH; heat 10 min at 90°.

*Step 5.* Transfer a 4.5 μl aliquot to 1 ml 6-P-gluconate reagent (tubes read before transfer to increase precision.)

*Kinetics*

The kinetics of glucose-6-P dehydrogenase from baker's yeast have been discussed in Chapter 4. To summarize, the high specific activity and low Michaelis constants, are exceptionally favorable for measuring low levels of either glucose-6-P or $TPN^+$. At pH 8 the $V_{max}$ is the order of 400 μmoles/mg/min (400 U/mg), the $K_m$ for glucose-6-P is 10–20 μ*M*, that for $TPN^+$ 2 μ*M*. This means that 0.1 μg/ml of crystalline enzyme will give a first-order rate constant as high as 4/min (0.1 × 400/10). Many assays using this enzyme as an auxiliary are conducted at pH 7 rather than 8. At pH 7 the lower $V_{max}$, about 200 μmoles/mg/min, is offset by even more favorable Michaelis constants, 2.5 μ*M* for glucose-6-P, 0.8 μ*M* for $TPN^+$.

Although the initial velocity at low glucose-6-P concentration is as rapid at pH 7 as at pH 8, the equilibrium is less favorable and the lactone is more slowly hydrolyzed; therefore a larger excess of $TPN^+$ is required to give stoichiometric reaction. A substantial $TPN^+$ excess is also desirable because of the strong inhibition by TPNH. This inhibition increases somewhat the amount of enzyme or time required for completion of the reaction.

## Glucose-1,6-Diphosphate

$$\text{P-Glucomutase} + \text{glucose-1,6-P}_2 \rightleftarrows \text{phosphorylated P-glucomutase} + \text{glucose-6-P} \quad (9\text{-}36)$$

$$\text{Glucose-1-P} \xrightarrow[\text{P-glucomutase}]{\text{phosphorylated}} \text{glucose-6-P} \quad (9\text{-}37)$$

$$\text{Glucose-6-P} + \text{TPN}^+ \xrightarrow[\text{dehydrogenase}]{\text{glucose-6-P}} \text{6-P-gluconolactone} + \text{TPNH} + \text{H}^+ \quad (9\text{-}38)$$

P-Glucomutase must be phosphorylated by glucose-1,6-$P_2$ to be active. A small amount of glucose-1,6-diphosphate will induce the formation of a large amount of TPNH. Although the method was originally used in the spectrophotometer (Paladini *et al.*, 1949), only a fluorometric method is presented here. It is suitable for measuring 1–8 × $10^{-12}$ mole of the coenzyme. Much greater sensitivity can be attained, as indicated at the end of this section.

*Reagent.* Imidazole buffer, pH 7 (30 m*M* imidazole base, 20 m*M* imidazole-HCl); $MgCl_2$, 2 m*M*; EDTA, 0.1 m*M*; bovine serum albumin, 0.01 %; glucose-1-P, 15 μ*M*; $TPN^+$, 50 μ*M*; P-glucomutase (rabbit muscle), 0.01 μg/ml (0.001 U/ml)*; glucose-6-P dehydrogenase (yeast), 0.125 U/ml, equivalent to 0.33 μg/ml of crystalline enzyme.

### *Conduct of the Assay*

The reagent is prepared complete except for the omission of glucose-1-P and is brought to room temperature. One milliliter of this reagent is pipetted

* Calculated from the activity as reported for commercial preparations. As observed under analytical conditions the activity is about twice this.

into each of a series of fluorometer tubes at room temperature. The fluorometer is adjusted so that 5 $\mu M$ TPNH will read approximately full scale. Standards are added to give 2, 5, and 8 n$M$ concentrations (2, 5, and 8 $\mu$l of 1 $\mu M$ standard glucose-1, 6-$P_2$). Samples are likewise prepared to make a set of 15 or 20 tubes. Several tubes are reserved as blanks. All tubes are read initially. The reaction is started by adding 5$\mu$l of 3 m$M$ glucose-1-P to each tube, including the blanks. The additions are made at intervals equal to the time it will take to read each tube. Readings are made at several exact intervals (e.g., 5, 10, 15, and 20 min) after glucose-1-P addition. Because the reaction has a high-temperature coefficient (the $Q_{10}$ is nearly 3) it is important that samples and standards be run at exactly the same temperature. The results are calculated from a standard curve.

### *Comment*

The overall blank should not exceed 0.2–0.3 n$M$ glucose-1,6-$P_2$. Part of the blank is contributed by the enzyme itself, which is present at about 0.1 n$M$ concentration and is partially phosphorylated as purchased. It is for this reason that P-glucomutase concentration is kept low. The glucose-1-P level is also kept low because preparations usually contain some of the coenzyme. Some preparations are so badly contaminated as to be unsuitable.

### *Increased Sensitivity*

Much smaller quantities of glucose-1,6-$P_2$ could be easily measured by an indirect procedure. With the P-glucomutase level recommended above, the rate of TPNH formation is about 2000 moles per mole of coenzyme per hour. This could be increased three, or fourfold with four times the enzyme concentration without seriously increasing the blank. Glucose-1,6-$P_2$, $10^{-14}$ mole in 10 $\mu$l of reagent (1 n$M$ concentration) would yield at least $5 \times 10^{-11}$ mole of TPNH, an amount easily measurable with accuracy by the strong alkali method.

### *Kinetics*

The apparent dissociation constant for glucose-1,6-$P_2$ is very low, 60 n$M$. Therefore in the analytical range recommended (1–10 n$M$) P-glucomutase is operating at 15% or less of its $V_{max}$. The $K_m$ for glucose-1-P varies with coenzyme concentration (Ray and Roscelli, 1964). At the low coenzyme levels encountered here the $K_m$ is less than 1 $\mu M$. Thus the level of glucose-1-P recommended (15 $\mu M$) is nearly saturating. The $V_{max}$ for commercial P-glucomutase is almost 200 $\mu$moles/mg/min under analytical conditions. The net result of these constants is that at the recommended enzyme level (0.01 mg/liter) the velocity is 0.03–0.05 $\mu M$/min with 1 n$M$ glucose-1,6-$P_2$.

## Glutamate

$$\text{Glutamate} + \text{DPN}^+ \xrightarrow[\text{dehydrogenase}]{\text{glutamic}} \alpha\text{-ketoglutarate} + \text{NH}_4 + \text{DPNH} + \text{H}^+ \quad (9\text{-}39)$$

The equilibrium of the analytical reaction is concentration-dependent (see Kinetics). At glutamate levels less than 10 $\mu M$ (fluorometric procedures) the reaction proceeds essentially to completion. At levels encountered in the spectrophotometer (100 $\mu M$), the reaction would be incomplete with practicable levels of $DPN^+$. Therefore, $H_2O_2$ is added to oxidize the α-ketoglutarate and pull the reaction. (Another method for glutamate measurement is given in Chapter 8.)

### A. Spectrophotometer

*Reagent.* Tris-acetate buffer, pH 8.4 (33 m$M$ Tris base, 17 m$M$ Tris-acetate); $DPN^+$, 1 m$M$; ADP, 100 $\mu M$; $H_2O_2$ 10 m$M$; glutamic dehydrogenase (beef liver), 100 $\mu$g/ml (4.5 U/ml)* from a 50% glycerol solution.

*Reaction Time.* Fifteen to 20 min.

*Conduct of the Assay*

The $H_2O_2$ is added to the reagent shortly before use. Since the 10 m$M$ $H_2O_2$ slowly inactivates glutamic dehydrogenase, the enzyme is ordinarily added to the reagent in the spectrophotometer either before or after the sample. Both $DPN^+$ and the enzyme contribute significant blank absorption at 340 nm which may slowly change with time. Therefore parallel blank samples are required. There is an initial rapid reaction as the initial equilibrium is attained. This is followed by a slower reaction brought about by α-ketoglutarate removal.

*Comment*

$NH_4^+$ ion should be low or absent since it is one of the products and will therefore delay the reaction. The $H_2O_2$ will slowly destroy the DPNH formed (2 or 3% per hour). ADP is added to protect the enzyme from inactivation and to overcome DPNH inhibition (Frieden, 1959a).

### B. Fluorometer Direct Assay, 0.1–10 × $10^{-9}$ Mole

*Reagent.* The buffer and ADP concentrations are the same as in A. $DPN^+$, 400 $\mu M$ (150 $\mu M$ with less than 2 $\mu M$ glutamate). Glutamic dehydrogenase, 50 $\mu$g/ml.

* Measured in the opposite direction.

*Reaction Time.* Fifteen to 30 min.

*Conduct of the Assay*

The enzyme is added last, after the sample is in, and an initial reading has been made.

*Comment*

No $H_2O_2$ is necessary if $NH_4{}^+$ concentration is kept low (less than 10 $\mu M$). Otherwise $H_2O_2$ is added as for the spectrophotometer. Thus $H_2O_2$ is required when glutamate is measured in the fluorometer as the terminal step in the $DPN^+$ cycling method. $DPN^+$ is reduced for lowest glutamate concentrations because it contributes blank fluorescence. It can be reduced even further if longer reaction times are acceptable.

*Kinetics*

The kinetic constants are not very favorable. Under analytical conditions $V'_{max}$ for glutamate oxidation with 1 m$M$ $DPN^+$ is about 12 $\mu$moles/mg/min. The Michaelis constant for each substrate is affected (favorably) by the other (Frieden, 1959b). With 1 m$M$ $DPN^+$ the $K_m$ for glutamate is about 600 $\mu M$; with 10 $\mu M$ glutamate the $K_m$ for $DPN^+$ is considerably greater than 1 m$M$. Thus if the back-reaction could be ignored (which it cannot) the half-time with 1 $\mu$g/ml of enzyme would be $0.7 \times 600/12[DPN^+] = 35$ min/$[DPN^+]$, with $[DPN^+]$ expressed as millimolar concentration. For example, with 50 $\mu$g/ml of enzyme and 0.4 m$M$ $DPN^+$ the half-time would be 1.7 min $[35/(50 \times 0.4)]$. The back-reaction prolongs this in a complicated manner, so that roughly 10 half-times are required.

Under analytical conditions at equilibrium:

$$\frac{[\alpha\text{-KG}][\text{DPNH}][\text{NH}_4^+]}{[\text{glut}][\text{DPN}^+]} = 12.5\ \mu M \qquad (9\text{-}40)$$

This expresses the fact that because two reactants are forming three products, the lower the initial glutamate concentration the more nearly complete will be the reaction, *even if $DPN^+$ is reduced in proportion*. The dilution effect is even more marked if $DPN^+$ is held constant. For example, according to Equation (9-40) with 1 m$M$ $DPN^+$ and 100 $\mu M$ glutamate (protocol A) the reaction will be only 70% complete at equilibrium, whereas with 20 $\mu M$ glutamate it will be 97% complete. With 10 $\mu M$ glutamate and 400 $\mu M$ $DPN^+$ (protocol B) in spite of the lower $DPN^+$, the reaction will be 98% complete. This shows why $H_2O_2$ is required only for glutamate concentrations greater than 10 or 20 $\mu M$.

For the reaction between $\alpha$-ketoglutarate and $H_2O_2$ the first-order rate constant is $k = 0.06$ min$^{-1}$ $[H_2O_2]$, with $H_2O_2$ expressed as millimolar concentration; thus $k = 0.6$ min$^{-1}$ with 10 m$M$ $H_2O_2$ as recommended. This means a rate of 60%/min of the α-ketoglutarate present at any instant, not of the glutamate present. Consequently, no matter how much enzyme is added, the rate of reaction with $H_2O_2$ will depend on how far the reaction is pushed by $DPN^+$ in the first place. Thus, when $H_2O_2$ has to be used, a high $DPN^+$ level is required not only to satisfy kinetic requirements but also to shift the equilibrium toward α-ketoglutarate.

## α-Glycerophosphate: Method I

$$\text{Glycero-P} + \text{DPN}^+ \xrightarrow[\text{dehydrogenase}]{\text{glycero-P-}} \text{dihydroxyacetone-P} + \text{DPNH} + \text{H}^+$$

$$\text{Dihydroxyacetone-P} + \text{hydrazine} \longrightarrow \text{DHAP-hydrazone}$$

Because of its greater simplicity, this will usually be the prefered method for direct assays, both in the spectrophotometer and in the fluorometer. Only one enzyme is required. For indirect assays, the second method, which avoids the use of hydrazine, has distinct advantages.

### A. Spectrophotometer

*Reagent.* Hydrazine buffer, pH 9.2 (350 m*M* hydrazine: 50 m*M* hydrazine-HCl); $DPN^+$, 2 m*M*; glycero-P-dehydrogenase, 20 μg/ml.

*Reaction Time.* Ten to 20 min.

*Conduct of the Assay*

The reaction is ordinarily started with the enzyme after a preliminary reading.

*Comment*

A blank sample should be run concurrently because of a small increase in absorption with time.

### B. Fluorometer Direct Assay, $0.2–10 \times 10^{-9}$ Mole

*Reagent.* Hydrazine buffer as for spectrophotometer. $DPN^+$, at least 50 times the concentration of glycero-P expected; glycero-P-dehydrogenase, 5 μg/ml.

*Reaction Time.* Ten to 20 min.

*Conduct of the Assay*

The enzyme is added last after an initial reading.

*Comment*

There is significant fluorescence enhancement with this amount of glycero-P-dehydrogenase. Lower concentrations of DPNH are enhanced disproportionately; therefore standards should be included at several levels in the range of samples.

With 5 $\mu$g/ml of glycero-P-dehydrogenase the fluorescence enhancement at pH 9 is 18%, 14%, and 8%, respectively, with 2, 4, and 10 $\mu M$ DPNH.

The time required depends largely on the ratio of $DPN^+$ to glycero-P (see below).

*Kinetics*

The rate of reaction of dihydroxyacetone-P with hydrazine is rather slow and depends on the concentration of both reactants. With high enzyme levels the dihydroxyacetone-P concentration depends in turn on the equilibrium constant for the enzyme reaction and the ratio of $DPN^+$ to glycero-P. The reaction proceeds in the kinetically unfavorable direction. At pH 9.4,

$$K_{eq} = \frac{[\text{glycero-P}][\text{DPN}^+]}{[\text{dihydroxyacetone-P}][\text{DPNH}]} = 50$$

It is therefore desirable to provide a high ratio of $DPN^+$ to glycero-P. The $DPN^+$ levels chosen are a compromise between this consideration and the wish to avoid excessive initial blank values.

Approximate kinetic constants at pH 9 are $V_{max} = 80$ $\mu$moles/mg/min, $K_{glycero-P} = 250$ $\mu M$ (with 850 $\mu M$ $DPN^+$), and $K_{DPN^+} = 350$ $\mu M$ (with 40 $\mu M$ glycero-P). With these constants, in the presence of 500 $\mu M$ $DPN^+$, the *initial* rate would be about 20%/min (and essentially independent of dihydroxyacetone-P concentration) with only 1 $\mu$g/ml of enzyme [i.e., $k = (80/250)(500/850)$ $min^{-1}$]. However, because of the back-reaction and the low $K_m$ for DPNH, much more enzyme is needed.

## α-Glycerophosphate: Method II

$$\text{Glycero-P} + \text{DPN}^+ \xrightarrow[\text{dehydrogenase}]{\text{glycero-P-}} \text{dihydroxyacetone-P} + \text{DPNH} + \text{H}^+ \tag{9-41}$$

$$\text{Dihydroxyacetone-P} \xrightarrow[\text{isomerase}]{\text{triose}} \text{glyceraldehyde-P} \tag{9-42}$$

$$\text{Glyceraldehyde-P} + \text{DPN}^+ \xrightarrow[\text{arsenate}]{\text{GAPDH}} \text{3-P-glycerate} + \text{DPNH} + \text{H}^+ \tag{9-43}$$

This analytical sequence proposed by Matschinsky (1964) avoids the use of hydrazine. It is therefore of advantage for indirect procedures. Also in its favor is the fact that 2 moles of DPNH are produced per mole of glycero-P.

The method has the disadvantage that two additional enzymes are required and that triose phosphate must be removed, or measured first, or ignored. Fortunately, in most cases glycero-P levels are much greater than those of the triose phosphates.

A. SPECTROPHOTOMETER

*Reagent.* 2-Amino-2-methylpropanediol, pH 9.0 (30 m*M* as the free base, 20 m*M* as the hydrochloride); $DPN^+$ 2 m*M*; mercaptoethanol, 2 m*M*; $Na_2HAsO_4$, 1 m*M*; glycero-P-dehydrogenase,* 20 $\mu$g/ml; triose-P isomerase, 10 $\mu$g/ml; glyceraldehyde-P dehydrogenase,* 100 $\mu$g/ml.

*Reaction Time.* Fifteen to 20 min.

*Conduct of the Assay*

For standardization purposes the glycero-P can be added last. In this case the three enzymes can be incorporated in the reagent where they are sufficiently stable for at least an hour. For biological samples the reaction is started with glycero-P-dehydrogenase after first allowing 5 or 10 min for triose phosphates, if present, to react.

*Comment*

Sulfate is removed to a large extent because of its inhibitory effect on all three enzymes. The level should not exceed 2 or 3 m*M*.

B. FLUOROMETER DIRECT ASSAY, $0.1–5 \times 10^{-9}$ MOLE

*Reagent.* Buffer, mercaptoethanol and arsenate as for the spectrophotometer; $DPN^+$, 1 m*M* (100 $\mu M$ with 1 $\mu M$ glycero-P or less); glycero-P dehydrogenase, 10 $\mu$g/ml; triose-P isomerase, 5 $\mu$g/ml; glyceraldehyde-P dehydrogenase, 40 $\mu$g/ml.

*Reaction Time.* Fifteen to 20 min.

*Conduct of the Assay*

The sample is added. If triose phosphates are present, glyceraldehyde-P dehydrogenase and triose-P isomerase are added and a reading taken 5 or 10 min later, after which the glycero-P-dehydrogenase is added.

*Comment*

In principle glyceraldehyde-P, dihydroxyacetone-P, and glycero-P can all be measured on the same sample by stepwise addition of the three enzymes. In practice this will usually not be satisfactory because of the relatively low

* Stock suspensions in $(NH_4)_2SO_4$ are centrifuged and dissolved in Tris buffer, pH 8, to remove most of the sulfate.

levels of the triose phosphates, particularly glyceraldehyde-P. It is ordinarily much better to measure the triose phosphates on a separate sample with a somewhat different reagent (see Method II for dihydroxyacetone-P, etc.).

Because of fluorescence enhancement by glycero-P-dehydrogenase, standards at several levels may be required, particularly at the lowest concentrations. With 10 $\mu$g/ml of glycero-P-dehydrogenase, enhancement at this pH is 35, 22, and 12%, respectively, for 2, 4, and 10 $\mu M$ DPNH.

*Kinetics*

The reaction sequence involves two reversible steps followed by a nonreversible one. The equilibria of the reversible steps are unfavorable. The triose isomerase equilibrium constant favors dihydroxyacetone-P by a factor of 25:1. For glycero-P-dehydrogenase at pH 9,

$$K_{eq} = \frac{[\text{glycero-P}][\text{DPN}^+]}{[\text{dihydroxyacetone-P}][\text{DPNH}]} = 125$$

As a result of these two equilibria, with an excess of glycero-P-dehydrogenase, and triose isomerase and a 10-fold excess of $DPN^+$ only 1% of the substrate would be present as glyceraldehyde-P (2% with a 50-fold excess). It is this situation which demands the relatively high concentration of glyceraldehyde-P dehydrogenase. The kinetics of this enzyme were not thoroughly studied under analytical conditions. However, the essential parameter is the first-order rate constant for glyceraldehyde-P which is about 2/min with 1 $\mu$g/ml of enzyme. (The rate is essentially the same for 100–2000 $\mu M$ $DPN^+$.) This means that if only 1% of substrate is present as glyceraldehyde-P, the apparent rate constant for glycero-P could not exceed 0.02/min with 1 $\mu$g/ml of glyceraldehyde-P dehydrogenase. For the reaction to be complete in 20 min the minimal rate constant would need to be at least 10 times this, i.e., at least 10 $\mu$g/ml of this enzyme is needed, no matter how much of other enzymes is used. In practice even more is needed because reasonable levels of the other enzymes cannot keep the two unfavorable steps at equilibrium.

## Glycogen

$$\text{Glycogen} + \text{P}_i \xrightarrow[\text{debrancher complex}]{\text{phosphorylase}} \text{glucose-1-P (92\%)} + \text{glucose (8\%)} \quad (9\text{-}44)$$

$$\text{Glucose-1-P} \xrightarrow{\text{P-glucomutase}} \text{glucose-6-P} \quad (9\text{-}45)$$

$$\text{Glucose-6-P} + \text{TPN}^+ \xrightarrow[\text{dehydrogenase}]{\text{glucose-6-P}} \text{6-P-gluconolactone} + \text{TPNH} + \text{H}^+ \quad (9\text{-}46)$$

A. SPECTROPHOTOMETER

*Reagent.* Imidazole-HCl buffer, pH 7.0 (30 m$M$ imidazole base, 20 m$M$ imidazole-HCl); $MgCl_2$, 0.5 m$M$; EGTA [ethylene glycol-bis($\beta$-aminoethyl ether)-$N,N'$-tetraacetic acid), 1 m$M$; AMP, 100 $\mu M$; $K_2HPO_4$, 5 m$M$;

dithiothreitol, 0.5 m$M$; glucose-1,6-$P_2$, 0.5 $\mu M$; $TPN^+$, 500 $\mu M$; glycogen phosphorylase *a* (rabbit muscle), 20–200 $\mu$g/ml (0.3 U/ml), depending on the debrancher activity, which should be 15–20 $\mu$moles/liter/min (see Comment); P-glucomutase (rabbit muscle), 1 $\mu$g/ml (0.2 U/ml); glucose-6-P dehydrogenase (yeast), 0.15 U/ml.

*Reaction Time.* Twenty to 40 min. Old glycogen solutions may react much more slowly (see Comment below).

*Conduct of the Assay*

For standardization purposes glycogen is ordinarily added last.

*Comment*

Many lots of commercial preparations of twice-crystallized phosphorylase *a* have been found to contain sufficient glucosidase and transglucosylase (debrancher complex) to degrade glycogen completely. However, commercial preparations are also contaminated with amylase, which can reduce the yield of glucose-1-P. Because amylase requires $Ca^{2+}$ for activity it can be inhibited by the chelating agent, EGTA, which binds $Ca^{2+}$ much more tightly than $Mg^{2+}$. The residual free $Mg^{2+}$ is ample for P-glucomutase function since its $Mg^{2+}$ requirement is very small.

Because all lots of phosphorylase *a* may not contain sufficient debrancher for measurement of glycogen the following test is recommended. The enzymes are activated by diluting the original suspension (20 mg/ml) at least five-fold in 20 m$M$ imidazole-HCl buffer, pH 7.0, containing 0.02% bovine plasma albumin, 10 m$M$ dithiothreitol, and 200 $\mu M$ 5′-AMP, and heating 60 min at 38°. The debrancher activity is assayed spectrophotometrically at 340 nm in a reagent like that given above but with EGTA omitted. The phosphorylase to be tested is added to give a concentration of 10 $\mu$g/ml. After taking a reading, glycogen is added to give a concentration of 0.25 m$M$ (as glucosyl units). The reaction is followed for 15–20 min. The rate is rapid until the outer tiers are gone (optical density increase of about 0.6), after which the rate is slow and limited by the debrancher complex. Enzyme preparations giving rates not less than of 0.006 optical density change per minute (1 $\mu$mole/liter/min) at 25° during the second phase are satisfactory. This is equivalent to 0.1 $\mu$mole of glucose-1-P formed per milligram of phosphorylase per minute. The debrancher is not stable indefinitely and it may be necessary to repeat the assay if the preparation is stored for a month or longer.

The protocol calls for 15–20 $\mu$moles/liter/min of debrancher activity. This would be provided by 150–200 $\mu$g/ml of enzyme with the minimal activity described above, or correspondingly less with more active preparations.

Standards may be prepared from rabbit liver glycogen. It is well to check the concentration by analyzing them for glucose. This can be done by hydrolyzing a 2 m$M$ solution (glucosyl units) in a sealed tube (2 hr at 100° in 1 $N$ HCl), after which the glucose is assayed in the spectrophotometer (see Glucose method). On long storage, particularly with repeated freezing and thawing, glycogen solutions tend to aggregate, as evidenced by increasing opalescence. Such solutions assayed with phosphorylase react slowly and incompletely. If solutions are not too old they can be restored by heating 10 min at 100°. Ultimately however, the reactivity is not restored and fresh standard solutions should be provided.

### B. Fluorometer Direct Assay, 0.1–10 × $10^{-9}$ Mole (of Glucosyl Units)

*Reagent.* The reagent is the same as in A except as follows: $TPN^+$, 50 $\mu M$; bovine plasma albumin, 0.02%; glucose-6-P dehydrogenase, 0.06 U/ml; P-glucomutase, 0.5 $\mu$g/ml; phosphorylase *a*, 10–80 $\mu$g/ml (to provide debrancher activity of 5–10 $\mu$moles/liter/min; see Comment to protocol A above).

*Reaction Time.* Fifteen to 40 min. Old glycogen preparations may react slowly. See Comment in A.

*Conduct of the Assay*

The P-glucomutase and glucose-6-P dehydrogenase can be incorporated in the reagent with phosphorylase added last after the sample is in and a few moments allowed for any preformed glucose-6-P or glucose-1-P to react.

## Preparation of Tissues for Glycogen Analysis

Preparation for glycogen analysis is different than for most metabolites because of the very high molecular weight. For example, if the tissue is extracted with $HClO_4$, part or all of the glycogen may remain in the precipitate. How the glycogen will be distributed depends on the tissue and the way the extract is made. If it is essential to make an $HClO_4$ extract, glycogen can be assayed in both the neutralized extract and in the precipitate, and the results combined. In this case, the $HClO_4$ must be first removed from the precipitate, which is easily accomplished by extracting with ethanol, using at least 20 times the volume of the precipitate. The precipitate is then dispersed in 20–50 volumes of 0.02 $N$ HCl and heated 10 min at 100°. Aliquots are analyzed without centrifugation. If the $HClO_4$ is not removed there will be serious loss of glycogen as far as phosphorylase assay is concerned. The simplest way to prepare tissues for glycogen analysis is to homogenize the frozen tissue in 10 to 100 volumes of 0.02 $N$ HCl, followed by 10 min heating at 100°. Without the acid heat treatment results may be low, although this varies with different tissues. Aliquots of the homogenate are analyzed without removal of insoluble material.

C. CYCLING, 1–10 × $10^{-12}$ MOLE

*Glycogen Reagent.* The same reagent is used as in B except that $TPN^+$ is reduced to 30 $\mu M$.

*Cycling Reagent.* As for protocol C of glucose method.

*Reaction Vessels.* Tubes, 4 × 50 mm; 2–2.5 mm i.d.

*Procedure*

Except for the difference in the specific reagent, the analysis is carried out exactly as described in protocol C for glucose.

*Comment*

All of the cycling methods are described for the analysis of frozen dried tissues. For other material the first step in each case could be modified accordingly.

The procedure as described will include glucose-6-P, glucose-1-P, and part of the glucose-1,6-$P_2$ in the assay (the latter is easily hydrolyzed to glucose-6-P in acid). Ordinarily none of these compounds are significant in amount compared to glycogen.

D. CYCLING, 1–10 × $10^{-13}$ MOLE

*Glycogen Reagent.* The reagent is the same as in B except as follows: $TPN^+$, 40 $\mu M$; bovine plasma albumin, 0.05%.

*Cycling Reagent.* The same as in protocol D for glucose.

*Reaction Vessels.* Oil wells.

*Procedure*

Except for the change in the specific reagent, the steps are identical to those described in protocol D for glucose. The time of incubation with the specific reagent (Step 2) is increased to 40 min.

E. CYCLING, 1–5 × $10^{-14}$ MOLE

*Reagent.* The reagent is like that given in B, but all components are double strength. The $TPN^+$ concentration, however, is only 40 $\mu M$.

*Cycling Reagent.* The same as in protocol E for glucose.

*Reaction Vessels.* Oil wells.

*Procedure*

Except for the change in the specific reagent, the steps are identical to those described in protocol E for glucose. The time of incubation with the specific reagent (Step 2) is increased to 40 min.

### *Kinetics*

Phosphorylase *a* has a $V_{max}$ of about 12 $\mu$moles/mg/min. The Michaelis constant for $P_i$ and glycogen are each affected favorably by the increases in the concentration of the other. In addition, 5′-AMP has a very favorable affect in lowering the $K_m$'s for both substrates. Under assay conditions the $K_m$ for $P_i$ is 4 m$M$ with 100 $\mu M$ glycogen (as glucosyl units) and more than 20 m$M$ with 5 $\mu M$ glycogen. The $K_m$ for glycogen with 4 m$M$ $P_i$ is 60 $\mu M$. The velocity falls off markedly as the glycogen branches are shortened.

The real kinetic problem concerns the debrancher complex, because of the relatively low activity in phosphorylase preparations. As discussed above the activity may be only 0.1 $\mu$mole/mg/min or less.

The complete degradation of glycogen requires both glucosidase and transglucosylase activities which may reside in the same molecule. Purified preparations of this complex are not available. Michaelis constants are not available; however, they must be quite low since substantially less activity is required to degrade glycogen in the 1 to 5 $\mu M$ range than in the 50 to 100 $\mu M$ range.

## Isocitrate

$$\text{Isocitrate} + \text{TPN}^+ \xrightarrow[\text{dehydrogenase}]{\text{isocitrate}} \alpha\text{-ketoglutarate} + CO_2 + \text{TPNH} + H^+ \quad (9\text{-}47)$$

### A. Spectrophotometer

*Reagent*. Tris-HCl, pH 8.1 (25 m$M$ Tris base, 2.5 m$M$ Tris-HCl); $MnCl_2$, 100 $\mu M$; $TPN^+$, 300 $\mu M$; isocitric dehydrogenase (TPN-specific from pig heart) 10 $\mu$g/ml.

Stock enzyme dilutions are made in 50% glycerol in Tris buffer (see Comment).

*Reaction Time*. Ten minutes.

### *Conduct of the Assay*

The enzyme is added last after the sample is in and the initial reading made.

### *Comment*

Isocitric dehydrogenase is unstable when highly diluted in buffer, especially at dilutions greater than 5 $\mu$g/ml. It is stabilized by isocitrate, which protects it during the assay. It is also stabilized by bovine plasma albumin, although this does not appear to be due to the albumin itself. Therefore if the enzyme must be added to the reagent before the sample, at least 0.01% bovine plasma albumin should also be present. The enzyme is also stabilized by 50% glycerol. Therefore this is used for preparing dilute stock solutions.

### B. Fluorometer Direct Assay, 0.1–10 × $10^{-9}$ Mole

*Reagent.* The same buffer and $MnCl_2$ concentrations are used as in A. $TPN^+$, 35 $\mu M$; bovine plasma albumin, 0.01%; isocitric dehydrogenase, 2 $\mu$g/ml.

*Reaction Time.* Five to 10 min.

*Conduct of the Assay*

The enzyme is added last.

*Comment*

For isocitrate concentrations of 1 $\mu M$ or less, $TPN^+$ can be reduced to 5 $\mu M$ and the enzyme reduced to 1 $\mu$g/ml. Greater dilution of the enzyme is not recommended because of increased instability. For very low levels of isocitrate, the albumin can be omitted but it may be necessary to increase the amount of enzyme or prolong the assay time.

Isocitrate levels are usually very low in living cells because the equilibrium ratio between citrate and isocitrate favors citrate by a factor of 15. Indirect cycling procedures may be required to obtain accurate estimations. We have had no experience with such assays, but no difficulty is anticipated if the specific reagent is prepared with at least 0.02% bovine plasma albumin, the isocitric dehydrogenase concentration is not less than 1 or 2 $\mu$g/ml, and the reagent is used promptly after enzyme addition. The possibility of trace contamination with aconitase should be carefully checked.

*Kinetics*

The $V_{max}$ is low, 4.5 $\mu$moles/mg/min. This is compensated by very low Michaelis constants for the substrates, 0.3 $\mu M$ for $TPN^+$, 0.4 $\mu M$ for isocitrate. From this the calculated amount of enzyme needed for an assay time of 10 min with 1, 10, 100 $\mu M$ isocitrate would be 0.05, 0.25, and 2.5 $\mu$g/ml [Chapter 2, Eq. (2-27)]. Because of enzyme instability at high dilution, somewhat larger amounts of enzymes are required in practice.

The $K_m$ for $Mn^{2+}$ is 7 $\mu M$. Levels higher than 100 or 200 $\mu M$ should be avoided because of possible development of significant absorption in the region of TPNH fluorescence. $Mg^{2+}$ can be substituted for $Mn^{2+}$. Its $K_m$ is 48 $\mu M$. The $V_{max}$ is about the same but the $K_m$ for isocitrate is four times larger. Therefore $Mn^{2+}$ is distinctly preferable with low substrate levels.

## Lactate: Method I

$$\text{Lactate} + \text{DPN}^+ \xrightarrow[\text{dehydrogenase}]{\text{lactic}} \text{pyruvate} + \text{DPNH} + \text{H}^+ \quad (9\text{-}48)$$

$$\text{Pyruvate} + \text{glutamate} \xrightarrow[\text{transaminase}]{\text{glut-pyr}} \text{alanine} + \alpha\text{-ketoglutarate} \quad (9\text{-}49)$$

Although tissue lactate levels are relatively high, lactate measurement may cause difficulty because of the ease of contamination (see Final Comments) and because of the unfavorable equilibrium (see Kinetics below). This first method, initially described by Noll (1966), has several advantages over the alternative method with hydrazine which compensate for the fact that an additional enzyme is required. Particularly in the spectrophotometer, blanks are easier to control, and there are fewer complications in adapting to high sensitivity.

## A. Spectrophotometer

*Reagent.* 2-Amino-2-methylpropanol buffer, pH 9.9 (50 m*M* in the free base, 50 m*M* in the hydrochloride); $DPN^+$, 1.5 m*M*; glutamate, 50 m*M* (pH 9.8–10, see Comment); lactic dehydrogenase (beef heart), 100 μg/ml (20 U/ml); glutamic–pyruvic transaminase (pig heart), 100 μg/ml (8 U/ml).

*Reaction Time.* Eight to 12 min.

### *Conduct of the Assay*

For standardization purposes the lactate can be added last after making a preliminary reading. In this case both enzymes should be in the reagent for 10 or 15 min to permit any contaminating lactate to react. For analyzing biological samples the lactic dehydrogenase is ordinarily added last to permit an initial reading of the reagent plus sample. In this case it is necessary to conduct a separate blank. Because the enzymes are somewhat unstable at this alkaline pH, neither enzyme should be added to the reagent more than 30 min before starting the reaction.

### *Comment*

The glutamate required is conveniently prepared from a stock 1 M solution of monosodium glutamate which is brought to a pH of about 10 by incorporating 0.9 M NaOH. (The third $pK_a$ of glutamic acid is about 9.) The reaction is carried out at pH 9.8–10. The addition of $(NH_4)_2SO_4$ with the enzymes can lower the pH, otherwise a 10 m*M* concentration of this salt does not materially affect the rate or end point.

## B. Fluorometer Direct Assay, 0.2–10 × $10^{-9}$ Mole

*Reagent.* The buffer concentration is reduced to 50 m*M*. $DPN^+$, 500 μ*M* (200 μ*M* with 2 μ*M* lactate or less); glutamate, 2 m*M* (0.5 m*M* with 2 μ*M* lactate or less); lactic dehydrogenase, 50 μg/ml; glutamic–pyruvic transaminase, 50 μg/ml.

*Reaction Time.* Fifteen to 20 min.

*Conduct of the Assay*

The transaminase can be incorporated in the reagent but not more than 30 min ahead of time. The reaction is ordinarily started with lactic dehydrogenase.

*Comment*

Because of the lower lactate levels in the fluorometer, the glutamate concentration is reduced; less is needed to drive the reaction to completion. The level used in the spectrophotometer is actually inhibitory. As pyruvate levels are decreased the glutamate required for maximal activity is also decreased (see Kinetics below). Because the lactate concentrations are lower than in the spectrophotometer procedure, contamination with lactate is more of a problem. The tubes must be handled carefully to avoid contamination by lactate from the fingers. It is best not to touch them near the top. It may be necessary to wear rubber gloves or to pick up the tubes with forceps.

C. FLUOROMETER INDIRECT, 0.1–1 × $10^{-9}$ MOLE

Because of the danger of lactate contamination, there is advantage in carrying out the analysis in two steps with the first step at a reduced volume.

*Reagent.* Buffer and transaminase as in B; $DPN^+$, 100 $\mu M$; glutamate, 5 m$M$; lactic dehydrogenase, 100 $\mu$g/ml.

*Procedure*

*Step 1.* Fifty microliters of complete reagent plus 1–10 $\mu$l of sample incubated 30 min in a tube 7 × 70 mm.

*Step 2.* Add 50 $\mu$l of pH 12 phosphate buffer (0.25 $M$ $Na_3PO_4$, 0.25 $M$ $K_2HPO_4$); heat 15 min at 60°.

*Step 3.* Add whole sample to 1 ml of 6 $N$ NaOH containing 0.03% $H_2O_2$; heat 15 min at 60°; read fluorescence.

D. CYCLING, 1–10 × $10^{-12}$ MOLE

*Reagent for Lactate Reaction.* 2-Amino-2-methyl-1-propanol buffer, pH 9.9 (50 m$M$ free base, 50 m$M$ hydrochloride); $DPN^+$, 50 $\mu M$; glutamate, 0.5 m$M$; bovine plasma albumin, 0.02%; lactic dehydrogenase, 100 $\mu$g/ml; glutamic–pyruvic transaminase, 50 $\mu$g/ml.

*Cycling Reagent.* Basic DPN cycling reagent (Chapter 8) with 100 $\mu$g/ml (4.5 U/ml) of glutamic dehydrogenase and 4 $\mu$g/ml (0.8 U/ml) of beef heart lactic dehydrogenase (about 500-fold amplification per hour).

*Reaction Vessels.* Tubes, 4 × 50 mm, 2–2.5 mm i.d.

*Procedure for Frozen Dried Samples*

*Step 1.* NaOH, 1 μl of 0.02 *N*, for samples and blanks, 1 μl of 1–10 μ*M* lactate in 0.02 *N* NaOH for standards, heated 10 min at 60°.

*Step 2.* Reagent, 5 μl; 30 min at room temperature.

*Step 3.* NaOH, 5 μl of 0.2 *N*; 10 min at 60°.

*Step 4.* A 2 μl aliquot into 100 μl of cycling reagent in fluorometer tubes. Incubate 1 hr at 38°; heat 2 min at 100°.

*Step 5.* Reduction of pyruvate with 100 μl of DPNH: lactic dehydrogenase reagent (Chapter 8).

*Step 6.* Development of $DPN^+$ fluorescent product in 1 ml of 6 *N* NaOH.

E. CYCLING, 1–10 × $10^{-13}$ MOLE

*Reagent for Lactate Reaction.* The same as in protocol C.

*Cycling Reagent.* DPN cycling reagent of Chapter 8 with 400 μg/ml of glutamic dehydrogenase and 50 μg/ml of lactic dehydrogenase (about 5000 cycle/hr).

*Reaction Vessels.* Oil wells.

*Procedure*

*Step 1.* NaOH, 0.1 μl of 0.02 *N*, for samples and blanks; 0.1 μl of 2–10 μ*M* lactate in 0.02 *N* NaOH for standards; 20 min at 80°.

*Step 2.* Reagent, 0.5 μl; room temperature 30 min.

*Step 3.* NaOH, 5 μl of 0.05 *N*; 20 min at 80°.

*Step 4.* Transfer a 4 μl aliquot to 50 μl of cycling reagent in fluorometer tube. Incubate 1 hr at 38°; heat 5 min at 100°.

*Step 5.* Reduction of pyruvate with 50 μl of DPNH: lactic dehydrogenase reagent; 5 min at room temp; 12 μl 5 *N* HCl (see Chapter 8).

*Step 6.* Development of $DPN^+$ fluorescent product in 1 ml of 6 *N* NaOH.

F. CYCLING, 1–10 × $10^{-14}$ MOLE

*Reagent for Lactate Reaction.* The same as in protocol C but all components at double concentration.

*Cycling reagent.* DPN cycling reagent of Chapter 8 with 400 μg/ml glutamic dehydrogenase and 100 μg/ml lactic dehydrogenase (about 8000 cycles/hr).

*Reaction Vessels.* Oil wells.

*Procedure*

*Step 1.* NaOH, 0.015 μl of 0.02 *N*, for samples and blanks; 0.015 μl of 2–5 μ*M* lactate in 0.02 *N* NaOH for standards; 20 min at 80°.

*Step 2.* Reagent, 0.015 μl; 30 min at room temperature.

*Step 3.* NaOH, 0.2 μl of 0.1 *N*; 20 min at 80°.

*Step 4.* Cycling reagent, 5 μl (added to oil well). Incubate 1 hr at 38°; heat 10 min at 90°.

*Step 5.* Five microliters of DPNH: lactic dehydrogenase reagent, Chapter 8 (added to oil well); 10 min at room temperature, then 1 μl of 5 *N* HCl.

*Step 6.* An 8 μl aliquot transferred to 1 ml of 6 *N* NaOH and fluorescence developed by heating 10 min at 60°.

*Final Comments*

As indicated, lactate is an exceedingly troublesome contaminant in the laboratory. All glassware used for reagents and analysis should be repeatedly rinsed, and it is recommended that gloves be worn during the rinsing process. The lactate contamination of the reagent can be reduced to 0.5 μ*M* or less if proper precautions are observed.

Skeletal muscle lactic dehydrogenase is less satisfactory than the heart enzyme for the first enzyme step, because it is much less stable at alkaline pH.

*Kinetics*

The kinetics of lactic dehydrogenase are complicated and rather unfavorable. The Michaelis constants are large and the equilibrium lies on the side of pyruvate reduction. There is major kinetic interaction between the two substrates. Under analytical conditions the apparent $K_m$ for $DPN^+$ is about 200 μ*M* whereas that for lactate is nearly 2 m*M*. The $V_{max}$ at pH 10 is about 100 μmoles/mg/min or about half that in the opposite direction at pH 7. The analytical reaction is carried out at as high a pH as possible without enzyme inactivation, because this shifts the equilibrium in the desired direction. It is also important to keep the $DPN^+$ concentration high, not only because of the large Michaelis constant but because this also shifts the equilibrium toward pyruvate formation.

At pH 9.9 the lactic dehydrogenase equilibrium is

$$\frac{[\text{lac}][\text{DPN}^+]}{[\text{pyr}][\text{DPNH}]} = 60 \tag{9-50}$$

The transaminase equilibrium at pH 9.9 is

$$\frac{[\text{glut}][\text{pyr}]}{[\alpha\text{-KG}][\text{ala}]} = 1.6 \tag{9-51}$$

For the overall reaction [Eqs. (9-48) and (9-49)] to proceed to virtual completion, the final concentrations of α-ketoglutarate, alanine, and DPNH will be essentially equal to the initial level of lactate, $[\text{lac}]_0$. Making these substitutions and combining Eqs. (9-50) and (9-51)

$$[\text{lac}]_f = \frac{96[\text{lac}]_0{}^3}{[\text{glut}][\text{DPN}^+]} \tag{9-52}$$

where $[\text{lac}]_f$ is the final lactate concentration.

For the reaction to be 99% complete, $[\text{lac}]_f = 0.01[\text{lac}]_0$. Making this substitution in Eq. (9-52) and rearranging:

$$[\text{glut}][\text{DPN}^+] = 9600[\text{lac}]_0{}^2 \tag{9-53}$$

This equation defines the concentrations of glutamate and $\text{DPN}^+$ which will drive the reaction 99% to completion with different levels of lactate. For example, with 0.1, 0.01, and 0.001 m$M$ lactate the product of glutamate and $\text{DPN}^+$ concentrations (also millimolar) would need to be at least 100, 1, and 0.01 m$M^2$, respectively. This has governed in part the choice of levels for the reagents above.

The kinetics of glutamic–pyruvic transaminase have not been well studied at pH 10, but the following suffices for the present purpose. The apparent $K_m$ for glutamate is about 1.5 m$M$ with 40 μ$M$ pyruvate (i.e., at initial spectrophotometer levels) but is much lower (0.15 m$M$) with 5 μ$M$ pyruvate. The apparent Michaelis constant for pyruvate is similarly affected by the glutamate concentration. It is 50 μ$M$ or more with 40 m$M$ glutamate and less than 5 μ$M$ with 0.1 m$M$ glutamate. These kinetic properties make it unnecessary to use high glutamate concentrations except for high lactate levels in the spectrophotometer. High levels of glutamate are actually inhibitory. The $V_{max}$ at pH 10 was not measured, but with 1 to 5 m$M$ glutamate the apparent first-order rate constant with low pyruvate levels is about 0.04/min with an enzyme concentration of 1 μg/ml.

## Lactate: Method II

$$\text{Lactate} + \text{DPN}^+ \xrightarrow[\text{dehydrogenase}]{\text{lactic}} \text{pyruvate} + \text{DPNH} + \text{H}^+ \tag{9-54}$$

$$\text{Pyruvate} + \text{hydrazine} \longrightarrow \text{pyruvate hydrazone} \tag{9-55}$$

### A. Spectrophotometer

*Reagent.* Hydrazine buffer, pH 9.6 (190 m$M$ hydrazine, 10 m$M$ HCl); $DPN^+$, 2 m$M$; Lactic dehydrogenase (heart), 100 μg/ml (20 U/ml).

The lactic dehydrogenase stock suspension is centrifuged and dissolved in 0.02 $M$ Tris buffer, after removing as much of the original supernatant fluid as possible (to eliminate most of the $NH_4^+$).

*Reaction Time.* Ten to 20 min (depending on lactate level).

*Conduct of the Assay*

The enzyme is added separately to start the reaction after the sample has been added and an initial reading has been made. Concomitant blank samples are essential.

*Comment*

Hydrazine is the only buffer used because the combination of hydrazine, $DPN^+$, and buffers tried (glycine, aminomethylpropanol) increases the blank absorbence. Hydrazine is a poor buffer at the best pH for the reaction, 9.7–10. Care must be taken that the pH is not shifted by the sample or by other additions. This is the reason for removal of most of the $NH_4^+$ from the enzyme preparation.

### B. Fluorometer, Direct Assay 0.5–10 × $10^{-9}$ Mole

*Reagent.* 2-Amino-2-methyl-1-propanol buffer, 100 m$M$, pH 10 (50 m$M$ free base, 50 m$M$ hydrochloride); hydrazine, 100 m$M$; $DPN^+$, 200 μ$M$ (for 2–10 μ$M$ lactate), 50 μ$M$ (for 0.5–2 μ$M$ lactate); lactic dehydrogenase, 40 μg/ml (8 U/ml).

*Reaction Time.* Ten to 20 min.

*Conduct of the Assay*

As in the spectrophotometer.

*Comment*

The reagent is better buffered than for the spectrophotometric assay. It is therefore unnecessary to remove $NH_4^+$ from the enzyme preparation if the final $NH_4^+$ concentration does not exceed 10 m$M$. If the buffer adds an appreciable blank, as it may, glycine buffer, pH 10, can be used instead, or the hydrazine buffer used in the spectrophotometer can be substituted. It is important that the ratio of $DPN^+$ to lactate does not fall below 20 : 1. (For more sensitive procedures for lactate see Method I.)

*Kinetics*

For the kinetics of lactic dehydrogenase at pH 10 see the first lactate method. The hydrazine reaction is relatively slow and is accelerated in proportion to the concentration of free pyruvate and the hydrazine concentration. At pH 10 the equilibrium constant is about 50 for

$$\frac{[\text{lactate}][\text{DPN}^+]}{[\text{pyruvate}][\text{DPNH}]}$$

This means that with an initial lactate : $DPN^+$ ratio of 20 : 1, the reaction will proceed about 45% to equilibrium without the benefit of hydrazine.

**Malate: Method I**

$$\text{Malate} + \text{DPN}^+ \xrightarrow[\text{dehydrogenase}]{\text{malic}} \text{oxalacetate} + \text{DPNH} + \text{H}^+ \quad (9\text{-}56)$$

$$\text{Oxalacetate} + \text{glutamate} \xrightarrow[\text{transaminase}]{\text{glut-oxal}} \text{aspartate} + \alpha\text{-ketoglutarate} \quad (9\text{-}57)$$

This method is analogous to the first lactate method. It is superior in several respects to the alternative procedure with hydrazine, even though it requires an additional enzyme. It can be made faster, there is less problem with blanks, particularly in the spectrophotometer, and it is much easier to adapt to high-sensitivity procedures.

A. SPECTROPHOTOMETER

*Reagent.* 2-Amino-2-methylpropanol buffer, pH 9.9 (50 m*M* in free base, 50 m*M* in the hydrochloride); $DPN^+$, 2 m*M*; glutamate, 40 m*M*, pH 9.9 (10% monosodium salt, 90% disodium salt); malic dehydrogenase (pig heart), 5 μg/ml (3.5 U/ml); glutamic–oxalacetic transaminase (pig heart), 5 μg/ml (0.9 U/ml).

*Reaction Time.* Four to 8 min.

*Conduct of the Assay*

For standardization purposes the malate can be added last after making a preliminary reading. For other purposes, if the sample may add an appreciable absorption blank, it is more satisfactory to add malic dehydrogenase last. In neither case should the enzyme remain in the reagent for more than 10 or 15 min because of slight instability at this alkaline pH.

*Comment*

The $NH_4^+$ in the enzyme preparations will lower the pH slightly. If the amounts to be added would exceed 10 m*M* most of the $(NH_4)_2SO_4$ should be removed by centrifugation before use. Otherwise the assay will be prolonged and the reaction may be incomplete.

### B. Fluorometer Direct Analysis, 0.1–10 × $10^{-9}$ Mole

*Reagent.* The buffer concentration is reduced to 50 m*M*. $DPN^+$, 200 $\mu M$ (50 $\mu M$ with 1 $\mu M$ malate or less); glutamate, 10 m*M* (1 m*M* with 1 $\mu M$ malate or less ); malic dehydrogenase, 5 $\mu$g/ml (10 $\mu$g/ml with 50 $\mu M$ $DPN^+$); glutamic–oxalacetic transaminase, 2 $\mu$g/ml.

*Reaction Time.* Three to 6 min.

*Conduct of the Assay*

Ordinarily the sample is added first. After an initial reading the enzymes are added together as a 2.5 :1 mixture.

*Comment*

Because of the weaker buffer used to reduce the fluorescence blank, the enzymes will ordinarily need to be centrifuged to remove most of the $NH_4^+$ (see above). With malate levels of 1 $\mu M$ or less, glutamate and $DPN^+$ are decreased to further reduce the fluorescence blank. At lowest levels it may be desirable to reduce the buffer concentration to 20 m*M*.

### C. Analysis with Less than $10^{-10}$ Mole

No experience has been had with measuring malate at very low levels, but no difficulty is anticipated. The excess of $DPN^+$ should be kept as small as possible; a five-fold excess is adequate if sufficient malic dehydrogenase and glutamate are used (see Kinetics, below). After the specific reaction is complete it should merely be necessary to heat the sample with alkali in preparation for cycling.

*Kinetics*

The analytical reaction depends on two equilibria, one unfavorable, one slightly favorable:

$$\frac{[\text{oxalacetate}][\text{DPNH}][\text{H}^+]}{[\text{malate}][\text{DPN}^+]} = 8 \times 10^{-13} \tag{9-58}$$

At pH 9.9 this becomes $6.4 \times 10^{-3} = 1/156$. Under conditions of the assay it was observed that

$$\frac{[\alpha\text{-ketoglutarate}][\text{aspartate}]}{[\text{glutamate}][\text{oxalacetate}]} = 2.7 \tag{9-59}$$

Others have found at neutral pH a value of about 7 (Krebs, 1953). Combining the two equilibrium constants:

$$\frac{[\text{malate}][\text{glutamate}][\text{DPN}^+]}{[\alpha\text{-ketoglutarate}][\text{aspartate}][\text{DPNH}]} = 58 \tag{9-60}$$

If the overall reaction is to approach completeness, the final concentrations of DPNH, $\alpha$-ketoglutarate, and aspartate will have to be nearly equal to the initial malate concentration $[\text{malate}]_0$. Substituting and rearranging Eq. (9-60) (and ignoring the decrease in glutamate and $DPN^+$ concentrations):

$$\frac{[\text{malate}]}{[\text{malate}]_0} = \frac{58[\text{malate}]_0^{\,2}}{[\text{glutamate}][\text{DPN}^+]}$$

From this the minimal concentrations of glutamate and DPN can be calculated. For the reaction to be 99% complete, i.e., $[\text{malate}]/[\text{malate}]_0 = 0.01$, with 0.001, 0.01, and 0.1 m$M$ malate, requires the product of the millimolar concentrations of glutamate and $DPN^+$ to be 0.006, 0.6, and 58, respectively. Because $DPN^+$ has more absorption at 340 m$\mu$ and more fluorescence than glutamate, it is an advantage to keep glutamate high and $DPN^+$ low.

The kinetic constants for both enzymes are quite favorable. Malic dehydrogenase has a $V_{max}$ of about 100 $\mu$moles/mg/min at pH 10, with Michaelis constants of about 200 $\mu M$ for both malate and $DPN^+$. Consequently when $DPN^+$ is reduced to the extent it is in the fluorometric procedures there is a need to increase malic dehydrogenase. Transaminase kinetics have not been well studied under assay conditions, but the $V_{max}$ is at least 50 $\mu$moles/mg/min. The $K_m$ for oxalacetate is probably about 15 $\mu M$ (much greater than at pH 7) and the $K_m$ for glutamate is less than 100 $\mu M$ (much smaller than at pH 7). Consequently the required concentration of transaminase is decreased at low malate levels.

## Malate: Method II

$$\text{Malate} + \text{DPN}^+ \xrightarrow[\text{dehydrogenase}]{\text{malic}} \text{oxalacetate} + \text{DPNH} + \text{H}^+ \qquad (9\text{-}61)$$

$$\text{Oxalacetate} + \text{hydrazine} \longrightarrow \text{oxalacetate-hydrazone} \qquad (9\text{-}62)$$

### A. Spectrophotometer

*Reagent.* Hydrazine pH 9.3 (180 m$M$ hydrazine base, 20 m$M$ hydrazine-HCl); $DPN^+$, 2 m$M$; malic dehydrogenase (pig heart), 3 $\mu$g/ml (2.1 U/ml).

*Reaction Time.* Thirty to 50 min.

*Procedure*

The reaction can be started with either the enzyme or the sample after an initial reading. It is important to have concomitant blanks.

*Comment*

The hydrazine serves as the only buffer. Inclusion of other buffers have given higher blank values which tend to increase with time. It is important that neither the sample nor the enzyme lower the pH significantly. For this reason, $NH_4^+$ from the enzyme preparation should not be greater than 5 m*M*.

B. FLUOROMETER DIRECT ASSAY, $0.1–10 \times 10^{-9}$ MOLE

*Reagent.* 2-Amino-2-methylpropanol, pH 9.9 (50 m*M* free base, 50 m*M* hydrochloride); hydrazine, 200 m*M*; $DPN^+$, 100 times the highest expected malate concentration. Malic dehydrogenase, 1 μg/ml (2 μg/ml and 5 μg/ml with 200 μ*M* and 50 μ*M* $DPN^+$).

*Reaction Time.* Fifteen to 30 min.

*Procedure*

The reaction is started with the enzyme after a preliminary reading.

*Comment*

The rate is very dependent on the ratio of $DPN^+$ to malate. This governs the steady-state proportion of oxalacetate, which in turn governs the rate or reaction with hydrazine which is rate-limiting. At low absolute $DPN^+$ levels the malic dehydrogenase is increased to compensate for the fact that the $K_m$ is about 200 μ*M*.

C. INDIRECT ASSAYS

For malate amounts of less than $0.2 \times 10^{-9}$ mole, the alternative method with transaminase as auxiliary enzyme is strongly preferred.

*Kinetics*

See Method I.

**Oxalacetate**

$$\text{Oxalacetate} + \text{DPNH} + \text{H}^+ \xrightarrow[\text{dehydrogenase}]{\text{malate}} \text{malate} + \text{DPN}^+ \quad (9\text{-}63)$$

A. SPECTROPHOTOMETER

*Reagent.* Tris-HCl buffer, pH 7.5 (10 m*M* Tris base, 40 m*M* Tris-HCl); DPNH, 50–150 μ*M* (at least a 20% excess); malic dehydrogenase (pig heart), 0.1 μg/ml (0.07 U/ml).

*Reaction Time.* Two to 4 min.

*Conduct of the Assay*

Either the enzyme or sample can be added last.

*Comment*

Oxalacetate is unstable except in strong acid (0.5 *N* HCl or stronger), in which it can be stored frozen for long periods without loss. It is actually less stable in slightly acid solutions (pH 3–5) than it is in neutral or slightly alkaline solution. The loss at 25° and pH 7 is 10%/hr.

B. FLUOROMETER DIRECT ASSAY, $0.1–8 \times 10^{-9}$ MOLE

*Reagent.* The same buffer and enzyme level as in A. DPNH, 0.2–10 $\mu M$ (20 to 75% excess).

*Reaction Time.* Less than 3 min.

*Conduct of the Assay*

The enzyme is added last after a preliminary reading. The samples should be kept frozen until shortly before analysis and should not be allowed to remain more than 30 min in the fluorometer tube before enzyme addition.

*Comment*

Although only 0.1 $\mu$g/ml of enzyme is recommended because of the extremely favorable kinetics, it is a generous amount and could be reduced if necessary except with the very lowest levels. Oxalacetate has been found to be very low in tissues. It may be necessary to use as much as 500 $\mu$l of tissue extract to obtain reliable results. In this case, the reagent can be made at twice the stated concentration and used in a volume equal to the extract.

*Kinetics*

The $V_{max}$ is over 700 $\mu$moles/mg/min. Under assay conditions the $K_m$ for DPNH is about 8 $\mu M$, and that for oxalacetate about 2 $\mu M$. This means that with 0.1 $\mu$g/ml of enzyme ($V_{max} = 70$ $\mu$mole/min), 10 $\mu M$ DPNH and 5$\mu M$ oxalacetate, the reaction should be 98% complete in 1 min [Chapter 2, Eq. (2-27)].

**6-P-Gluconate**

$$\text{6-P-Gluconate} + \text{TPN}^+ \xrightarrow[\text{dehydrogenase}]{\text{6-P-gluconate}} \text{ribulose-5-P} + \text{CO}_2 + \text{TPNH} + \text{H}^+ \quad (9\text{-}64)$$

A. SPECTROPHOTOMETER

*Reagent.* Imidazole acetate, pH 7.0 (30 m$M$ imidazole base, 20 m$M$ imidazole acetate); $MgCl_2$, 5 m$M$; dithiothreitol, 0.5 m$M$; ammonium acetate, 30 m$M$; EDTA, 1 m$M$; $TPN^+$, 500 $\mu M$; 6-P-gluconate dehydrogenase (yeast), 5 $\mu$g/ml (0.06 U/ml).

*Reaction Time.* Eight to 10 min.

*Conduct of the Assay*

It may be preferable to add the enzyme last after an initial reading has been made. The enzyme tends to lose part of its activity at high dilution.

*Comment*

The reaction is faster with higher $TPN^+$ levels, but, due to impurities usually present, extra $TPN^+$ will increase the initial blank reading more than is desirable.

B. Fluorometric Direct Analysis, $0.1\text{–}10 \times 10^{-9}$ Mole

*Reagent.* The ingredients are the same as in A except as follows: $TPN^+$, 100 $\mu M$ (may be reduced to 20 $\mu M$ for 1 $\mu M$ 6-P-gluconate or less); 6-P-gluconate dehydrogenase, 2 $\mu$g/ml.

*Reaction time.* Five to 10 min.

*Conduct of the Assay*

The enzyme is added last.

*Comment*

Tissue levels are apt to be very low. It may be desirable to increase the imidazole buffer strength to 0.1 $M$ to quench flavin fluorescence. We have not had experience with the application of cycling amplification to 6-P-gluconate measurement, but there would be a distinct advantage even when the amount of tissue is not limiting. No difficulty is forseen in such adaptation. See, for example, the glucose-6-P methods.

*Kinetics*

The kinetics are favorable, in spite of the low $V_{max}$ (12 $\mu$mole/mg/min) because of the low Michaelis constants for both substrates. The $K_m$ for 6-P-gluconate is affected by reagent composition. As shown by Pontremoli *et al.* (1961), $NH_4^+$ and $Mg^{2+}$ each have a favorable affect by lowering this $K_m$, which under assay conditions is about 20 $\mu M$. The additives do not have much effect on the $V_{max}$. The $K_m$ for $TPN^+$ is 2 to 4 $\mu M$. In spite of this low value, when measuring higher levels of 6-P-gluconate it is necessary to add high levels of $TPN^+$ because TPNH is an exceedingly potent inhibitor, competitive with $TPN^+$. In the spectrophotometer under assay conditions, because of this inhibition, the reaction follows almost a first-order rate curve, whereas one would expect an almost linear course with both subtrates so much higher than their Michaelis constants.

## 3-P-Glycerate: Method I

$$\text{3-P-Glycerate} + \text{ATP} \xrightarrow[\text{kinase}]{\text{P-glycerate}} \text{1,3-diphosphoglycerate} + \text{ADP} \quad (9\text{-}65)$$

$$\text{1,3-Diphosphoglycerate} + \text{DPNH} + \text{H}^+ \xrightarrow[\text{dehydrogenase}]{\text{glyceraldehyde-P}} \text{glyceraldehyde-P} + \text{DPN}^+ + \text{P}_i \quad (9\text{-}66)$$

This is the simpler of the two methods offered, but has the disadvantage for direct analyses that it is based on DPNH disappearance. Therefore, the DPNH concentration must be chosen with care. For indirect analyses $DPN^+$ appearance can be measured; this method is therefore preferable in this case, except possibly with extremely small volumes (see below).

### A. Spectrophotometer

*Reagent.* Imidazole-HCl buffer, pH 7.1 (30 m$M$ imidazole base, 20 m$M$ imidazole-HCl); mercaptoethanol, 2 m$M$; $MgCl_2$, 1 m$M$; NaCl, 20 m$M$; ATP, 1 m$M$; DPNH, 100 to 150 $\mu M$ (at least a 20% excess); P-glycerate kinase (rabbit muscle), 2 $\mu$g/ml (0.36 U/ml); glyceraldehyde-P dehydrogenase (rabbit muscle), 5 $\mu$g/ml (0.18 U/ml).*

*Reaction Time.* Three to 5 min.

*Conduct on Assay*

For standardization purposes the 3-P-glycerate can be added last, eliminating the need for a separate blank. Otherwise the sample is added first. The enzymes can be added in either order or together as a mixture. In this case 1,3-diphosphoglycerate would be included in the assay, however, the levels in biological samples are exceedingly low and, moreover, 1,3-diphosphoglycerate is extremely unstable and would disappear from extracts without special precautions.

### B. Fluorometer Direct Assay, 0.2–8 × $10^{-9}$ Mole

*Reagent.* This is the same as for the spectrophotometer except for the following changes: DPNH, 0.5 to 10 $\mu M$ (a 20–80% excess); ATP, 300 $\mu M$.

*Reaction Time.* Five to 10 min.

* Most of the $(NH_4)_2SO_4$ should be removed from this enzyme by centrifugation.

*Conduct of the Assay*

The enzymes are ordinarily added last after the sample. Otherwise, if glyceraldehyde-P dehydrogenase is added to the reagent before the sample, it may be desirable to add 0.01% bovine serum albumin to protect it from partial inactivation.

*Comment*

Sulfate increases the $K_m$ for glyceraldehyde-P and should be kept below 1 m*M*. Indirect procedures have not been presented, but would be of advantage at low levels of substrate. After allowing the reaction to proceed in a volume of 200 $\mu$l or less, excess DPNH would be destroyed and the $DPN^+$ measured by the strong alkali procedure. This would permit greater leeway in regard to excess DPNH, and would diminish the significance of the blank contribution from the biological material. For very small volumes, where cycling is required, DPNH oxidation in the blanks becomes a problem. If ascorbic acid does not provide a solution (Chapter 4) it may be preferable to use Method II, in which DPNH rather than $DPN^+$ is the product to be measured.

*Kinetics*

The kinetics of P-glycerate kinase are quite favorable. Under analytical conditions the $V_{max}$ is about 400 $\mu$moles/mg/min. The kinetic relationships are somewhat complicated. Half-maximal velocity is obtained with about 500 $\mu M$ ATP and 500 $\mu M$ 3-P-glycerate; however, the rate with 20 $\mu M$ 3-P-glycerate is 20% of the maximum rather than 4% as expected from "normal" kinetics. NaCl, as well as other salts, has a favorable effect by lowering the $K_m$ for 3-P-glycerate.. With 300 $\mu M$ ATP, as recommended for the fluorometric method, the half-time with low levels of 3-P-glycerate would be 15 sec for 1 $\mu$g/ml of enzyme.

The kinetics of glyceraldehyde-P dehydrogenase are not quite so favorable. The $V_{max}$ is moderately high, probably about 150 $\mu$moles/mg/min, the $K_m$ for DPNH is about 2 $\mu M$ and that for 1,3-diphosphoglycerate is also low, 15 or 20 $\mu M$. The problem is that the equilibrium constant for [ATP][3-PGA]/[ADP][DPGA] is about 2000 at pH 7, i.e., very unfavorable. With 1 m*M* ATP, the 3-P-glycerate level required to give half-maximal activity (glyceraldehyde-P dehydrogenase rate-limiting) is about 200 $\mu M$. With 300 $\mu M$ ATP it is nearer 1 m*M*. Offsetting this is the fact that with ATP held constant, the diphosphoglycerate equilibrium level is roughly the square root of the 3-P-glycerate level. Consequently with 300 $\mu M$ ATP, reduction of 3-P-glycerate from 800 $\mu M$ to 3$\mu M$ (266-fold) only reduces the rate about 30-fold.

## 3-P-Glycerate: Method II

$$\text{3-P-Glycerate} + \text{ATP} \xrightarrow[\text{kinase}]{\text{P-glycerate}} \text{1,3-diphosphoglycerate} + \text{ADP} \tag{9-67}$$

$$\text{1,3-Diphosphoglycerate} + \text{DPNH} + \text{H}^+ \xrightarrow[\text{dehydrogenase}]{\text{glyceraldehyde-P}} \text{glyceraldehyde-P} + \text{DPN}^+ + \text{P}_i \tag{9-68}$$

Destruction of DPNH with acid (9-69)

$$\text{Dihydroxyacetone-P} \xrightarrow[\text{isomerase}]{\text{triose}} \text{glyceraldehyde-P} \tag{9-70}$$

$$\text{Glyceraldehyde-P} + \text{DPN}^+ \xrightarrow[\text{GAP dehydrogenase}]{\text{arsenate}} \text{3-P-glycerate} + \text{DPNH} + \text{H}^+ \tag{9-71}$$

This is more complicated than the alternative method, which consists of reactions (9-67) and (9-68) alone. It has the advantage that readings increase with increased substrate, and it is consequently more flexible and is more accurate with lower substrate levels.

Current preparations of P-glycerate kinase are badly contaminated with triose-P dehydrogenase. This means that at the time of acidification, most of the 3-P-glycerate has been converted to dihydroxyacetone-P. Consequently it is necessary to include triose-P isomerase in the reagent after acidification. In spite of the complexity of this method it has given highly reproducible results without difficulty.

No spectrophotometric assay is presented because it is simpler and completely satisfactory to standardize 3-P-glycerate solutions with the first method.

### A. Fluorometer, Direct Measurement of Final Steps, 0.1–10 × $10^{-9}$ mole

*Reagent before Acidification.* Imidazole-HCl buffer, pH 7 (18 m*M* imidazole base, 12 m*M* imidazole-HCl); $MgCl_2$, 1 m*M*; mercaptoethanol, 5 m*M*; ATP, 100 $\mu M$; NaCl, 20 m*M*; DPNH, 20 $\mu M$ (5 $\mu M$ with 1 $\mu M$ 3-P-glycerate or less); P-glycerate kinase (rabbit muscle), 1 $\mu$g/ml (0.18 U/ml); glyceraldehyde-P-dehydrogenase* (rabbit muscle), 10 $\mu$g/ml (0.36 U/ml).

*Reagent Composition during Final Step.* Imidazole-HCl buffer, pH 7.0 (78 m*M* imidazole base, 52 m*M* imidazole-HCl); $MgCl_2$, mercaptoethanol, ATP, and NaCl from the first step essentially unchanged; $DPN^+$, 100 $\mu M$; $Na_2HAsO_4$, 1 m*M*; triose-P isomerase (rabbit muscle), 1 $\mu$g/ml (2.4 U/ml); glyceraldehyde-P dehydrogenase*, 20 $\mu$g/ml.

* Centrifuged to remove most of the $(NH_4)_2SO_4$.

*Procedure*

*Step 1.* The sample is added to 1 ml of reagent in a fluorometer tube. The P-glycerate kinase can be incorporated in the reagent, however it is preferable to add the dehydrogenase last after the samples, because otherwise it loses part of its activity. After 5 min at room temperature 10 $\mu$l of 4 $N$ HCl are added and the tube is allowed to stand for at least 10 min at room temperature. Twenty microliters of 5 $M$ imidazole base are then added, followed by 10 $\mu$l of a mixture of 10 m$M$ $DPN^+$ in 100 m$M$ $Na_2HAsO_4$.

*Step 2.* After making a reading, 4 $\mu$l of a mixture containing 0.5% glyceraldehyde-P dehydrogenase and 0.02% triose-P isomerase are added. A final reading is made when the reaction is over (5–10 min).

*Comment*

Most of the $(NH_4)_2SO_4$ is removed from the glyceraldehyde-P dehydrogenase because it is inhibitory to triose-P isomerase and the glyceraldehyde-3-P dehydrogenase. Imidazole-HCl rather than the acetate is used as the buffer to decrease the amount of HCl required for acidification. NaCl is added because of its favorable effect on P-glycerate kinase kinetics (see Kinetics in Method I). The first-step reaction is expected to be finished in 1 or 2 min. It is desirable to check this directly in the fluorometer each day with an extra tube to which 3-P-glycerate has been added equal to 50–70% of the DPNH level. Otherwise, no readings are necessary in Step 1; however, readings on one tube in the set can be used to check on the time required for complete DPNH destruction (expected to be less than 5 min).

The assay as described measures the sum of 3-P-glycerate and the triose phosphates. As a rule, 3-P-glycerate is likely to be present in greater amounts than the triose phosphates. When necessary correction can be made by a separate measurement of the triose phosphates. Alternatively they can be destroyed beforehand by heating 10 min at 60° in 0.02 $N$ NaOH; 3-P glycerate is extremely stable in alkali.

*Kinetics*

The kinetics of the reactions in Step 1 are given in Method I. The kinetics of the reactions in Step 2 are described in Method II for dihydroxyacetone-P.

**P-Pyruvate**

$$\text{P-Pyruvate} + \text{ADP} \xrightarrow[\text{kinase}]{\text{pyruvate}} \text{pyruvate} + \text{ATP} \quad (9\text{-}72)$$

$$\text{Pyruvate} + \text{DPNH} + \text{H}^+ \xrightarrow[\text{dehydrogenase}]{\text{lactate}} \text{lactate} + \text{DPN}^+ \quad (9\text{-}73)$$

Levels of P-pyruvate are low in most tissues. Because lactate dehydrogenase reacts with pyruvate and other keto acids present in tissue extracts, precision is greatly improved if these keto acids are removed with hydrazine prior to the assay. During the assay lactic dehydrogenase is used in sufficient amount to reduce pyruvate before it has a chance to react with the hydrazine. Because the spectrophotometric assay will ordinarily be used only for standardization, the hydrazine step is omitted.

A. SPECTROPHOTOMETER

*Reagent.* Phosphate buffer, pH 7 (30 m*M* $K_2HPO_4$ : 20 m*M* $NaH_2PO_4$); $MgCl_2$, 2 m*M*; ADP, 200 $\mu M$; DPNH, 50–150 $\mu M$ (at least a 20% excess); pyruvate kinase, 1 $\mu$g/ml (0.15 U/ml); lactic dehydrogenase (beef heart), 1 $\mu$g/ml (0.2 U/ml).

*Reaction Time.* Two to 4 min.

*Conduct of the Assay*

Unless pyruvate is also to be measured, the reagent may be prepared with lactic dehydrogenase added but without the pyruvate kinase. The sample is added. A reading is made after 1 or 2 min to allow time for possible presence of pyruvate. Pyruvate kinase is then added.

*Comment*

For standardization of P-pyruvate solutions there is no reason to pretreat the sample with hydrazine. If, however, P-pyruvate is to be measured spectrophotometrically in the presence of a larger amount of pyruvate the pretreatment may be carried out exactly as described in the next protocol and with the same reagent (enzymes omitted). Notice the increase in lactic dehydrogenase. Because the hydrazine pretreatment takes 20 or 30 min, this would be conveniently carried out in test tubes ahead of time. The sample would then be transferred to the spectrophotometer cell, lactic dehydrogenase added, a reading made, and the reaction started with pyruvate kinase.

B. FLUOROMETER DIRECT ASSAY, 0.1–8 × $10^{-9}$ MOLE

*Reagent.* The same as in A except: DPNH, 0.2–10 $\mu M$ (to provide a 20–75% excess); hydrazine-HCl, 10 m*M*; lactic dehydrogenase, 2.5 $\mu$g/ml (increased to 5 $\mu$g/ml with DPNH less than 1 $\mu M$).

*Reaction Time.* Two to 5 min.

*Conduct of the Assay*

The reagent is prepared without either enzyme. The sample is added to reagent in the fluorometer tube. After 20–30 min at room temperature, lactic dehydrogenase is added and a reading made (1 or 2 min delay is sufficient to make sure that any trace of pyruvate in sample or reagent has reacted).

The analytical reaction is then started with pyruvate kinase. The final reading should be made within 30 min (5 min should be sufficient) because the pyruvate reaction with hydrazine is slowly reversible.

Imidazole can be substituted for phosphate to diminish fluorescence of tissue flavins. With 200 m*M* imidazole the pyruvate kinase and lactate dehydrogenase should both be doubled. If imidazole is used KCl is added at a concentration of 75 m*M* to take care of the $K^+$ requirement of pyruvate kinase.

*Kinetics*

The approximate kinetic constants for pyruvate kinase at pH 7 in 50 m*M* phsophate buffer are $V_{max}$, 100 μmoles/mg/min (100 U/mg); $K_m$ for P-pyruvate, 30 μ*M*; $K_m$ for ADP, 200 μ*M*. From these data, the calculated half-time for 10 μ*M* substrate or less with 200 μ*M* ADP and 1 μg/ml of enzyme would be 0.4 min [0.7(30/100)(400/200)]. In 200 m*M* imidazole at the same pH, the velocity with low P-pyruvate is a third to half as fast, probably due to an increase in the Michaelis constant.

Pyruvate kinase requires both $Mg^{2+}$ and $K^+$.

The reaction of pyruvate with the large excess of hydrazine proceeds at a rate which can be considered as a first-order reaction for pyruvate. The rate constant for this reaction at 25° with 10 m*M* hydrazine is about 0.2/min (0.02/min multiplied by the millimolar concentration of hydrazine), giving a half-time of 3 or 4 min. For 99% of liberated pyruvate to react with DPNH, the lactic dehydrogenase apparent first-order rate constant for pyruvate must be 100 times faster than that with hydrazine, or 20/min. Even with 100 μ*M* P-pyruvate the free pyruvate concentration will never exceed first-order concentrations, therefore the same lactic dehydrogenase concentration is appropriate for high and low P-pyruvate concentrations. However, at very low levels if the DPNH concentration is reduced below 1 μ*M*, an increase in lactic dehydrogenase is indicated.

**Pyruvate**

$$\text{Pyruvate} + \text{DPNH} + \text{H}^+ \xrightarrow[\text{dehydrogenase}]{\text{lactic}} \text{lactate} + \text{DPN}^+ \quad (9\text{-}74)$$

Although pyruvate is not as abundant a contaminant as lactate, it is nevertheless widespread, and because tissue levels are much lower than in the case of lactate, the danger from contamination is almost as great.

A. SPECTROPHOTOMETER

*Reagent.* Phosphate buffer, pH 7 (30 m*M* $K_2HPO_4$ : 20 m*M* $NaH_2PO_4$); DPNH, 50–150 μ*M* (at least a 20% excess); lactic dehydrogenase (beef heart), 0.5 μg/ml (0.1 U/ml).

*Reaction Time.* Two to 4 min.

*Conduct of the Assay*

Ordinarily the sample is added first. After an initial reading the reaction is started with enzyme added in small volume. For standardizing pure solutions the enzyme can be incorporated in the reagent and the reaction started with the sample, eliminating the need for a blank.

B. FLUOROMETER DIRECT ANALYSIS, 0.1 TO $\times 10^{-9}$ MOLE

*Reagent.* Phosphate buffer as in spectrophotometer; DPNH, 0.2–10 $\mu M$ (20–75% excess over expected pyruvate); lactic dehydrogenase, 0.2 $\mu$g/ml (1 $\mu$g/ml with less than 1 $\mu M$ DPNH).

*Reaction Time.* Three to 10 min.

*Conduct of the Assay*

The reaction is ordinarily started by addition of the enzyme after the sample is in and an initial reading made.

*Comment*

Pyruvate stock solutions are not indefinitely stable even frozen, due to a tendency to polymerize. Therefore, when used as fluorometer standards they should be occasionally restandardized in the spectrophotometer. Heart lactate dehydrogenase can strongly enhance the fluorescence of DPNH. At the enzyme level recommended this is scarcely noticeable. This is handled by running suitable standards and blanks. This is one reason for not using more enzyme than necessary. A more important reason is that lactic dehydrogenase reacts with many keto acids besides pyruvate, although at a much slower rate. If a large excess of enzyme is used, certain tissues, brain for example, will give an erroneously high result. If tissue blank fluorescence from flavins is a problem, this can be greatly reduced by substituting 100 or 200 m$M$ imidazole buffer, pH 7, for the phosphate. With the higher imidazole concentration lactic dehydrogenase should be increased threefold or the reaction time prolonged (see Kinetics below).

C. FLUOROMETER INDIRECT ANALYSIS, 1–10 $\times 10^{-11}$ MOLE

*Reagent.* Imidazole buffer, pH 7 (30 m$M$ imidazole, 20 m$M$ imidazole-HCl, hydrochloride); DPNH, 2 $\mu M$; bovine plasma albumin, 0.01%; lactic dehydrogenase, 0.2 $\mu$g/ml.

The reagent should be prepared within an hour of use and kept on ice. The DPNH is added after the ascorbic acid and should be as free as possible of $DPN^+$ (see Chapter 1). The ascorbic acid is added from a 100 m$M$ stock solution stored at $-20°$ or below.

*Reaction Vessels.* Fluorometer tubes.

*Procedure*

*Step 1.* One hundred microliters of reagent at 0°. Standards, 10 $\mu$l of 1 to 10 $\mu M$ pyruvate. Samples, in volumes of 20 $\mu$l or less.

*Step 2.* Twenty minutes at room temperature (water bath) and return to ice bath.

*Step 3.* HCl, 20 $\mu$l of 2 $N$; room temperature at least 5 min.

*Step 4.* NaOH, 1 ml of 6 $N$; 10 min 60°.

*Step 5.* Fluorescence reading exactly at room temperature.

*Comment*

This indirect assay is based on measurement of $DPN^+$. In general, living cells contain more $DPN^+$ than pyruvate, often much more. Therefore in analyzing tissues or microorganisms with this indirect analysis it would usually be necessary to destroy most of the $DPN^+$. This can be accomplished by heating in alkali as in D below. Pyruvate is not indefinitely stable in alkali, therefore the alkaline treatment should not be excessive (see Chapter 1 for minimal conditions for $DPN^+$ destruction). The buffer is changed from phosphate to imidazole because phosphate buffers cause appreciable DPNH oxidation.

D. Indirect Assay, 1–5 × $10^{-12}$ Mole (Cycling)

*Pyruvate Reagent.* The same buffer and DPNH concentration as in C. Ascorbic acid, 2 m$M$; bovine plasma albumin, 0.02%; lactic dehydrogenase 0.1 $\mu$g/ml. The ascorbic acid is added before the DPNH from a stock 9% (50 m$M$) solution which is stored at −20°.

*Cycling Reagent.* $DPN^+$ cycling reagent (Chapter 8) designed for final glutamate measurement, i.e., $(NH_4)_2SO_4$ concentration reduced to 1 m$M$. Glutamic dehydrogenase and lactic dehydrogenase are 400 $\mu$g/ml and 50 $\mu$g/ml, respectively. [The latter is centrifuged to remove most of the $(NH_4)_2SO_4$.]

For glutamate reagent see Chapter 8.

*Reaction Vessels.* Oil wells.

*Procedure for Frozen Dried Tissue Samples*

*Step 1.* Use 1 $\mu$l of 0.02 $N$ NaOH for samples and blanks, 1 $\mu$l of 2 to 5 $\mu M$ pyruvate in 0.02 $N$ NaOH for standards; heat 20 min at 80°.

*Step 2.* Add 5 $\mu$l of pyruvate reagent; incubate 20 min at room temperature.

*Step 3.* Add 2 $\mu$l of 0.5 $N$ HCl. Allow to stand at least 30 min.

*Step 4.* Transfer 2 $\mu$l into 50 $\mu$l of cycling reagent in a fluorometer tube. Incubate 1 hr at 38°; heat 10 min at 95°.

*Step 5.* Add 1 ml of glutamate reagent. After 20 min add 5 $\mu$l of 1% glutamic dehydrogenase in 50% glycerol. Read after 20–30 min.

### E. Indirect Assay, 1–5 × $10^{-13}$ Mole

The reagents are the same as in D.

*Reaction Vessels.* Oil wells.

*Procedure*

*Steps 1 and 2.* The same as in D but all volumes reduced 10-fold.

*Step 3.* Ten microliters of cycling reagent are added to the sample in the oil. Incubation 2 hr at 38°; heat 10 min at 95°.

*Step 4.* Add 1 $\mu$l of 1 *N* NaOH to each sample; heat 5 min at 95°.

*Step 5.* Transfer 10 $\mu$l aliquots to fluorometer tubes containing 100 $\mu$l of glutamate reagent. Incubate 20 min then add 1 $\mu$l of 0.5% glutamic dehydrogenase. After 30 min add 1 ml of $H_2O$; mix and read.

*Kinetics*

The kinetic situation is favorable. The $V_{max}$ is large, 200–250 $\mu$moles/mg/min, and the Michaelis constants are small. Increase in one substrate increases the apparent Michaelis constant for the other. At pH 7 in 50 m*M* phosphate buffer the apparent $K_m$ for pyruvate is 30 $\mu M$ with 100 $\mu M$ DPNH, but only 6 $\mu M$ with 0.25 $\mu M$ DPNH. Conversely, the $K_m$ for DPNH is about 2.5 $\mu M$ with 400 $\mu M$ pyruvate and only 1 $\mu M$ with 3 $\mu M$ pyruvate. The practical result is that the amount of lactic dehydrogenase does not need to be increased as much as otherwise expected to measure small concentrations of pyruvate with low levels of DPNH.

The equilibrium is very favorable at pH 7. Starting with equal concentrations of pyruvate and DPNH the reaction would proceed until only 0.5% of the initial pyruvate remained.

Because of the use of lactic dehydrogenase as an auxiliary enzyme in a number of analyses, its kinetics are of interest under several conditions. At pH 8 in 50 m*M* Tris or glycylglycine buffers, the $V_{max}$ is at least as great as at pH 7, and the $K_m$ for DPNH is if anything more favorable, but the $K_m$ for pyruvate is increased 8- or 10-fold. This means that more enzyme must be used. Dilute imidazole (50 m*M*) at pH 7 is about as favorable as phosphate, but if the concentration is increased to 200 m*M*, three times as much enzyme is required because of an increase in the $K_m$ for pyruvate.

Skeletal muscle lactic dehydrogenase has less favorable kinetics than the heart enzyme for measuring pyruvate. Although the $V_{max}$ is 50% greater, the $K_m$ for pyruvate is five or six times larger. The $K_m$ for DPNH is about the same for both enzymes.

**Total Nucleotide Triphosphates**

$$\text{XTP} + \text{Fructose-6-P} \xrightarrow{\text{P-fructokinase}} \text{XDP} + \text{fructose-1,6-P}_2 \quad (9\text{-}75)$$

$$\text{Fructose-1,6-P}_2 \xrightarrow{\text{aldolase}} \text{dihydroxyacetone-P} + \text{glyceraldehyde-P} \quad (9\text{-}76)$$

$$\text{Glyceraldehyde-P} \xrightarrow[\text{isomerase}]{\text{triose}} \text{dihydroxyacetone-P} \quad (9\text{-}77)$$

$$2\ \text{Dihydroxyacetone-P} + 2\ \text{DPNH} \xrightarrow[\text{dehydrogenase}]{\text{glycero-P-}} 2\ \text{2-glycero-P} + 2\ \text{DPN}^+ + 2\text{H}^+ \quad (9\text{-}78)$$

The procedure is adapted from an unpublished method of Dr. Stanley R. Nelson; we wish to thank him for allowing us to include it. P-Fructokinase reacts with all purine and pyrimidine nucleotide triphosphates tested (ATP, ITP, GTP, CTP, UTP, TTP).

A. SPECTROPHOTOMETER

*Reagent*. Imidazole-HCl buffer, pH 7.0 (25 m*M* imidazole base, 25 m*M* imidazole-HCl); fructose-6-P, 300 $\mu M$; $MgCl_2$, 2 m*M*; $K_2HPO_4$, 5 m*M*; DPNH, 50–150 m*M* (20–50% excess); P-fructokinase (rabbit muscle), 0.12 U/ml (2 $\mu$g/ml of enzyme of about 50% purity); aldolase (rabbit muscle), 1 $\mu$g/ml (0.1 U/ml); triose-P-isomerase (rabbit muscle), 0.5 $\mu$g/ml (1.2 U/ml); glycero-P-dehydrogenase (rabbit muscle), 4 $\mu$g/ml (0.16 U/ml).

*Reaction Time*. Four to 10 min, depending on the nucleotide.

*Conduct of the Assay*

Unless it is necessary to measure fructose-1,6-$P_2$ in the same sample, all enzymes except P-fructokinase can be incorporated in the reagent. The sample is added, and after taking an initial reading, P-fructokinase is added.

*Comment*

It is desirable to test the system with UTP, which is one of the slower reacting ribonucleotides. In the case of desoxyribonucleotides it may be necessary to increase the amount of P-fructokinase. TTP, at least, reacts rather slowly. P-Fructokinase is very unstable in simple buffers, particularly at pH 7 or below. Dilutions of stock enzyme preparations should be made in pH 8 Tris buffer containing 10 m*M* $K_2HPO_4$. Phosphate is a potent stabilizer and is included in the reagent itself for this reason.

### B. Fluorometer Direct Assay, 0.05–4 × $10^{-9}$ Mole

*Reagent.* The reagent is the same as in A except as follows: DPNH, 0.2–10 $\mu M$ (20–75% excess); P-fructokinase, 0.06 U/ml; aldolase, 1 $\mu$g/ml; glycero-P-dehydrogenase, 1 $\mu$g/ml.

*Reaction Time.* Four to 8 min.

*Conduct of the Assay*

See A above.

*Comment*

In the case of animal tissues it will seldom be desirable to measure fructose-1,6-$P_2$ in the same sample as the total triphosphates because of the large discrepancy in levels. In bacterial extracts this may not always be true, in which case a stepwise addition of the enzymes can be made. (see also Comments under A.)

*Kinetics*

For P-fructokinase with specific activity of 60 $\mu$moles/mg/min (60 U/mg) the half-times for substrate concentrations of 10 $\mu M$ or less, and 1 $\mu$g/ml of P-fructokinase, would be 0.5–1 min for ATP and GTP and 1–2 min for ITP, CTP, and UTP. This refers to the P-fructokinase step only. The $K_m$ for fructose-6-P is less than 50 $\mu M$.

## Uridine-5′-Diphosphoglucose

$$\text{UDP-glucose} + 2\ \text{DPN}^+ \xrightarrow[\text{dehydrogenase}]{\text{UDPG}} \text{UDP-glucuronate} + 2\ \text{DPNH} + 2\ \text{H}^+ \qquad (9\text{-}79)$$

### A. Spectrophotometer

*Reagent.* Tris-HCl buffer, pH 8.1 (25 m$M$ Tris base, 25 m$M$ Tris-HCl); $MgCl_2$, 2 m$M$; $DPN^+$, 1 m$M$; UDP-glucose dehydrogenase 0.008 U/ml (200 units/ml as given by Sigma Chemical Co.)

*Reaction Time.* Ten to 20 min.

*Conduct of the Assay*

For standardization purposes it is preferable to add the UDP-glucose last because the enzyme adds appreciably to the absorption at 340 nm.

*Comment*

The enzyme solution loses activity at 4°, but keeps well frozen at −40°.

### B. Fluorometer Direct Analysis, 0.05–5 × $10^{-9}$ Mole

*Reagent*. The buffer and $MgCl_2$ are the same as in A. $DPN^+$, 100 $\mu M$ (20 $\mu M$ with 1 $\mu M$ UDPG or less); UDP-glucose dehydrogenase, 0.004 U/ml (0.002 U/ml with 1 $\mu M$ UDPG or less).

*Reaction Time*. Ten to 20 min.

#### *Conduct of the Assay*

Ordinarily the enzyme is added last. Because of the amount of enzyme needed with current preparations, there is likely to be an appreciable contribution of fluorescence. In some instances a small drift in the blank reading has also been observed.

#### *Kinetics*

Currently available preparations of UDP-glucose are far from pure, but because of the low Michaelis constants are quite satisfactory. The $K_m$ for UDP-glucose is of the order of 5 $\mu M$, that of $DPN^+$ about 10 $\mu M$ under analytical conditions. DPNH is a very strong competitive inhibitor of $DPN^+$. Consequently the reaction rate falls off even when both substrates are far above their $K_m$'s. When UDP-glucose is in excess and $DPN^+$ is in the 50 $\mu M$ range, it is virtually impossible to drive the reaction to completion with $DPN^+$ because of this inhibition. In the measurement of UDP-glucose it is advisable to provide a 10-fold excess of $DPN^+$.

PART II

# QUANTITATIVE HISTOCHEMISTRY

# INTRODUCTION

The following portion of this book represents a coherent system of quantitative analysis for histochemical purposes. It derives from the original system of Linderstrøm-Lang and Holter, but differs in several significant details. In the original methodology of the Carlsberg Laboratory, serial frozen sections were cut from a uniform tissue block. Alternate sections were either analyzed, or fixed and stained for histological control. This procedure has been further developed and exploited with great success by David Glick and is fully described in book form (Glick, 1961, 1963).

One essential change in the system to be described is that before analysis the sections are dried from the frozen state and then cleanly dissected into defined histological portions. This permits direct histological control of the actual sample analyzed and the study of very small structures which might not extend into adjacent sections. One of the virtues of this modification is that most biological materials are relatively stable in the frozen dried state. Consequently a given set of sections, perhaps from a tissue in an otherwise transient physiological state, can serve for years as a source of material for analysis.

A second essential modification from the Carlsberg procedure is that sample size is determined by dry weight rather than by the dimensions of the tissue section. This permits analysis of irregular pieces and of much smaller specimens.

Finally, sensitivity has been increased almost without limit by the introduction of enzymatic cycling and the oil well technique. These changes now permit single cells or even parts of single cells to be analyzed for enzymes and metabolites of all kinds, even when present at very low concentrations.

The complete analytical system is designed for study of solid tissues. It is, of course, possible to use the tools and methods for other purposes. For example, the starting material might be a frozen dried ovum or protozoan. The fishpole balance might be used for weighing single pollen grains. The oil well technique combined with enzymatic cycling could be used for measurement of the lactate production of single bacteria or the activity of single enzyme molecules.

CHAPTER 10

# PREPARATION OF TISSUES AND SECTIONS

## Freezing

The procedure used in preparation of tissues for histochemical analysis will vary somewhat with the intended purpose. If the material is to be used for the analysis of biologically labile metabolites the tissue is frozen *in situ* or as rapidly as possible after removal, to preserve the *in vivo* state (see Chapter 7). When the tissue is intended for the measurement of enzyme activity, rapid changes following removal of the tissue are less likely to interfere. In either case the freezing process itself is made as rapid as possible and an effort is made to keep ice crystals small and thereby maintain the structural integrity of the tissue.

Samples larger than a gram or two are probably best frozen directly in liquid $N_2$ (Chapter 7). If the sample is a gram or less it is plunged with vigorous, stirring, into Freon-12 ($CCl_2F_2$) brought to its freezing point (−150°) with liquid $N_2$. With small samples the freezing is faster than with liquid $N_2$ itself. The use of chilled propane or isopentane is avoided because of the fire hazard. This hazard is enhanced by the danger of condensation of $O_2$ from the air, even though the chilling is accomplished with liquid $N_2$.

For small samples another means of freezing is available, which was originally suggested by Dr. Fernandez Moran. It is particularly convenient for situations such as may occur in the operating room, when samples are presented on short notice. This is to freeze by immersion in liquid nitrogen which has been chilled from its boiling point (−196°) to its freezing point (−210°) by rapid evaporation. Ordinarily the gas which is evolved at the surface of liquid $N_2$ at its boiling point delays heat transfer and is thus not satisfactory for rapid freezing. The use of $N_2$ chilled below its boiling point prevents the formation of gas. To chill the $N_2$ the Dewar flask is connected to a vacuum pump through a rubber hose and rubber stopper. A portion of the liquid $N_2$ is pumped off. Freezing of the liquid $N_2$ begins when about 10% of the volume has been evaporated. In the Dewar flask, the liquid $N_2$ will remain below its boiling point for an hour or two. If a substantial part of the liquid $N_2$ is frozen by the evaporation process it will stay below its boiling point for even longer periods. This method is not suitable for large samples because the heat capacity of liquid $N_2$ is low.

When identification is needed, and time permits, tissue samples may be placed on a piece of labeled hardened filter paper before freezing.

### Storage of Frozen Tissues

The frozen tissue samples can be stored before cutting sections but the temperature must be kept very low to prevent ice crystal growth, enzyme action, or possible migration of constituents. As discussed in Chapter 7, labile metabolites are drastically changed in a few hours at −20°, and a few minutes at −5°. Many enzymes are unstable at −20° during storage for a few weeks. On the other hand, after several months at −80° tissues have been cut without noticeable crystal growth or loss of enzyme activity, and after a week at −35° there is no discernible loss of ATP or P-creatine. In general, however, it is recommended that sections be cut and dried as soon after preparation as convenient, since the dangers of loss or migration of materials are greatly reduced after drying.

### Mounting the Tissue for Cutting

Experience has shown that mounting the block of tissue for cutting is one of the most critical steps. It is important to mount the tissue without thawing, since this would result in rapid diffusion of soluble compounds and formation of large ice crystals on refreezing.

The frozen tissue is softened somewhat by allowing it to warm up to −15° to −20°. Blocks suitable for mounting are cut out with a razor blade. If labile metabolites are to be measured, the time at this warmer temperature level should be kept to a minimum (see Chapter 7). In this case, until ready to mount, the blocks should be kept at dry ice temperature. Wooden dowel rods, 5–6 cm long, 5–8 mm in diameter, and drilled at one end to form a shallow

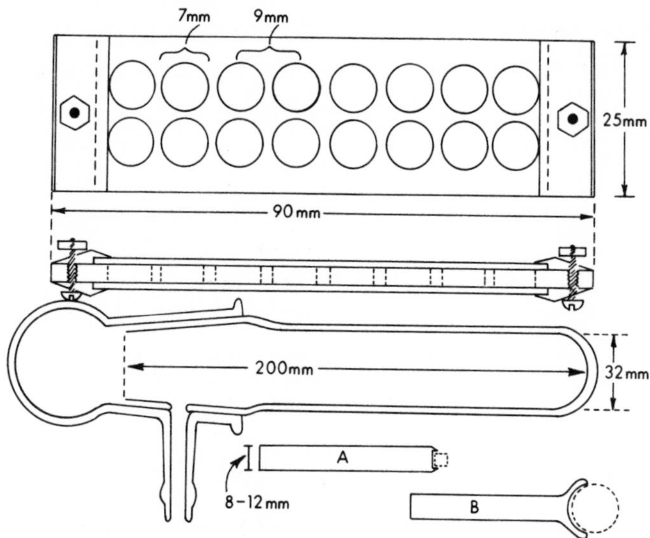


***Fig 10-1.*** Aluminum holder for tissue sections, glass vacuum tube for drying and storing sections (in aluminum holders), and wooden holders for tissue blocks. Wooden holder A is for routine use with tissue blocks that can be trimmed to fit the slight depression. Holder B is an example of a special holder for use with an eyeball for cutting retinal sections.

cup are used for mounting (Fig. 10-1). The diameter of the dowel and depth of cup will vary with the tissue to be mounted. A paste of brain tissue, prepared for this purpose, is applied in generous amount to the dowel cup. (Adherence to the tissue may be improved if the paste is diluted with one-third volume of $H_2O$.) With the paste just above freezing, the tissue block is picked up with insulated forceps and pushed into the paste with suitable orientation and at once plunged into the freezing medium. This medium consists of hexane or heptane in a small beaker which is maintained at a low temperature by placing it inside a larger beaker containing dry ice. Liquid $N_2$ or Freon-12 chilled to $-150°$, are not suitable for this purpose because the temperature is too cold and the tissue blocks may crack. To mount successfully the block should be cold enough to freeze the brain paste rather than itself be melted. To make sure this is the case, the tissue block, if not already at dry ice temperature, can be placed briefly on a small piece of dry ice.

As soon as safely frozen in place in the hexane or heptane bath, the block and holder are removed and excess solvent is blotted off. If difficulty is encountered in having the block crack loose during sectioning, it can be further anchored in place by painting on a little more of the paste after the solvent has been completely removed. The mounted tissue is kept at $-50°$ or below until just before cutting, particularly if labile metabolites are to be measured.

**Sectioning the Tissue**

The holder and block of tissue are placed in a microtome in a cryostat. We strongly recommend the use of the closed type of cryostat (Appendix) because of the much better temperature control. The cryostat temperature is maintained at −18° to −30°, depending on the thickness of sample desired (see below). The angle of the blade of the microtome should be set so that the leading edge of the bevel is nearly parallel to the plane of motion of the tissue block. The knife can be properly positioned with a mirror, using as a guideline the metal support affixed to the knife as an aid to sharpening. (See Appendix.)

The thicker the section, the greater the tendency for it to crack as it comes off the blade and therefore the higher the temperature must be. Sections of 5 $\mu$ can usually be cut at −25° to −30°, whereas at 25 $\mu$ it may be necessary to raise the temperature to −18° to −20°. If the temperature is too warm, the section will be compressed. The sections are cut with a slow steady motion and kept flat by holding a small camel's hair brush against the surface of the block as the section is cut. If the sections tend to fragment, in spite of a sharp clean knife and suitable temperature, this can usually be controlled by cutting very slowly and smoothly and increasing the pressure from the brush.

The sections are transferred serially as they are cut to a suitable holder with a sharp pointed wooden or plastic stick. Metal instruments are not suitable because of heat conduction. Satisfactory tissue holders are made of drilled aluminum blocks sandwiched between glass slides (Fig. 10-1). (See Appendix for a commercial source). If the cut sections are nearly as large as the hole, several sections may be placed in the holder without mixing the order. Moreover, the exact order is seldom important. It is usually not wise to crowd more than five or six sections in a hole. As soon as the holder is filled, the cover is replaced to prevent drying at the temperature of the cryostat. Great care should be exercised not to warm the holder with the gloved hand. Whenever possible the holder should be handled with rubber-tipped forceps. If the sections are to be used for the measurement of labile constituents, they should not remain longer than necessary at the comparatively warm temperature of cutting.

**Drying of Samples**

The section holders are placed in a large, round-bottomed tube fitted with a combined glass cap and stopcock (Fig. 10-1). This improved model was designed by Wenger (1955). The cap is greased with silicone stopcock grease with care not to get grease into the airway. The tube is transferred without warming to a constant temperature box at −35° to −40°. After the tube and contents are cooled to the lower temperature, the sections are dried by applying a vacuum equivalent to 0.01 mm of Hg or less (vacuum gauge, Appendix). A dry ice trap is used to receive the moisture. Most of the water can be removed in 1 to 6 hr, depending on the number and thickness of the sections, but it is usually safer to continue evacuation overnight. Care is exercised not to get crystals of ice into the evacuation tube since these, being solid, are slow to evaporate. A minute crystal of ice (1 mg) if left in the tube could ruin the samples. (Air saturated with moisture contains only 0.5 mg of $H_2O$ per 100 ml at 0°, and 2.5 mg at 24°.) As a precaution against possible residual moisture or vagrant ice crystals the tube is taken from the cold box and evacuation is continued outside until the contents reach room temperature.

The evacuation temperature recommended, −35° to −40°, avoids dangers encountered at higher temperatures and difficulties in dessication at lower temperatures. At temperatures above −25° there is some shrinkage on drying and the possibility exists that salts present might result in a small fluid phase permitting diffusion or enzyme action. Below −40° the vapor tension of ice falls to very low pressures. At −40°, −60°, and −80°, respectively, the vapor tension is 0.1, 0.008, and 0.0004 mm of Hg. Thus it would take 250 times longer to dry tissue at −80° than at −40°.

At −40° there is little advantage in using a vacuum of less than 0.01 mm Hg since this gives a pressure gradient between the tissue and the trap of 0.09 mm of Hg out of a possible 0.1 mm. A pressure of 0.01 mm can be easily maintained with a mechanical pump (and a dry ice trap). The most important element in achieving this pressure is not the pump but the prevention of air leaks. It pays to use a minimum of rubber tubing. For a given pressure at the pump, the rate of drying is governed by the diameter and length of the pathway from sections to trap. A narrow or partially plugged stopcock or small-bore connecting tubing will greatly delay drying.

## Storage of Dry Sections

In contrast to the fresh frozen tissue, the dried sections can be stored for long periods in the evacuated tubes at −20° or below without loss of enzyme activities or breakdown of labile metabolites such as ATP or P-creatine. Even if air leaks in, enzyme activities tested have not been noticeably affected by many years of storage. As a precaution against getting ice crystals into the stopcock orifice, it is plugged with a cork or covered with a rubber cap.

The time of exposure to room temperature and room air is more critical. A test in dry brain sections of three enzymes, hexokinase, 6-P-gluconate dehydrogenase, and glucose-6-P dehydrogenase, showed that the first two lost 40% activity in 3 days at 25° in room air, but lost little or no activity under vacuum at 25° during the same period. Glucose-6-P dehydrogenase was more unstable and under the same conditions lost 65% of its activity in air and 45% under vacuum.

The stability of several metabolites in dry sections has also been tested at room temperature. Under vacuum at 20° glucose, glycogen, and ATP were not significantly changed in 10 days. In air at 20° with 50% humidity, glucose fell 25% in 10 days, and ATP fell 35% in 16 days. In 12 days glycogen fell less than 10%. Lactate during even brief storage at room temperature increased markedly, in brain tissue at least, both in air and under vacuum. The changes far exceed the changes in glucose and glycogen, showing that the lactate must come from another source. In some cases the humidity at room temperature is very critical. For example, oxidized and reduced pyridine nucleotides are stable for several hours at 25° if the relative humidity is 50% or less. However, if the humidity reaches 80%, DPNH and TPNH are rapidly oxidized and $DPN^+$ and $TPN^+$ are destroyed.

CHAPTER 11

# DISSECTION AND HISTOLOGICAL CONTROL

## Dissection and Balance Room

A room in which temperature and humidity can be controlled facilitates dissection. As noted above, humidity greater than 50% increases the danger of changes in tissue constituents. It also seems wise to keep the temperature as cool as comfort permits (20° is suggested) to further increase the stability of sensitive components. Some tissues, specifically myelinated portions of the the nervous system, become sticky and difficult to handle if the temperature rises much above 25°.

Since the tissue sections and dissected fragments are very light, they are easily blown about. It is almost essential to surround the dissection bench and balance area with lightweight curtains (preferably of fiber glass to avoid a fire hazard). A most satisfactory arrangement is to place metal screen 3 or 4 feet above the bench top with curtains on the sides.

## Removal of Sections from Evacuation Tube

The tube is taken from the deep freeze and the vacuum is immediately checked, since the sections should warm up under vacuum. Because heat transfer is slow in the evacuated tube warming will require an hour ordinarily, or the time can be reduced to 20 or 30 min by warming the outside with a wet towel. During the warm-up the stopcock outlet must continue to be plugged or capped to prevent condensation of moisture inside. It cannot be too strongly emphasized that admission of even a few microliters of $H_2O$ would ruin the sections. Opening the vacuum tube while the sample holders are cold would be fatal to the sections as the result of condensation of moisture.

## Dissection

As the previously cited tests of enzyme and metabolite stability show, to minimize changes in air at room temperature it is desirable to keep dry sections under vacuum as much as possible. A few sections are removed for dissection; the tissue holder is returned to the large storage tube and the vacuum re-established at once. This prevents exposure of the bulk of the samples for long periods to warm air. In the case of a few exceptionally sensitive enzymes and certain metabolites it may be necessary to strictly control the period of exposure to room temperature even under vacuum.

The dissection is done freehand under a dissecting microscope. The sections on the stage are protected from drafts and the operators breath by a piece of lint-free toweling taped to the front of the miscroscope. A hard plastic surface such as Plexiglas (Lucite, Perspex) is fastened to the microscope stage. The plastic chosen is translucent (opalescent) so that the illumination provided from below is diffused. This gives excellent visibility of structures within the frozen dried material. In addition, cutting the tissue on a plastic surface rather than on glass prolongs the life of the knife edge. The plastic surface also facilitates dissection of very small samples, as will be described below.

To control static electricity it is necessary to have a source of radiation on the microscope stage or suspended over it. A small amount of radium such as used in the fishpole balance (Chapter 12) is satisfactory or a polonium source can be used (see Appendix).

The frozen dried material is held in place during dissection with a hair loop (Fig. 11-1) or in some cases with a hair point (see below). Dissection is accomplished by a series of short vertical cuts through the section. The knives are made of splinters of razor blades mounted on a flexible bristle (Fig. 11-1). The cutting edge is ordinarily 0.5–1 mm long and is broken away at a sharp angle so as to leave a pointed tip. The blade is mounted with epoxy resin to a short bristle and this in turn is fastened with the same resin to a piece of copper wire in a wooden dowel. The flexibility results in easier dissection than with an ordinary scalpel, since the knife blade adjusts itself to the plane of the cutting surface. At the same time it prevents dulling or breaking of the blade. (See Appendix for construction.)

The limiting factor in freehand dissection is tremor. For larger structures this is sufficiently minimized by grasping the tools firmly with not more than

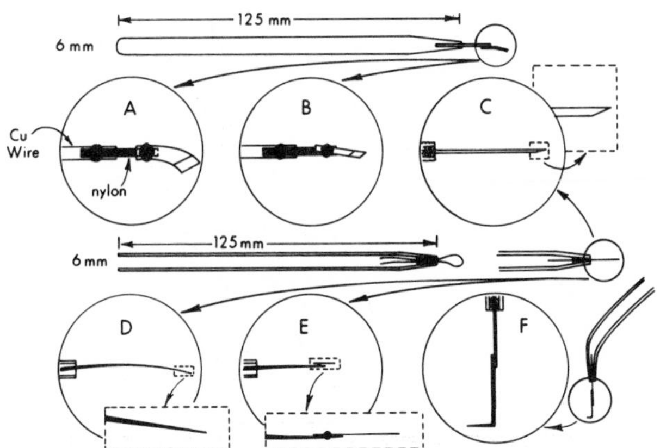


***Fig. 11-1.*** Dissecting tools: razor blade knives, hair points, hair loop. A and B illustrate knives of two sizes. Hair point C is made from a human hair that is sharpened with a razor blade. Hair point D is a fine sable or camel's hair with a natural taper. E is a quartz-tipped hair point made from a fine camel's hair to which a 3–5 $\mu$ diameter quartz fiber is attached with epoxy cement. F is a special hair point for loading samples through oil in an oil well. A sufficiently stiff quartz fiber is attached with epoxy to a stiff hair (to prevent breakage). Note that the quartz fiber is drawn down to a tip beyond the bend.

20 mm extending beyond the fingers. The side of the hand, the last three fingers, and if possible even the thumb, are pressed firmly against the dissecting surface. The flexibility of the bristle mounting of the knife blade permits a dissection technique for further reduction of tremor and freehand dissection of very small objects. The tip of the razor blade is touched to the plastic surface just in front of the place to be cut (Fig. 11-2). The tip cuts into the surface enough to anchor it and stop the vibration. The blade is exactly positioned by a slight lateral motion of the hand with the anchored tip as a pivot. When the position is correct a slight downward pressure is exerted, causing the blade to rotate vertically, again with the tip as a pivot, making the desired cut in the dry tissue. In this manner, structures 15 or 20 $\mu$ in diameter may be readily dissected freehand.

When discrete layers from a structure such as cerebellum or retina are to be analyzed, the original microtome sections are usually cut at a small angle to the plane of the layers in order to make the layers appear as wide as possible. Consequently, unless the section is exceedingly thin, the edge of a given layer may be in a slightly different position on one side of the tissue than the other. It may therefore be necessary to examine both sides of the section and discard the zone where adjacent layers overlap.

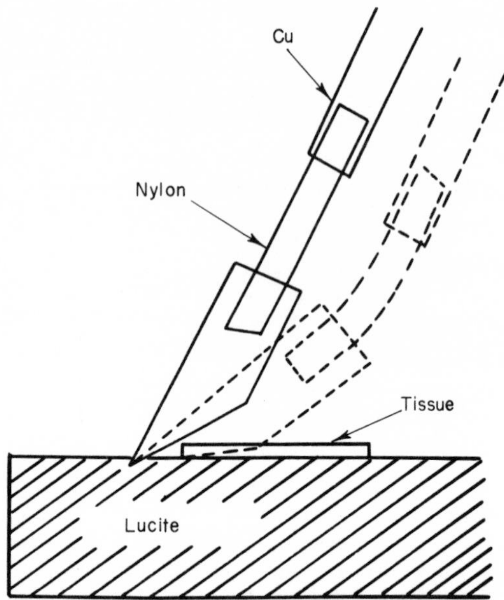


*Fig. 11-2.* Dissection technique to avoid tremor of the hand. See text.

## Handling Dissected Samples

After dissection the samples are picked up and transferred with a short piece of hair mounted with Duco cement or similar adhesive in a pencil-shaped glass holder (Fig. 11-1). Such "hair points" are made from hair of different degrees of stiffness for different samples sizes or types of tissue. Human hair, extending 4–6 mm and sharpened by cutting at an acute angle, is suitable for samples of 1–5 μg. The flexibility depends on the length of the hair, and may be varied to suit the toughness of the tissue. A better hair point for smaller samples, down to 0.1 μg, is a fine hair with a natural taper, such as those in a sable or camel's hair paint brush. The tip of such hairs taper to a diameter 10 or 20 μ. A still better tip is provided by sealing to the end of a fine hair point a piece of quartz fiber 2–5 μ in diameter (Fig. 11-1). This is convenient for use with 0.1 μg samples and is almost essential with samples weighing 0.01 μg or less. The tip of the hair is dipped in epoxy resin, leaving a very small amount on the hair. This is touched to a fine quartz fiber 1 or 2 cm long. Surface tension helps align the two fibers. When the resin has set, the quartz tip is cut off with scissors to a length of 0.5–1 mm.

Relatively large samples (over 2 or 3 μg) are lifted by pressing into the sample with a stiff sharp hair point to which the tissue then adheres. For smaller samples it suffices to just touch the surface with the hair point or better to lift up the sample from underneath. Sometimes it is found easier to pick up the specimen than it is to disengage it subsequently, perhaps on a balance pan. This is a situation in which the fine quartz tip has great virtue, since a sample is easier to remove from a small tip than a large one. Static charge must be controlled with radium or polonium for satisfactory transfer of samples, particularly in the case of small sizes.

## Identification and Histological Control

In most tissues some study is required to recognize sufficient histological detail in frozen dried material. With familiarity and a good light, however, a surprising amount of detail can be visualized. If the initial freezing is too slow, ice crystals will distort the material and make identification difficult. The thickness of the section may affect the ability to see and identify small structures. Slight differences in color are often helpful, and for this reason white light is usually preferred for illumination.

As needed, sections are appropriately stained as a guide. The sections are placed in recorded order on a single slide lightly coated with histological glycerine–egg albumin. The sections can be flattened if necessary with hair loops, and pressed down with the finger tip. The sections are then fixed and stained as desired. The accuracy of dissection may also be checked by mounting and staining the unused parts of the dissected section. Selected dry unfixed sections may be stored for study and future reference. The sections are placed on a slide, a cover slip sealed at the edges with paraffin, and the slides stored at 4° to prevent deterioration.

## Sample Carriers

Once the desired structures are dissected, the fragments are lined up on a tissue carrier for transfer to the balance and for temporary storage. It is convenient to design the carrier to hold 10–12 samples of the anticipated size (Fig. 11-3).

A piece of hardwood cut and shaped to be easily handled is used as the handle. To this is glued a segment of a glass microscope slide or, for smaller balances, a hemocytometer cover glass cut to the appropriate size. The glass is mounted at an angle which will facilitate insertion of the samples into the balance case for weighing. The hardwood handle is painted with black ink to reduce glare during weighing. The carrier with sections is placed in a petri dish so the sections will not be disturbed by air currents when moved. Handling at this stage is made easier if the Petri dish is turned upside down and if the lid (now the bottom) is shallower than the other part (Fig. 11-3). This permits the petri dish and contents to be lifted without danger of pulling the two

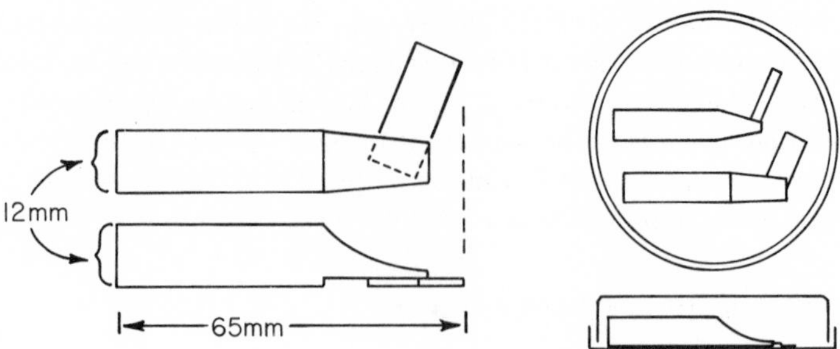


*Fig. 11-3.* Sample carriers. Two sizes are shown as well as the inverted petri dish for protecting the samples from air currents. The body of the carrier is made of hard wood, the carrier surface is made of a glass strip cut from a microscope slide. For smallest carriers the glass strip is cut from a hemocytometer cover.

halves apart, which would suck in air and dislodge the samples. The petri dishes should be set down gently, because if there is an electrostatic charge on the cover a sharp jar may cause some of the samples to fly to the top. If this becomes a problem it may be desirable to keep a small amount of radium on metal in each dish.

## Defatting Frozen Dried Samples

In many tissues there are regional differences in lipid content. It may be desirable to measure the lipid present, if for no other reason than to permit calculation of data on a fat-free dry weight basis. This is usually a truer basis physiologically than the total dry weight basis. An extreme example is presented by the nervous system, in which lipid can range in different regions from 20–75% of dry weight.

Samples weighing 1 μg or more are placed in a tube 4 or 5 mm i.d. and extracted with absolute ethanol by simply filling the tube a third full with the solvent and tapping just enough to dislodge the sample. After 10 min the ethanol is carefully decanted in such a way as to leave the sample on the wall. The tube is then inverted for 30–60 sec and any fluid collected at the mouth is blotted off. The tube is immediately filled part way with hexane, the tube is tapped gently to be sure the sample is free and the solvent is decanted as before. Hexane is added a second time to be sure all of the ethanol has been removed and the last traces of solvent are aspirated carefully with a fine pipette. The section should come loose spontaneously or with a sharp tap. The sample can be removed from the tube by tapping onto a clean surface with the usual precaution to dissipate electrostatic charge.

Sticking will be caused by failure to use residue-free hexane, failure to wash away all of the ethanol, failure to use anhydrous ethanol or leaving the ethanol on so long that it absorbs moisture from the air. Moisture can be easily absorbed if hexane is not added promptly after removal of the bulk of the ethanol. If the ethanol should dry, the section will be inevitably and irretreivably stuck. Sections ranging from 0.02 to 0.2 $\mu$g, because they are more difficult to see, are more easily extracted in an 8–10 mm glass tube of 2 or 3 mm bore sealed to a 50–60 mm length of tubing for a handle. The extraction procedure is the same as for larger samples except that the solvents are removed by aspiration with a fine-tipped pipette. Good visibility is maintained during aspiration to be sure the sample is not lost. Low-power magnification is helpful. These smaller samples may not fall out on tapping but can be removed with a hair point.

## Sample Volume

The sample volume can be measured provided the sections have been cut without serious compression. The volume is calculated from the thickness multiplied by the area. The fragment is placed on a microscope slide on a microscope stage, using a coverslip to keep the sample flat. The monocular eye piece is replaced with a right-angled prism. With a low-power objective and bright illumination the image is projected on the wall and traced on a piece of paper. An ocular micrometer disk is projected at the same time and marked on the same paper to establish the magnification. The area of the fragment is measured with a planimeter.

CHAPTER 12

# THE QUARTZ FIBER FISHPOLE BALANCE

The fishpole balance is simply a quartz fiber mounted horizontally by one end. For larger sizes a light glass or quartz pan is attached to the free end. The sample is placed on the pan or the free end and the displacement measured (Fig. 12-1).

This is an inelegant instrument compared to other balances. As ordinarily used it is not capable of great precision. Any given balance is limited in usefulness to a narrow range. Nevertheless the quartz fishpole balance is ideal for weighing small dry tissue samples and it is capable of far greater useful sensitivity than any other mechanical balance presently available.

The great advantage of this simple balance is that it can be placed when necessary in a very small chamber and thereby be protected from the chief enemy of reproducibility: air currents. A balance of whatever design with sensitivity, for example, of $\pm 10$ picograms, would be useless because of air currents if mounted in a balance case of 50 ml volume. However, a fishpole balance with this sensitivity can be mounted in a chamber of 0.2 ml volume where air currents are easily controlled.

Quartz is preferred to glass because of its greater strength and resistance to breakage and because unlike most glasses it does not flow under tension. It will therefore retain its position indefinitely. A quartz fishpole balance that has been in continuous use for nearly 20 years shows no detectable change in position.

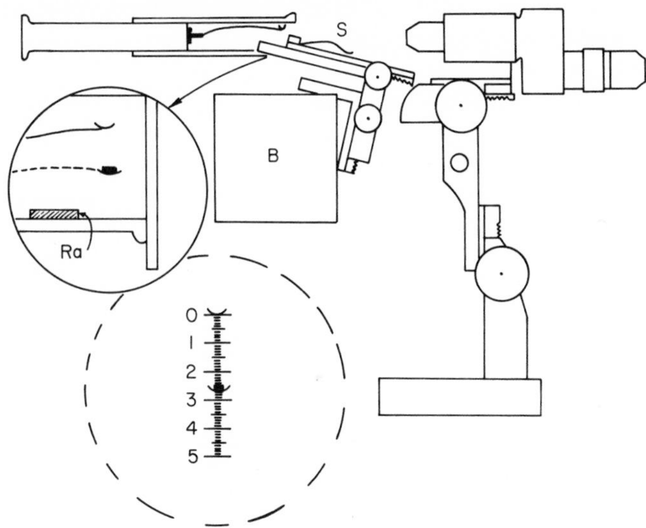


***Fig. 12-1.*** Arrangement for loading and reading quartz fiber balance. The wooden sample holder is held by spring S to the double rack and pinion loading device shown (a modified stage micrometer). The loader is mounted on a 4 × 4 inch beam (B) which also provides a steady rest for the right hand with which samples are transferred to the balance tip. Transfers are supervised through the wide-angle microscope. The displacement of the balance tip is measured on the micrometer ocular in one eyepiece of the same microscope (see inset). The zero point is adjusted by means of the vertical rack and pionion of the microscope mount.

## Measurement of Displacement and Useful Range

For most purposes it is convenient to measure displacement of the balance tip on a micrometer ocular of a wide-angle stereomicroscope. The same microscope is used to supervise loading and unloading. A 5 mm ocular divided in 0.1 mm divisions has been found most satisfactory. (With a 10 mm ocular, loading is less convenient, since during loading the pan is near the zero point, which would be located at the extreme top of the field of view.)

A 5 mm ocular scale represents a real distance which varies with the magnification of the objective, not with the magnification of the ocular. A 5 mm scale represents real distances of 5 mm, 2.5 mm, and 1.25 mm with 1 ×, 2 ×, and 4 × objectives, etc. For most purposes a 1 × or 2 × objective is most satisfactory. Higher powers usually have inconveniently short working distances.

The scale of the 5 mm ocular can be read to a one-tenth of 1 small division or 0.01 mm. Since there is a 0.01 mm uncertainty at the zero point as well as at the point of deflection, the smallest deflection that can be measured to

$\pm 2\%$ is about 1 mm or one-fifth of the full scale. This limits the useful range to about fivefold for any given balance. Consequently, several balances may be needed if a wide range of sample sizes are to be analyzed. Nevertheless, to satisfactorily cover the range from 0.001 to 10 $\mu$g only six balances would be required.

## Choice of Suitable Fiber. Sensitivity as a Function of Length and Diameter of the Fiber

For quartz fibers the sensitivity can be adequately expressed by the formula

$$S = \frac{L^3}{d^4} \text{ mm}/\mu\text{g} \tag{12-1}$$

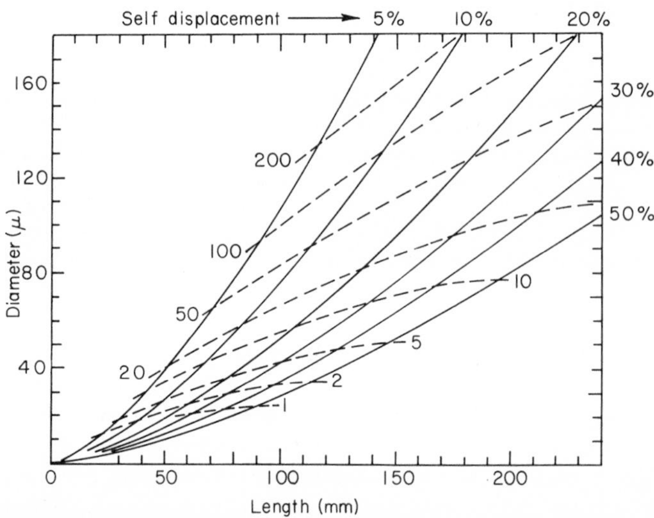


***Fig. 12-2.*** Self-displacement (droop) and sensitivity of quartz fibers as a function of diameter and length [corrected for deviations from formulas (12-1) and 12-2)]. Each solid line represents the relationship between diameter and length for a given percentage droop. For example, a fiber 80 $\mu$ in diameter at a length of about 135 mm would droop 20% or 27 mm. The droops are calculated for the condition in which the mounting is adjusted to bring the tip into line horizontally with the point of attachment. The dotted lines represent the sensitivity expressed as micrograms per millimeter of displacement. For example, a fiber 80 $\mu$ in diameter and 135 mm long would have a sensitivity of 20 $\mu$g/mm.

This figure is also applicable to fibers under 0.18 $\mu$ in diameter. In this case the ordinate scale would be 0–0.18 $\mu$, the abscissa scale 0–2.4 mm, and the numerals on the sensitivity lines would represent picograms per millimeter of displacement. For example, a fiber 0.15 $\mu$ in diameter and 2.4 mm long would droop 30% of its length (0.7 mm) and have a sensitivity of 50 pg/mm of displacement.

where $L$ is the length in millimeters and $d$ is the diameter in microns. Thus the sensitivity increases as the cube of the length and inversely as the fourth power of the diameter. For example, a fiber 100 $\mu$ in diameter and 100 mm long would have a sensitivity of $10^6/10^8 = 0.01$ mm/$\mu$g or 100 $\mu$g/mm.

Formula (12-1) only applies strictly to small displacements, since as the fiber bends the horizontal distance from the point of attachment decreases.

Although the sensitivity increases as the cube of the length, there is for every fiber a length beyond which it would droop so far under its own weight as to be useless. For most purposes this limit is reached when the droop (self-displacement of the tip) is greater than 40 or 50% of its length. If the fiber balance requires a pan it is usually desirable to keep the droop to 20 to 30%.

Figures 12-2 to 12-4 show the relationships between diameter, droop, and fiber length. These figures also show the sensitivity as a function of these three parameters and can serve as a guide to the choice of a suitable fiber for the sensitivity desired. For example, according to Fig. 12-2, a sensitivity of 2 $\mu$g/mm could be obtained with fibers ranging from 15 to 35 $\mu$ in diameter. A 15 $\mu$ fiber would be 25 mm long with droop equal to 5% of its length. A 35 $\mu$ fiber would be 115 mm long with droop equal to 50% of its length.

Although a suitable fiber could be chosen on the basis of its diameter, this becomes difficult to do accurately with the smaller sizes. A much more convenient method for fibers of all diameters is offered by the relationship seen in Figs. 12-2 to 12-4 between droop, length, and sensitivity. The fiber is

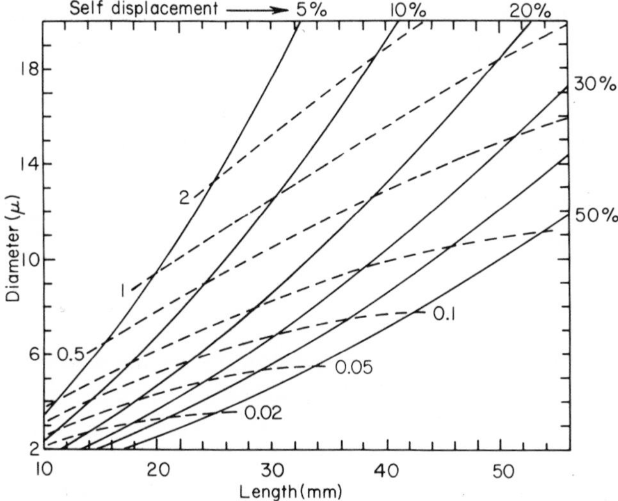


***Fig. 12-3.*** Self-displacement–sensitivity chart for fibers ranging from 2 to 20 $\mu$ in diameter (see Fig. 12-2).

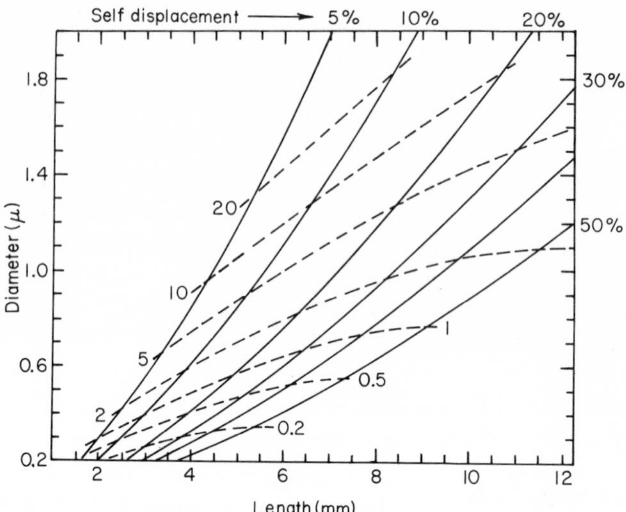


*Fig. 12-4.* Self-displacement-sensitivity chart for fibers ranging from 0.2 to 2 $\mu$ in diameter (see Fig. 12-2). In this case dotted lines represent sensitivity expressed as nanograms per millimeter of displacement.

suspended on a temporary mounting in a suitable glass container (to curtail air currents) and the displacement of the tip is estimated when the vessel is turned to shift the fiber axis from vertical to horizontal. Or, with the fiber horizontal, the vessel can be rotated 180° along the fiber axis, in which case the tip will move a distance equal to twice the droop. As an example, if a fiber 200 mm long has a droop of 50%, it is about 80 $\mu$ in diameter and would be suitable for balances with sensitivity of 10 $\mu$g/mm or less (Fig. 12-2). If the droop is greater than 50% of the length, a shorter segment of the fiber can be tested by mounting it so that only a portion of the total length projects past the support.

Instead of using the figures, the sensitivity can be calculated from the droop by the following formula:

$$S = \frac{3 \times 10^6 D^2}{L^5} = 3 \times 10^6 \left(\frac{D}{L}\right)^2 \left(\frac{1}{L^3}\right) \text{ mm}/\mu\text{g} \qquad (12\text{-}2)$$

where $D$ is the droop in millimeters and $L$ is also in millimeters. This applies strictly only to small droops, but the discrepancy is not serious with droops up to 50%. The error in using formula (12-2) with large droops tends to underestimate the sensitivity of the fiber.

With the use of formula (12-2) it is unnecessary to calculate the fiber diameter. For example, if a fiber 50 mm long droops 25 mm under its own weight, the sensitivity is approximately

$$3 \times 10^6 \times \left(\frac{1}{2}\right)^2 \times \frac{1}{50^3} = 6 \text{ mm}/\mu\text{g or } 0.17\ \mu\text{g/mm}$$

To make from this fiber a balance with a sensitivity of 0.4 $\mu$g/mm, since the sensitivity varies as the cube of the length [formula (12-1)], the fiber is shortened to $50\sqrt[3]{0.17/0.4} = 38$ mm.

Note the paradox that because of the limitation on useful length set by droop, higher sensitivity can only be achieved ultimately by using *shorter* (and of course finer) fibers.

If 50% droop is set as a practical limit ($D/L = 0.5$), the maximum useful sensitivity for quartz fiber balances in general can be calculated as a function of the length. Substituting $D/L = 0.5$ in Eq. (12-2):

$$S = \frac{7.5 \times 10^5}{L^3} \text{ mm}/\mu\text{g} \qquad \text{or} \qquad L = \frac{90}{\sqrt[3]{S}} \tag{12-3}$$

i.e., ultimate useful sensitivity for quartz fiber balances varies *inversely* as the cube of the length. This is illustrated in Figs. 12-2 to 12-4. A balance with a sensitivity of 10 $\mu$g/mm can be 200 mm long, but a balance with a sensitivity 0.2 ng/mm can (for practical purposes) be no longer than 5.5 mm.

## Balance Fibers

Commercially produced quartz fibers are available in sizes ranging down to 3 $\mu$ (Appendix). They may also be easily made with a gas–oxygen torch, and at present this is necessary for the smallest sizes. The method of preparing the fibers will depend on the size of fiber desired. In the first method, one end of a piece of quartz tubing is fastened with adhesive tape to a 2 to 4 meter length of rubber band or thin rubber tubing which in turn is tied under tension to a stationary object so as to stretch the rubber when the quartz is held in the flame. The quartz is heated in the center until the lumen disappears. A small portion of the quartz is narrowed somewhat, then heated hotter until it is fluid, at which moment the tension on the rubber tubing is released. The heated quartz will be drawn into a long fiber. The fiber diameter depends on the amount of melted quartz, its temperature, and the tension and length of the rubber tubing. Fibers can easily be made in this way with diameters down to 10 or 15 $\mu$, i.e., of a size suitable for balances capable of weighing $1 \pm 0.01$ $\mu$g or more.

A second method which will permit preparation of fibers of smaller diameter utilizes the draft produced by the flame to pull the fiber. The quartz is first heated in a hot flame and pulled by hand to form a rod 0.5–1 mm diameter and 10–20 mm long. The narrow portion is then held nearly lengthwise in a large cool flame where the rush of air will gradually thin the rod and then suddenly whip it out into a long fine filament. This will float in the air and can be trapped on a wooden club wrapped with black cloth. By varying the flame, smaller fibers of a wide range of sizes can be produced down to less than 0.2 $\mu$ in diameter.

A third method uses the weight of the quartz to pull the fiber. A rod of quartz is prepared from tubing as above. The tubing is then suspended vertically 4 or 5 feet above the working surface. The narrow rod portion is heated just enough to soften the quartz and let it drop of its own weight, pulling a very fine fiber. This method is perhaps most useful for making balance fibers of 1–10 $\mu$ diameter.

The portion of the fiber selected for the balance ought to be of nearly uniform diameter. However, a slight uniform curvature is permissible since the fiber can then be mounted with the concave edge upward. The result is to provide a more nearly straight fiber, when mounted, than if the fiber itself were perfectly straight.

### Pans

Balances to weigh samples of 0.05 $\mu$g or more need pans to hold the samples. The pans are made from very thin fragments of Pyrex or quartz bubbles. Quartz is stronger than Pyrex but more difficult to blow. A compromise is Vycor, which is almost pure quartz but melts at a somewhat lower temperature. To make the thin bubble a piece of tubing is first closed at one end with heat, and then a fairly large portion of the closed end is nearly melted in the flame. The tubing is removed from the flame and quickly blown to produce a bubble thin enough to show interference colors. The bubble usually bursts, and the fragments are collected. The cuts necessary to reduce a fragment to an appropriate size and shape are made by placing it on a plastic surface and pressing straight down with the edge of a razor blade. The weight of the pan itself should not cause the fiber to bend by a distance of more than 20 or 30% of its length. This usually means that the pan should not weigh more than the load limit of the balance. No pan is needed with balances used to weigh samples of less than 0.05 $\mu$g, since such samples adhere well to the bare fiber tip.

### Balance Case

The fiber is permanently mounted in a case to protect it from air currents and dust. The chamber must be of proper size in relationship to the balance to facilitate weighing and minimize air currents within the case. A balance case made from a glass syringe has been found most convenient (Fig. 12-1). The size must be chosen according to the sensitivity of the balance, in order to keep internal air currents within tolerable limits. The proper sizes have been determined empirically. A 50 ml syringe is satisfactory for a balance of 5 μg capacity (1–5 μg range, ±0.01 μg). For balances of 0.5 μg, 0.05 μg, 5 ng, and 0.5 ng capacities the chamber volume should not exceed 10 ml, 2 ml, 0.5 ml, and 0.2 ml, respectively. (Since the two smaller balances will be very short they can be accommodated in 2 ml and 1 ml syringes respectively.)

To make the balance case the closed end of the syringe barrel is cut off and the plunger reversed. The barrel may need to be ground out slightly with fine carborundum and water to admit the plunger. The flared end of the barrel is ground flat against a glass plate with carborundum and water to permit a nearly air-tight closure by a glass cover.

A short piece of copper wire of suitable flexibility (20 gauge, 0.75 mm diameter) is sealed to the end of the plunger with epoxy resin. The wire is mounted in such a way that 5–15 mm (depending on the size of the balance) extends straight out from the center (Fig. 12-5). For the more sensitive

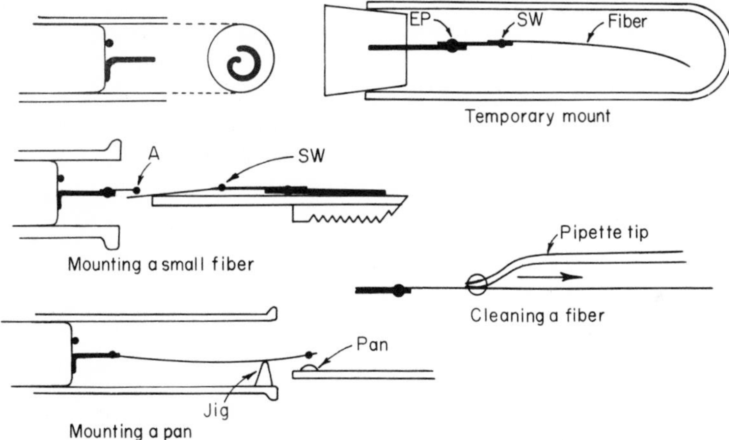


***Fig. 12-5.*** Mounting quartz fibers and balance pans, and cleaning the fiber. At upper left is shown how a copper wire is bent to permit secure fastening (with epoxy cement) to the end of the syringe plunger. At upper right is a temporary mount for a quartz fiber. A fine steel wire is sealed to a stiff wire with epoxy (ep). This in turn is sealed to the fiber with a minute amount of sealing wax (sw) or similar material. (The sealing wax is first affixed to the steel wire; the wax is touched to the fiber tip, which is then trapped by momentarily melting the wax with an electrically heated wire held nearby.)

A small fiber is mounted on the syringe barrel as shown. In the case illustrated a short piece of fine steel wire has been fastened to the heavier copper wire with epoxy. A very small droplet of epoxy is placed on the tip (A). The quartz fiber tip is now raised until it touches A, at which time the epoxy is polymerized with heat (see text). The other end of the fiber is released from its temporary mount by applying heat to the sealing wax.

A pan is mounted in a similar manner as shown at lower left (see text).

Note the special bend near the tip of the pipette used to clean a fiber. This is to prevent the fiber from being drawn to the pipette shaft by capillarity. The diameter of the pipette is exaggerated in the figure and the walls of the balance case are not shown.

balances, to facilitate attachment of the fiber, a 5 mm length of fine steel wire (e.g., 38 gauge) may be fastened to the copper, or in the case of the most sensitive balances a short piece of stiff quartz fiber about 20 $\mu$ in diameter is used.

The balance is mounted rigidly on a heavy ringstand, and once the fiber is in place and calibrated, the orientation of the balance should not be changed without checking the calibration. Just under the open end of the balance a firm support is provided on which to rest the hand when loading the sample. This support should not touch the balance support, to make sure that the balance position is not disturbed during loading. Loading samples is made easier by attaching to this same support a simple rack and pinion manipulator to hold the sample holder and move the samples in and out of the balance. This can be constructed from a microscope stage micrometer which is turned on edge and provided with a suitable spring clip (Fig. 12-1).

The balance closure is made with a piece of good quality flat glass. A large or small microscope slide or portion thereof will usually suffice. Wire hooks or loops are sealed to the glass with epoxy cement to provide for attachment of long rubber bands to hold the glass plate in place. The rubber bands should be strong enough to provide firm closure but at the same time permit easy removal of the plate without the slightest displacement of the balance. (Very weak long springs would probably be preferable to the rubber bands, which must be replaced at intervals.)

## Mounting the Fiber

The quartz fiber is sealed to the copper or steel wire or stiff quartz with epoxy resin. Handling a fine fiber is made easier by the use of a temporary support. Convenient for this is a 5 mm length of fine wire (36–38 gauge) which is permanently attached with epoxy cement to a larger wire (Fig. 12-5). A cement which softens with heat, such as De Khotinsky cement or sealing wax, is used to seal the fiber to the tip of the fine wire. The heat is provided by an electrically heated short piece of platinum or Nichrome wire brought nearby. The larger wire (above) can be inserted in a cork for easy manipulation of the fiber or for storage in a tube and for testing its sensitivity.

The fiber is permanently mounted with the aid of a dissecting microscope. The plunger is extended through the syringe barrel until the wire or quartz support emerges. A small amount of epoxy resin is placed on the tip of the support. The free end of the quartz fiber is brought into contact with the resin with a hair point. For finer fibers a rack and pinion device mounted on a ring stand may be used (Fig. 12-5). With the fiber properly aligned the resin is polymerized by heat from the electrically heated wire held close by. The resin is approached with the hot wire until bubbles begin to form. The heating is continued until the bubbles do not shrink when the hot wire is temporarily withdrawn. The other end of the fiber is then released with heat from the temporary mounting. If necessary any residual cement on the free end can be either burned off or removed with fuming $HNO_3$. The fiber is positioned by bending the wire support and if necessary the length is reduced with scissors to provide the desired sensitivity. The plunger should be positioned so that the natural curvature of the fiber, if any, is concave in the upward direction.

If a pan is needed, it is now attached upside down. A small amount of epoxy resin is placed on the tip of the fiber supported by a paper jig above the pan (Fig. 12-5). The pan is placed on a glass slide with its concave side down. The fiber is lowered to the pan by removing the paper jig or raising the glass slide, and the resin polymerized with heat from the hot wire.

The final adjustment of the balance is made so that the tip is 2–3 mm above the point of support to favor linearity of response. (Mounted in this way, as the fiber bends under a load the horizontal lever arm will first lengthen slightly, then shorten slightly.) To minimize disturbance from air currents the tip should be drawn into the syringe a distance equal to the radius of the syringe and the syringe plunger taped in position. A piece of metal bearing 5 to 20 $\mu$g of radium (Appendix) is placed in the bottom of the balance case to dissipate static electricity.

The distance of the tip from the top of the syringe should be recorded, as fibers may get dirty and droop further. This is particularly noticeable with very sensitive balances. When this occurs, the fiber should be cleaned (see below) so that the original position and therefore the original calibration is regained.

## Lighting and Viewing

Good lighting is required with a minimum of heat. With larger balances this can easily be accomplished with fluorescent lamps. With smaller balances the fiber, especially when new and clean, may be difficult to see clearly. Sufficient illumination can be provided with a focused light beam, such as that from a microscope lamp (e.g., American Instrument Co. No. 653). In this case it is usually necessary to pass the light through glass filters chosen to transmit the best ratio of visible light to heat. A satisfactory filter, which transmits maximally where the eye is most sensitive (560 nm), is provided by a combination of Corning glass No. 3962 with either No. 9780 or No. 4600. The fiber can be most easily visualized if the light comes at a grazing angle from behind and above.

## Weighing

The zero point is adjusted. The balance case is opened. With the rack and pinion the samples on the holder are moved into the balance case just under the fiber tip. A sample is loaded onto the balance pan or tip with a hair point and the holder withdrawn. The glass cover is replaced and a reading made.

Loading a balance fiber which does not have a pan is not as difficult as it might seem. The sample is picked up from the carrier with a fine hair point, preferably quartz-tipped (see Chapter 11). The sample is touched to the fiber tip as it floats freely in the air. The touching is repeated several times if necessary, perhaps with slightly different orientation, till the transfer is accomplished. The process is repeated to remove the sample after the weighing has been made. Samples of 0.02 $\mu$g or less will adhere either to the tip of the hair or the tip of the balance and will almost never fall off. A sharp tap of the balance case might, however, cause a sample to fall. Good control of electrostatic charge makes for easier transfer.

In general, it is easier to load and unload samples of all sizes if air currents in the neighborhood are also well controlled so that the balance tip remains reasonably stationary when the case is open. (See Chapter 11 for suggestions on screening the room.) On this account it is useful to shield the balance from the operator's breath by draping the front of the microscope head with a piece of cloth. Ordinarily readings can be made within 10–20 sec after closing the case and should be reproducible to 0.01 mm or at most 0.02 mm on the micrometer ocular (0.005–0.01 mm real distance with a $2\times$ objective). If there is a slow drift to a stable position there may be insufficient radium in the balance, or the operator may be wearing clothing particularly likely to generate static electricity. Other sources of instability could be too much heat from the lighting, poor closure of the chamber and a drafty room, unstable mounting of the balance, or wobble in the microscope. In any event the zero point should be checked frequently. Often there is a slight shift in the zero point during the first few weighings—possibly due to heat from the hand.

A good balance will give a nearly linear relationship between weight and deflection, although there must, because of the nature of the fiber balance, always be some departure from linearity, since the horizontal distance between the two ends must change as the fiber bends. It is easily possible to keep the nonlinearity to within 2–5%. Often the difference can be ignored, but for some purposes it may be worthwhile to make a correction. In any event, a new balance should be checked for linearity (see below).

If the pan is large in relation to the length of the fiber, reproducibility will be increased slightly if the sample is always placed near the center of the pan. Since the sample actually sticks to the surface of the pan, the length of the lever arm is determined by the center of gravity of the sample, not by the point at which the sample happens to touch the pan. (The displacement of the tip varies as the square of the horizontal distance of the sample from the mounting. Thus a 1% difference in this distance will make a 2% difference in tip displacement.)

## Correction of Weight for Adsorption of Gases and Moisture

Porous frozen dried tissue has a large surface area which adsorbs $O_2$, $N_2$, and moisture. When the sample is taken out of a vacuum this adsorption reaches equilibrium in a few seconds. In the case of animal tissue, adsorbed $O_2$ and $N_2$ adds 2% to the weight. The amount of moisture adsorbed is proportional to the humidity and is equal to about 1% for each 10% humidity. With 50% humidity, therefore, a tissue sample will weigh approximately 7% above its true dry weight. This correction can be incorporated into the balance factor. (Dry tissue rich in lipids will increase somewhat in weight if left in room air for several days, probably as the result of lipid oxidation.)

## Calibration of Balances and Testing Linearity

Balances of 0.5 μg capacity or larger can be calibrated colorimetrically using *p*-nitrophenol crystals. The deflections due to single crystals of suitable size are measured. Each is then dissolved in a measured volume of carbonate buffer (50 m*M* $Na_2CO_3$, 50 m*M* $NaHCO_3$) in the proportion of 1 ml for a 1–6 μg crystal. Readings are made at 400 nm in the spectrophotometer and the weights are calculated from readings made with appropriate standards. For example, crystals ranging from 0.5 to 3 μg in size are dissolved in 500 μl to give optical densities of 0.13–0.8 at 400 nm. A stock solution containing 1 mg/ml is prepared from the same bottle of *p*-nitrophenol. This is diluted 200-fold with the carbonate buffer (giving a 5 μg/ml) solution and the reading obtained is used to calculate the weight of the single crystal samples.

Balances of smaller capacity can be calibrated fluorometrically with crystals of quinine hydrobromide. (Quinine sulfate crystals are unsuitable because they are flat and tend to stick to flat surfaces including that of the balance pan.) The fluorescence of the quinine crystal dissolved in 1 ml of 0.1 *N* $H_2SO_4$ is read against an appropriate standard. Quinine at a concentration of 0.001 μg/ml will yield measurable fluorescence. However, it is difficult to manipulate crystals smaller than about 0.01 μg.

[Small crystals are somewhat difficult to pick up and transfer. For this reason McCann (1966) has described a calibration procedure using small pieces of gold leaf which are easier to handle than crystals. After the deflection is determined the gold leaf is dissolved and measured colorimetrically.]

Frozen dried tissue samples, being light and soft, are much easier than crystals to pick up and to manipulate on and off the balance pan or fiber. Therefore, once a relatively large balance has been calibrated the easiest procedure is to calibrate a new balance by comparison with the old one. For example, suppose a balance with a capacity of 5 $\mu$g has been calibrated and it is desired to calibrate a balance with a capacity of about 1 $\mu$g. A piece of tissue weighing 3 or 4 $\mu$g is cut into four pieces of nearly equal size (see Chapter 11). The four pieces are weighed together on the larger calibrated balance and each piece is then weighed separately on the smaller balance. From the sum of the weights on the smaller balance its sensitivity is readily calculated. This second balance can in turn be used in the same manner, to calibrate a still smaller balance and so on.

Frozen dried tissue samples can also be used to assess the linearity of a balance. A sample which gives a deflection slightly less than full scale is divided into four or five nearly equal parts. These are weighed all together, then singly and in combinations of two and three. For example, suppose there are four pieces which weighed together give a deflection of 4.00 divisions and this has been determined to equal 4.00 $\mu$g (i.e., the sensitivity in this region is 1.00 $\mu$g/division). The individual samples a, b, c, and d give deflections of 0.90, 1.10, 1.15, and 1.05 divisions which total 4.20 divisions. Therefore in the region of the first division on the balance the sensitivity is 4.00 $\mu$g/4.20 divisions or 0.95 $\mu$g/division. When samples a and b are weighed together they give a deflection of 1.98 divisions. Therefore in the region of two divisions the sensitivity is $0.95\ (0.90 + 1.10)/1.98 = 0.96$ $\mu$g/division, etc.

Once standardized, the sensitivity of a quartz fiber balance has not been found to change even after years unless the fiber becomes coated with dirt. Nevertheless it may be wise to check the sensitivity occasionally. For this purpose permanent balance standards can be prepared with gold leaf (Appendix). These standards can also be used to calibrate new balances. The piece of gold leaf of suitable size is crumpled slightly to make it easier to pick up and less likely to stick to a smooth surface. The weight is determined on a calibrated balance and it is then stored where it is unlikely to collect a film of dirt. A convenient storage vessel is one shaped like the sample holders for dissected tissue samples (Fig. 11-3) except that a piece of aluminum replaces the glass. A shallow depression somewhat larger than the piece of gold leaf is drilled in the metal toward the front. The gold leaf is stored in this hole covered with a small piece of glass which is held in place with a clip. Only weights of 1–3 $\mu$g have been used, but it is probable that smaller weights could be handled and stored satisfactorily.

### Cleaning the Balance Fiber

In spite of nearly airtight closure of a balance, the fiber will ultimately become covered with a dirt film and droop further. To clean the fiber a small droplet of cleaning fluid is formed at the end of a very fine-tipped pipette. This droplet is carefully touched to the fiber near the mounting (Fig. 12-5) and then drawn forward the length of the fiber until the fiber escapes from the droplet. This may be repeated if necessary. The dirt film can usually be removed by successive rinsings with 20% ethanol in $H_2O$, 95% ethanol, and finally hexane. With care this can be done without dislodging or breaking the pan. The ethanol is added to the $H_2O$ for the first rinse to lower the surface tension, which might cause a delicate fiber to be pulled off.

If these solutions do not clean the fiber, fuming $HNO_3$ may be used as a last resort, but this will usually take off the pan. Cleaning is therefore easier (and required more often) with balances small enough not to need a pan. This is true except with exceedingly sensitive balances (capacities under 0.001 $\mu$g), for which great care must be exercised not to remove the fiber along with the film of dirt.

CHAPTER 13

# HISTOCHEMICAL ANALYSIS

## Introduction

Analyses of histological samples are made by the same basic principles described for larger specimens. Samples of course will generally be small and may become very minute as the analyst's ambition grows to study smaller and smaller structures. This implies a demand for an enormous range of sensitivity for studies all of which could be denoted as quantitative histochemistry or cytochemistry. Consequently, in adapting methods to histochemical purposes certain modifications must be introduced depending on the sample size and the substance concerned. To measure lactic dehydrogenase in a 2 $\mu$g sample from the molecular layer of cerebellum requires a million times less sensitivity than to measure lactate in a 0.002 $\mu$g portion of an anterior horn cell. To measure the lactic dehydrogenase it is easy to provide sufficient sensitivity, but for reasons of convenience the reaction is carried out at high tissue dilution. This makes it necessary to be concerned about the effects of dilution on enzyme stability. To measure the lactate requires not only great sensitivity but also the use of very small volumes for the reaction in order to circumvent blank problems.

## Enzyme Stability at High Dilution

Primarily because of surface denaturation, most enzymes become unstable at high dilution. This is particularly significant when the ratio of surface to volume is large, as it is when the volume is small. The addition of an excess of inert protein crowds out most of the enzyme from the surface and protects it. A rule of thumb is that total protein content should be 0.02% in volumes down to 2 or 3 μl, and 0.05–0.1% in the much smaller volumes sometimes used in oil wells. A 1 μg dry tissue sample in a 10 μl volume constitutes a 0.005–0.01% protein solution.

## Addition of Sample and Reagent

The introduction of the dry sample into the reagent in a small test tube is a critical step, particularly in the case of a small sample in an enzyme assay. When the sample becomes wet it becomes more difficult to visualize and it could be left stranded unnoticed on the wall above the meniscus. Every effort should be made to see the sample after it is in the reagent. It is important to have good lighting and clear vision. The operator may benefit from special glasses that permit close examination, or a wide-field microscope may be used. Another more serious problem with small samples is discussed below.

### Dry Loading

Relatively large samples (1 μg or more) are transferred to the bottom of a dry tube. Usually the sample is placed inside the mouth of the tube and then dislodged with sharp taps while the tube is in a vertical position. Sometimes electrostatic charges make it difficult to make the sample fall to the bottom. In this event it is helpful not to handle the tubes more than necessary and to keep a radiation source near the samples being loaded.

When reagent is added to the sample, in order to prevent the possibility of the sample attaching itself to the pipette tip, it is helpful to keep the tip away from the sample when delivery is started and to keep the tip in the extreme upper part of the meniscus as delivery continues.

In an enzyme assay it is unnecessary and undesirable to mix the sample after reagent has been added. In addition to the danger of washing the sample up onto the wall, violent mixing could result in surface denaturation. Repeated tests have shown that with volumes of reagent even as large as 40 μl, diffusion is adequate to bring substrate to the enzyme. This is only true for reagent added to the porous frozen dried material. If an enzyme *solution* were to be added to a reagent, mixing would, of course, be required. If in spite of precautions a section should be stranded above the reagent, gentle tapping with the finger can be used to swirl the liquid gently and wash the sample down.

In a typical enzyme assay, with 1 μg samples or larger, the tubes with sections are lined up in a rack at *room temperature*. The tubes are identified by their position in the rack. (Until reagent is added the tubes must not be chilled because of the danger of condensation of moisture from the air.) As each tube receives the reagent it is transferred to the same position in a second rack in an ice bath. When all tubes, standards, and blanks are ready, this second rack is transferred to the incubation bath (usually 38°) for the required period, and then returned to the ice bath. If a second solution is used to stop the reaction this is added at about the same pace as that used for the first reagent addition. Consequently all samples are exposed to the first reagent for the same total time at 0° (as well as at 38°) and therefore the enzyme will be given the same opportunity for action in all samples. If the reaction is instead to be stopped by heat, allowance may have to be made for the fact that the first samples were exposed to reagent for a longer time at 0° than the last samples. (It is highly desirable to know the relative rate at 0° for the enzyme concerned. This can vary from 2 to 40% of that at 38°.)

## WET LOADING

For samples that weigh 0.1 μg or less the addition of reagent to a dry sample carries high risk. For one reason, it becomes more difficult than with larger samples to see the sample after reagent has been added and to be sure it has not been washed up on the wall (or carried out of the tube on the pipette tip!). In addition, for reasons that are not entirely clear, there is a strong likelihood of partial inactivation of the enzyme. The best explanation is that, when the reagent first hits the tube, surface tension causes a thin film to flash up the tube past the visible meniscus, carrying with it some of the tissue elements and perhaps denaturing some of the proteins.

One solution found for these problems is to reverse the loading procedure and add the sample to the reagent already present in the tube (wet-loading). The reagent is pipetted into two or three tubes at a time (it is less desirable to fill all the tubes with reagent ahead of time). The sample (0.01–0.2 μg) is placed on a fine glass (or better quartz) needle (Fig. 13-1). The tube with reagent is held horizontally on a manipulator made from a rack and pinion. The tube is moved over the needle and the sample is discharged into the reagent. The whole process is observed through a low-power wide-field dissecting microscope and the dried sample can be introduced into the reagent without ambiguity. If the samples are smaller than 0.01 μg, the tissue may adhere to the tip of the quartz needle. In this case a piece of fiber glass about a millimeter in length is placed at the end of the quartz needle and the sample placed on the fiber glass (Fig. 13-1). This comes off together with the sample as the reagent arrives at the tip.

In many cases it will prove an easier solution to the loading problem to substitute oil wells for the micro test tubes (see below).

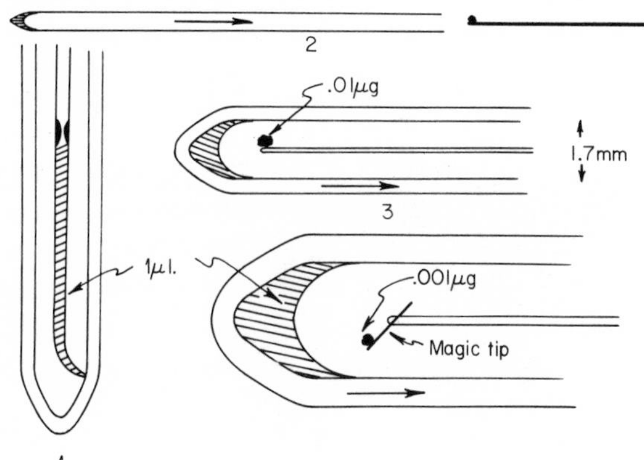


***Fig. 13-1.*** Wet loading in tubes (see text).

## Oil Well Technique

To utilize effectively the sensitivity of enzymatic cycling it is usually necessary to carry out the first analytical steps in very small volumes. It is difficult to work with volumes of less than 1 μl in open test tubes without serious evaporation. By working under oil in holes drilled in a block of Teflon, aqueous volumes as small as 0.001 μl can be employed with little or no evaporation Lowry (1963), Matschinsky *et al.* (1968). According to the demands of the particular analysis, successive additions can be made to the sample, as in any test tube, except that there is greater latitude in regard to volume. For example, an initial volume of 0.005 μl may be allowed to grow to 10 μl within the oil well. Usually all but the last analytical steps are carried out in the oil well. Heating steps can be employed, with temperatures approaching 100° without evaporation.

### Composition of Oil Well Rack

It is essential that the oil well be hydrophobic, otherwise droplets will wet the surface and spread, making mixing with subsequently added reagents difficult or impossible. Teflon is nearly ideal in that it is both hydrophobic and chemically inert. Its one disadvantage is its opalescence, which decreases visibility. Polyethylene is better in this respect and may be used, but destructive reagents must not be employed for cleaning. Siliconed glass has been tried but alkali will remove the silicone coating. Therefore no analytical step requiring alkalinization can be employed. This is a great disadvantage. At the present time Teflon is recommended for most purposes.

*Teflon Oil Well Racks*

The size of the Teflon rack and of the oil wells may be varied as needed. A convenient rack consists of a 20 × 120 mm block 5 mm ($\frac{3}{16}$ inch) thick drilled with 60 holes of 3 mm ($\frac{1}{8}$ inch) diameter (Fig. 13-2). A flat-ended drill

***Fig. 13-2.*** An oil rack. The piece of metal held above the rack contains a radioactive source to dispel electrostatic charges.

is used to create a flat-bottomed well deep enough so that only a very thin translucent layer remains. The visibility of very small volumes or samples is thus enhanced. To remove the threads or burrs that may remain from drilling, the inside of the well is polished with a piece of fine sandpaper wrapped around a metal rod slightly smaller in diameter than the well.

For very small tissue samples, there is an advantage in using thinner Teflon (e.g., $\frac{3}{32}$ inch) with correspondingly shallower wells.

*Oil*

A mixture of 40% hexadecane and 60% U.S.P. light mineral oil (specific gravity, 0.83–0.86) has been found to be satisfactory. The viscosity of oil used must be great enough to protect the fluid droplet from access of $CO_2$ from the air and still permit the droplet to fall rapidly to the bottom of the well when delivered from a pipette. Too viscous an oil adheres to the tip of the pipette to a disturbing degree.

In order to ensure that the oil is free of disturbing impurities the following washing procedure is recommended. The oil mixture is shaken vigorously in a separatory funnel with two volumes of 1 *N* NaOH and the oil and water phase separated. The oil is then washed with $H_2O$, 0.02 *N* HCl, and repeated changes of water until the washings are neutral. The oil is finally centrifuged at 10,000 rpm and the oil phase removed from the water phase. The last traces of water are removed by drying the oil in a vacuum dessicator. Unless the oil is cleaned, alkaline droplets may react with the oil during heating steps and the droplets spread out and be difficult to see.

If it is necessary to work at 0°, light mineral oil alone is substituted for the hexadecane mixture, which solidifies below 10°.

*Heating and Cooling of Teflon Racks*

Because Teflon is a good thermal insulator, the time of heating and cooling of the racks must be made longer than that for tubes immersed in water. It takes 5–10 min for the rack to reach the ambient temperature. Heating is most easily done in an oven of the desired temperature, or the rack may be placed in a metal pan covered with foil, and floated in a water bath. The foil is to protect the rack from any condensation of water. Too high temperatures must be avoided as convection currents may bring droplets to the surface where they will evaporate. If it is necessary to make additions or remove samples at 0°, the racks may be cooled by placing them on a plastic bag filled with crushed ice in a petri dish. As said above, when such a low temperature is needed, mineral oil alone is substituted, for the hexadecane–oil mixture solidifies.

*Cleaning Oil Well Racks*

The bulk of the oil is removed by inverting the rack and tapping it out. The remaining oil is rinsed out with acetone, and the racks boiled for 15 min in 0.5 *N* ethanolic KOH or NaOH. The racks are carefully rinsed with tap water and deionized water and dried in an oven at 100°. They are, of course, protected from dust.

## Reagent Wells

During the analytical procedure it is convenient to have the reagents in a container that can be viewed in the microscope along with the rack. A convenient holder is another Teflon rack with larger wells in which the reagent can be stored under oil. However when using very small volumes, pipetting the reagent from the Teflon rack may be somewhat difficult to visualize. Glass reagent vessels of 0.2–1 ml capacity with a very small orifice (2 mm) to retard evaporation can be substituted in this case. The vessels are made from Pyrex tubing, with a glass rod affixed for stability (Fig. 13-3).

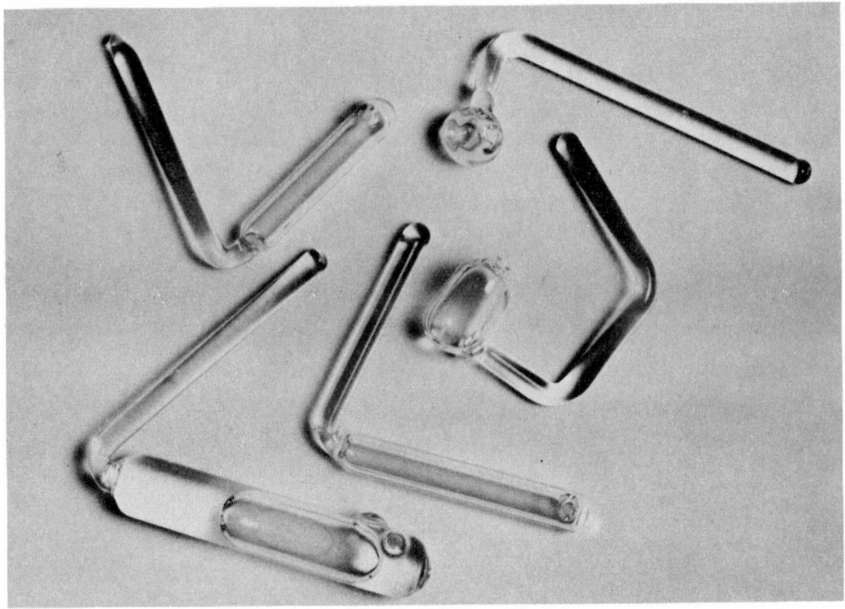

*Fig. 13-3.* Vessels for reagents to be added to oil wells.

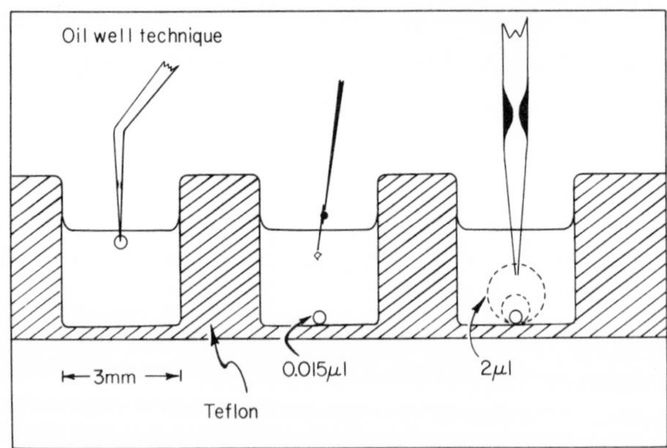


*Fig. 13-4.* Steps in an analysis in oil wells. The hair point shown would be improved if the tip had a short foot (see Fig. 11-1). The volumes indicated are completely arbitrary.

## Pipettes

Constriction pipettes of a special design are made for use with oil wells. (Figs. 3-1, 13-4). A finer tip than usual is needed since the volume must be removed by scraping it off at the oil surface rather than by transferring it with the aid of surface tension to a glass surface. If the tip is too blunt none or only part of the volume will be removed. Quartz rather than glass is used because its better heat conductivity makes it easier to make a uniform constriction, and its greater strength permits a sturdier sharp tip. The tip is drawn out in a slender taper to a fine point and bent at an angle of 45°, 5–6 mm above the orifice. The bend makes it possible to pipette into the oil with the hand steadied by resting on the working surface. The constrictions must be well balanced (Chapter 3) because if air should be delivered after the liquid the bubble will ruin the sample. If delivery is too rapid the liquid may break up into several droplets.

As with the usual constriction pipette, the tip should not be dipped further than necessary into the fluid to be pipetted. Since there is no hydrophilic surface against which to deliver the fluid, pipetting is made directly into the oil. The droplet is dislodged by withdrawing the pipette from the oil. The droplet is stripped from the tip and sinks to the bottom of the well. If a second volume of reagent is added, the pipette is touched to the surface of the previous droplet, and the fluid delivered into it (Fig. 13-4).

There is frequently a tendency for droplets to remain suspended from the oil surface. It is essential that they be dislodged, otherwise disastrous evaporation will occur through the point of contact.

## Microscope and Working Stage

All the operations are carried out under low magnification (0.5 × or 1 × objective, 10 × oculars) of a stereo dissecting microscope. A variable magnification (Zoom) microscope with long focal distance is especially useful. The microscope should be mounted on an arm projecting over the working surface (see Appendix).

The rack is illuminated by transmitted light from below. A suitable working surface and light is provided by a wooden stand in which a 90° glass prism is fitted to project the light through the rack (Fig. 13-5). Visibility is further improved if the illumination is restricted to a single well, thereby reducing glare from the rest of the rack. The stage is covered with a piece of sheet metal drilled with a hole the size of an oil well (Figs. 13-2, 13-5). The light is focused on this opening. [Better than the sheet metal is a thin piece of dark-colored or smoky (but not opaque) Plexiglas (Fig. 13-5).]

Static electricity is eliminated by suspending a piece of metal with radium or polonium on a stand above the rack during analyses (Fig. 13-2). The static must be dissipated or the aqueous droplets may migrate capriciously or even disintegrate into smaller droplets.

Dust must be prevented from falling into the oil wells. Except when additions are being made the oil rack should be continuously covered. Except for the least demanding analyses it is advisable to suspend a shelf of clear

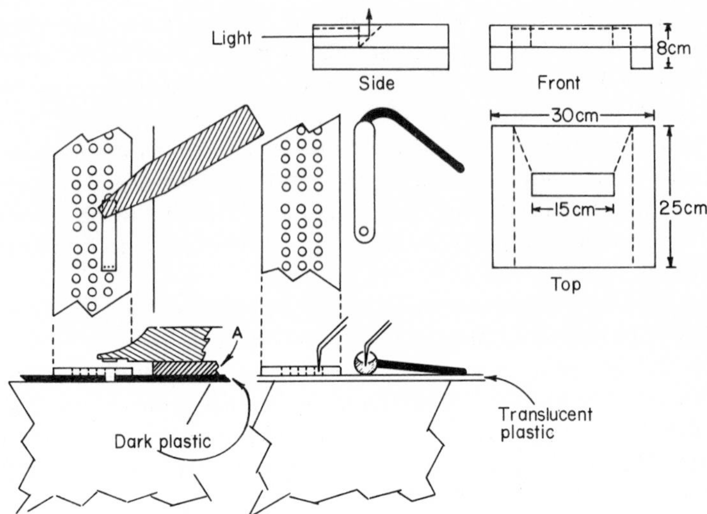


*Fig. 13-5.* Platform for oil well work. Light is transmitted from below by means of a right angle prism, or mirror. When loading samples, as shown on the left, the surface is completely covered by a sheet of dark plastic to absorb most of the light except where a hole is drilled with a diameter about equal to that of the oil wells. The right half of the platform is covered by smooth plate (A) somewhat thicker than the oil well rack. This is fixed firmly in place with tape and permits the oil well rack to slide under the sample holder. When adding reagent, as shown in the middle, plate A is not needed and better visibility is provided by covering the platform with a sheet of translucent plastic to illuminate the reagent vessel as well as the oil wells.

Plexiglas over the wooden stand just below the microscope. This protects the oil wells from dust (and dandruff!) during analytical manipulations.

### Addition of Sample and Reagent

Two different procedures have been used to bring the sample and reagent together. If the volume of the initial step is greater than 0.1 $\mu$l, the tissue sample can be placed in an empty well, the reagent added, and the droplet immediately covered with oil from an eye dropper.

An alternative better procedure of general applicability and almost essential with smaller volumes was kindly suggested to us by Dr. J.-E. Edström. Oil is added first, followed by the reagent droplet, and then the sample is introduced into the droplet through the oil (Fig. 13-4). It is guided or pushed through the oil with a special quartz-tipped hair point until it makes contact with the reagent. During the passage through the oil the porous tissue becomes soaked with the oil but this is displaced upon contact with the aqueous reagent, upon which the sample rapidly spreads.

In the case of very small samples, the introduction of the sample through the oil can be difficult at first. Therefore this step deserves elaboration. The hair point used to guide the sample through the oil is constructed with a fine foot (Fig. 11-1; notice the angle of the shaft). The sample is picked up from the carrier with the top of this foot and pushed into the oil where actual contact is lost. Thereafter, the foot acts as a kind of fan to wave the sample down to the droplet. Once real contact is made with the droplet the sample will spread over the surface. If this sign is not observed the sample can be gently pushed into closer contact. If possible, however, the hair point is kept free of the droplet, otherwise there is danger that part of the sample will adhere to the hair point. As the sample is pushed or fanned through the oil it is followed by focusing down with the microscope. If the sample falls to the bottom of the oil well without making contact with the droplet it may become invisible.

The sample, while in the oil, becomes translucent because of impregnation with oil and may be somewhat difficult to visualize. This is made worse by the fact that the oil meniscus acts as a negative lens, causing the sample to appear to shrink, more and more, the further it gets from the surface. If the volume of oil is kept to a minimum, however, this lens effect is decreased and the distance through which the sample must be pushed is diminished. (The substitution of shallower oil wells was mentioned above.) If too much difficulty is experienced the oil volume can be reduced until it does not quite cover the aqueous droplet, leaving a minute portal through which the sample is introduced. After sample addition more oil is added. (Reducing the percentage of hexadecane or eliminating it altogether may improve visibility by increasing the difference in refractive index between sample and oil.)

Loading either directly or through the oil appears to give the same results. Glucose, glucose-6-P, glycogen, and a number of enzymes have been specifically tested in this regard. Nevertheless, the possibility should be borne in mind that some compounds or certain enzymes might be affected by direct exposure to the oil. Cleaning the oil as recommended above should help eliminate possible harmful impurities.

## Special Considerations

The presence of the oil–water interface creates certain problems. Inactivation of enzymes may be more likely to occur during incubation; high concentrations of bovine plasma albumin (1 mg/ml) have been found to prevent this in situations tested. If a cycling step is carried out in the oil well, the high concentrations of enzymes used form a precipitate during heating which makes the samples difficult to transfer. In this case the reaction may be stopped by adding NaOH to give a 0.1 or 0.2 *N* concentration followed by heating 10 min at 90°. The alkali keeps the protein in solution and the sample can be easily pipetted.

Sample volumes greater than 10 $\mu$l so fill the standard-sized wells that there is danger of breaking through the oil surface with consequent evaporation. When the analysis demands that the volume be expanded beyond 10 $\mu$l the samples can be transferred to ordinary glass test tubes for subsequent steps. In the majority of cases, however, the sample can be transferred directly from the oil well to the fluorometer tube for the final step and for reading in a volume of 1 ml.

PART III

# APPENDIX

# APPENDIX

## Acid–Base Made Easy

### Some Definitions

*Logarithm* (*to the base 10*). A logarithm is an exponent. It is a number which if used as an exponent to 10 will give the number whose log it is.

*Examples*

$$10^3 = 1000 \qquad \log 1000 = 3$$
$$10^{-3} = 0.001 \qquad \log 0.001 = -3$$
$$10^{0.3} = 2 \qquad \log 2 = 0.3$$
$$10^{-0.3} = \tfrac{1}{2} = 0.5 \qquad \log 0.5 = -0.3$$

Remember, $\log N + \log M = \log N \times M$, $\log N - \log M = \log N/M$, $\log 10^{-N} = \log 1/10^N$, $\log \sqrt{N} = \frac{1}{2} \log N$.

(See also *A Homemade Log Table* at the end of this section.)

*pH.* pH is the negative logarithm of the hydrogen ion *concentration*, i.e., $\text{pH} = -\log[H^+]$ or $10^{-\text{pH}} = [H^+]$. (Note *concentration* is always *molar*.)

*Examples*

| | | |
|---|---|---|
| $[H^+] = 0.1\ M$ | $0.1 = 10^{-1}$ | $pH = 1$ |
| $[H^+] = 0.0001\ M$ | $0.0001 = 10^{-4}$ | $pH = 4$ |
| $[H^+] = 0.5\ M$ | $0.5 = 10^{-0.3}$ | $pH = 0.3$ |
| $[H^+] = 0.0002\ M$ | $0.0002 = 2 \times 10^{-4}$ | $pH = -(0.3 - 4) = 3.7$ |
| $[H^+] = 5 \times 10^{-9}$ | | $pH = -(0.7 - 9) = 8.3$ |

*Acid.* Any compound that can lose a hydrogen ion.

*Examples*

$$CH_3COOH,\ CH_3CH_2OH,\ CH_3NH_3^+,\ NH_4^+$$

(Note "strong acids" such as HCl really don't fit this definition, see under Conjugate Base.)

*Base.* Any compound that can gain a hydrogen ion.

*Examples*

$$CH_3COO^-,\ CH_3CH_2O^-,\ CH_3NH_2,\ NH_3$$

*Conjugate base* (*or acid*). The base (or acid) in the following general equation:

$$\text{acid} \rightleftharpoons \text{base} + H^+ \tag{A-1}$$

i.e., every acid has its conjugate base (and every base has its conjugate acid). Strong acids and bases, such as HCl and NaOH, really do not fit this formulation. HCl in solution has already lost its $H^+$, i.e., it is completely dissociated and can be written $H^+Cl^-$. NaOH is really $Na^+OH^-$, and the real base is $OH^-$ which qualifies because it *can* gain a $H^+$:

$$OH^- + H^+ \rightleftharpoons H_2O$$

*Buffer.* A mixture of a conjugate acid and base. It is a buffer (i.e., a pH buffer) because if $H^+$ is added most of the $H^+$ will combine with the base [Eq. (A-1)], lessening the increase in $H^+$ concentration which would otherwise result. And, similarly, if $H^+$ is removed from such a mixture more will be formed from the acid [Eq. (A-1)], lessening the decrease in $H^+$ concentration which would otherwise result.

## *pK's* AND THE HENDERSON-HASSELBALCH EQUATION

The more readily an acid loses $H^+$ the stronger an acid it is. This strength is measured by its *dissociation constant*.

$$K = \frac{[\text{base}][H^+]}{[\text{acid}]} = \frac{[B][H^+]}{[A]} \tag{A-2}$$

When the acid is 50% dissociated, base and acid are equal and $K = [H^+]$.

For example, the dissociation constant of acetic acid is $2.5 \times 10^{-5}$ *M*. Therefore, in a mixture of equal molar concentrations of acetic acid and sodium acetate $[H^+] = 2.5 \times 10^{-5}$ *M* or pH $= -(0.4 - 5) = 4.6$.

Similarly, the dissociation constant of $NH_4^+$ is $5 \times 10^{-10}$. Therefore in a mixture of equal molar concentrations of ammonium hydroxide ($NH_3 \cdot H_2O$) and $NH_4^+ Cl^-$, $[H^+] = 5 \times 10^{-10}$ or pH $= -(0.7 - 10) = 9.3$.

Because of this relationship between $K$ and $[H^+]$, and by analogy with pH, the *negative log of K is called pK*. It is convenient to replace $K$ and $[H^+]$ in Eq. (A-2) with pH and p$K$ by taking the log of each side.

$$\log K = \log \frac{[B][H^+]}{[A]} = \log \frac{[B]}{[A]} + \log [H^+]$$

rearranging:

$$-\log [H^+] = -\log K + \log \frac{[B]}{[A]} \text{ or pH} = \text{p}K + \log \frac{[B]}{[A]} \qquad \text{(A-3)}$$

This is the famous Henderson-Hasselbalch equation. This equation was originally written:

$$\text{pH} = \text{p}K + \log \frac{[\text{Salt}]}{[\text{Acid}]} \qquad \text{(A-3a)}$$

This was before Brönsted proposed the definition of acids and bases given above. Equation (A-3a) is harder to use than Eq. (A-3), and is not quite exact.

From the p$K$ of an acid and the Henderson-Hasselbalch equation it is easy to calculate the pH for any ratio of acid to base.

| Base/ acid | pH | Base/ acid | pH |
|---|---|---|---|
| 100 | p$K$ + 2 | 2 | p$K$ + 0.3 |
| 1/100 | p$K$ − 2 | 1/2 | p$K$ − 0.3 |
| 20 | p$K$ + 1.3 | 1 | p$K$ |
| 1/20 | p$K$ − 1.3 | 200 | p$K$ + 2.3 |
| 10 | p$K$ + 1 | 5000 | p$K$ + 3.7 |
| 1/10 | p$K$ − 1 | 10,000 | p$K$ + 4 |

It is easy to make this calculation in your head, as accurately as there is usually any need, especially if you memorize the log of 2 and 1.25 (0.3 and 0.1). (See A HomeMade Log Table below.) You can even forget whether it is base/acid or acid/base in the H-H equation if you simply remember that if *base* predominates the pH is on the *basic* side of the p$K$, and if *acid* predominates, the pH is on the *acid* side of the p$K$.

However, if you prefer, Table A-1 makes the calculation for you and gives

**TABLE A-1**

BASE: ACID RATIOS TO GIVE DESIRED pH

| pH | Acid/ base | Base as % of A + B | pH | Base/ acid | Base as % of A + B |
|---|---|---|---|---|---|
| p*K* + 1.0 | 10 | 9 | p*K* − 1.0 | 10 | 91 |
| p*K* + 0.9 | 8 | 11 | p*K* − 0.9 | 8 | 89 |
| p*K* + 0.8 | 6.3 | 14 | p*K* − 0.8 | 6.3 | 86 |
| p*K* + 0.7 | 5 | 17 | p*K* − 0.7 | 5 | 83 |
| p*K* + 0.6 | 4 | 20 | p*K* − 0.6 | 4 | 80 |
| p*K* + 0.5 | 3.2 | 24 | p*K* − 0.5 | 3.2 | 76 |
| p*K* + 0.4 | 2.5 | 29 | p*K* − 0.4 | 2.5 | 71 |
| p*K* + 0.3 | 2 | 33 | p*K* − 0.3 | 2 | 67 |
| p*K* + 0.2 | 1.6 | 39 | p*K* − 0.2 | 1.6 | 61 |
| p*K* + 0.1 | 1.25 | 44 | p*K* − 0.1 | 1.25 | 56 |
| p*K* + 0.0 | 1.0 | 50 | p*K* − 0.0 | 1.0 | 50 |

the percentage of the buffer that is in the basic form. Notice it takes three times more acid or base to make a pH change of 0.1 unit near the p*K* than it does 1 pH unit away. At 1.3 units from p*K* a 1% shift from acid to base makes a 0.1 unit pH change. Thus the useful pH range is limited to about one pH unit on either side of the p*K*.

## MAKING A BUFFER

A buffer mixture can be made by either (a) mixing the free acid or base with its salt, or (b) adding a calculated amount of strong acid or strong base to the free base or free acid. For example, a pH 4.6 buffer can be made by making the solution 50 m*M* in acetic acid and 50 m*M* in sodium acetate. Or, it can be made by making the solution 100 m*M* in acetic acid and 50 m*M* in NaOH. This ignores the fact that some $H^+$ had to be produced ($10^{-4.6}$ *M*) from the acetic acid. Therefore the base (acetate) is in fact slightly greater than the added sodium acetate or NaOH. In this case the difference is only 1 part in 2000. Only below pH 3 and above pH 11, and with low buffer concentrations, does this ever become significant.

## MIXTURE OF TWO OR MORE ACID–BASE PAIRS

The H-H equation must apply simultaneously to all acid–base pairs in a given solution, i.e.,

$$\mathrm{pH} = \mathrm{p}K_1 + \log\frac{[\mathrm{B}_1]}{[\mathrm{A}_1]} = \mathrm{p}K_2 + \log\frac{[\mathrm{B}_2]}{[\mathrm{A}_2]} \qquad \text{etc.}$$

Take as an example a mixture of acetic acid, acetate (p$K$ = 4.6), Tris$^+$ and Tris base (p$K$ = 8.1). At pH 7,

$$\log [B_1]/[A_1] = \log [Ac^-]/[HAc] = 2.4 = \log 500/1 \quad \text{and}$$
$$\log [B_2]/[A_2] = \log [Tris]/[Tris^+] = -1.1 = \log 1/12.5$$

Dissociation Constant of $H_2O$

There is one acid that must not be forgotten; this is $H_2O$. It is a very weak acid; its p$K$ is 15.7. This means that

$$\frac{[H^+][OH^-]}{[H_2O]} = 10^{-15.7} \quad \text{and} \quad pH = 15.7 + \log \frac{[OH^-]}{[H_2O]}$$

Because the concentration of $H_2O$ is practically constant in most aqueous solutions (55.5 *M*), it has become customary to merge its concentration with the dissociation constant:

$$[H^+][OH] = 55.5 \times 10^{-15.7} = 10^{-14}$$

In the logarithmic form this becomes,

$$pH + pOH = 14 \quad \text{or} \quad pOH = 14 - pH \tag{A-4}$$

where pOH is the negative log of ($OH^-$).

pH of Free Acids and Bases

The pH of a weak acid (or base) in the absence of added conjugate base (or acid) is easily calculated in all practical situations. When an acid is dissolved in $H_2O$, $A \rightleftharpoons B + H^+$, i.e., $B = H^+$. Equation (A-2) rearranged becomes $[B][H^+] = K[A]$, or in this case $[H^+]^2 = K(A)$, and $[H^+] = \sqrt{K[A]}$. Or,

$$pH = \tfrac{1}{2}\,(pK - \log [A]) \tag{A-5}$$

For example, 0.1 *M* acetic acid has pH = $\frac{1}{2}(4.6 + 1) = 2.8$. A 0.01 *M* solution ($-\log\ [A] = 2$) would have a pH of 3.3. The p$K$ of $NH_4^+$ is 9.3. A 50 m*M* solution of $NH_4Cl$ has pH = $\frac{1}{2}(9.3 + 1.3) = 5.3$. The pH of a 1 *M* solution would be 4.65.

When a free base is dissolved in $H_2O$ it removes $H^+$, forming hydroxyl ion: $B + H_2O \rightleftharpoons A + OH^-$. Here $[A] = [OH^-] = 10^{-14}/[H^+]$ (see above). Equation (A-2) rearranged becomes $[H^+]/[A] = K/[B]$. Substituting for [A],

$$[H^+]^2/10^{-14} = K/[B],\ [H^+]^2 = 10^{-14}\,K/[B], \text{ and } [H^+] = \sqrt{10^{-14}K/[B]},$$

or

$$pH = \tfrac{1}{2}(14 + pK + \log [B]) \tag{A-6}$$

For example, 0.2 *M* Tris base has $pH = \frac{1}{2}(14 + 8.1 - 0.7) = 10.7$. A 0.1 *M* sodium acetate solution has $pH = \frac{1}{2}(14 + 4.6 - 1) = 8.8$.

[An easy way to remember Eqs. (A-5) and (A-6) is that an acid has a pH half way between its p*K* and the pH it would have if it were a strong acid, and a base has a pH half way between its p*K* and the pH it would have if it were a strong base.]

Equations (A-5) and (A-6) are not strictly true because they ignore the decrease in A or in B required to form $H^+$ in one case and $OH^-$ in the other. The difference is exceedingly small except with low concentrations of acids having p*K*'s below 3 or bases with p*K*'s over 11.

### Some Complications with p*K*'s

We have been talking about p*K*'s and dissociation constants as though they were invariant. Actually they are affected by temperature and salt concentration (ionic strength). Increasing temperature and increasing ionic strength increase the dissociation constants (decrease the p*K*'s). Tables of p*K*'s are often calculated for 25° for "infinite dilution" (zero ionic strength). An observed p*K* is therefore always lower than these "true" p*K*'s and should properly be called an apparent p*K* and written p*K*′.

Ionic strength has a much greater affect if the dissociation involves a separation of charges than if it does not. Consider the three dissociations:

$$RNH_3^+ \rightleftharpoons RNH_2 + H^+$$
$$RCOOH \rightleftharpoons RCOO^- + H^+$$
$$RPO_2H^- \rightleftharpoons RPO_2^{2-} + H^+$$

Ionic strength will have little affect in the first case, a moderate affect in the second, and a large affect in the third. (Increasing ionic strength, by increasing the dielectric constant of the solution, shields the charges from each other, making it easier for them to come apart.)

As examples, the true p*K* of acetic acid is given as 4.76, but a 0.1 *M* acetate buffer has a p*K*′ near 4.6. The $pK_2$ of a phosphate buffer is 7.2, a 0.1 *M* phosphate buffer has a $pK_2'$ of 6.8, and in a 1 *M* phosphate buffer the $pK_2'$ is around 6. (The $pK_3$ is even more sensitive to ionic strength.) "True" p*K*'s are given in Table A-2 and in some cases ionic strength effects may need to be taken into account.

There is one final complication in regard to p*K*'s. We have been considering $NH_4^+$, for example, as an acid with $pK = 9.3$, i.e., a dissociation constant of $5 \times 10^{-10}$. Unfortunately, in pre-Brönsted days $NH_3$ was regarded in solution as $NH_4OH$ which dissociated as $NH_4OH \rightleftharpoons NH_4^+ + OH^-$. Measurements of $OH^-$ concentration indicated a dissociation constant of $2 \times 10^{-5}$, or $pK = 4.7$. To distinguish these, one is written $pK_a$ (for the acid dissociation constant); the other is written $pK_b$ (for the "base dissociation

**TABLE A-2**

DISSOCIATION CONSTANTS FOR ACIDS AND BASES USED IN MAKING BUFFERS[a,b]

| Acid or base | pK | Acid or base | pK |
|---|---|---|---|
| Phosphoric acid, $K_1$ | 1.96 | HEPES | 7.55[c] |
| Arsenic acid, $K_1$ | 2.25 | Triethanolamine | 7.9 |
| Glycine, $K_1$ | 2.35 | Glycylglycine | 8.07 |
| Citric acid, $K_1$ | 3.08 | Tris | 8.1 |
| Glycylglycine, $K_1$ | 3.14 | Tricine | 8.15[c] |
| Formic acid, $K_1$ | 3.76 | Hydrazine, $K_2$ | 8.23 |
| Citric acid, $K_2$ | 4.75 | Pyrophosphoric acid, $K_4$ | 8.22 |
| Acetic acid | 4.76 | 2-amino-2-methyl-1,3-propanediol | 8.78 |
| Pyrophosphoric acid, $K_3$ | 5.77 | Histidine, $K_3$ | 9.18 |
| Citric acid, $K_3$ | 6.00 | Ammonium hydroxide | 9.3 |
| Histidine, $K_2$ | 6.10 | Glycine, $K_2$ | 9.78 |
| MES | 6.15[c] | 2-Amino-2-methyl-1-propanol | 9.9 |
| Cacodylic acid | 6.19 | Carbonic acid, $K_2$ | 10.36 |
| Arsenic acid, $K_2$ | 6.77 | Arsenic acid, $K_3$ | 11.6 |
| Imidazole | 7.07 | Phosphoric acid, $K_3$ | 12.3 |
| Phosphoric acid, $K_2$ | 7.12 | | |
| BES | 7.15[c] | | |

[a] All p*K*'s are for acid dissociation constants at zero ionic strength and 25° except as noted. Where there is more than one dissociation constant, $K_1$ refers to the most acidic p*K*.

[b] Abbreviations: MES, 2-(*N*-morpholino)ethanesulfonic acid; BES, *N*,*N*-bis(2- hydroxyethyl)-2-aminoethanesulfonic acid; HEPES, *N*-2-hydroxyethylpiperazine-*N*′-2-ethanesulfonic acid; Tricine, *N*-tris(hydroxymethyl)methylglycine.

[c] At 20°, from Good *et al.*, who introduced MES, BES, HEPES (1966), and Tricine (Good, 1962).

constant"). It is clear that since pOH + pH = 14, $pK_a + pK_b = 14$. $pK_b$'s are still to be found in handbooks, often written simply as p*K*, and it is not always clear whether $pK_a$ or $pK_b$ is meant.

## INDICATORS

Indicators are weak acids that have conjugate bases of a different color. In the case of phenolphthalein, a commonly used indicator, the acid form is colorless and the conjugate base is red. The $pK_a$ of this indicator is 9.7. From the Henderson-Hasselbalch equation, it is seen that at pH 7.7 only 1% of the indicator is in the red form, and the solution will be nearly colorless. At pH 8.7 there will be 9% of the red form (B/A = 1/10) and as the pH is shifted further to the alkaline side, the dissociated base form will predominate and the solution will turn red. It is apparent that phenolphthalein would not be a good indicator to detect pH changes lower than 8.7 or greater than 10.7. It is clear that a given indicator is useful in only a narrow pH range, at most 2 pH units, and often less because of insensitivity of the eye to changes in intensity of the particular hue.

## TITRATIONS

When an indicator is used for an acid–base titration, the first decision is to select the proper pH for the end point. If a strong acid such as HCl is titrated with a strong base such as NaOH, there is great latitude. If the solution is 0.01*M* in HCl ($10^{-2}$ *M*) it will be 99.9% neutralized at pH 5 ($10^{-5}$ *M* $OH^-$). In this case any indicator with detectable color change between 5 and 9 would be satisfactory. If, however, 0.01 *M* acetic acid is to be titrated with NaOH, the correct end point (ignoring dilution due to the NaOH) would be at a pH half way between the p*K* (4.6) and that of 0.01 *N* NaOH (pH 12), or pH 8.3 (±0.7 for 0.1% accuracy). The acid would be 1% undertitrated at pH 6.6 (B/A = 100), and 1% overtitrated at pH 10 ($[OH^-] = 0.0001$ *M*).

Similarly for titrations of 0.01 Tris base with HCl, the correct end point (ignoring dilution due to the HCl) would be half way between the p*K* (8.1) and the pH of 0.01 *M* HCl (2) or pH 5.0 (±0.1 for 0.1% accuracy). The base would be 1% undertitrated at pH 6.1 (B/A = 1/100) and 1% overtitrated at pH 4 (0.0001 *M* HCl). Often the simplest solution is to use a pH meter.

## A HOMEMADE LOG TABLE

All the logarithms needed for practical acid–base aspects of most analytical work can be carried in your head, or figured out quicker than you can look up a log table. All you need to remember is that the log of 2 = 0.3 (actually 0.30103) and the log of 1.25 = 0.1 (actually 0.0969). Knowing the log of 2, it is clear that log 4 = 0.6, and log 8 = 0.9. Similarly log 5 = 0.7 (i.e., log 10 − log 2), and log 2.5 = 0.4 (i.e., log 5 − log 2). We are only missing 0.2, 0.5, and 0.8, which we obtain as shown in the following tabulation:

| | | |
|---|---|---|
| log 1 | = 0.0 | |
| log 1.25 | = 0.1 | (memory) |
| log 1.6 | = 0.2 | (log 2 − log 1.25) |
| log 2 | = 0.3 | (memory) |
| log 2.5 | = 0.4 | (log 10 − log 4) |
| log 3.2 | = 0.5 | (log 4 − log 1.25) |
| log 4 | = 0.6 | (log 2 + log 2) |
| log 5 | = 0.7 | (log 10 − log 2) |
| log 6.3 | = 0.8 | (log 5 + log 1.25) |
| log 8 | = 0.9 | (log 4 + log 2) |
| log 10 | = 1.0 | |

If you need to be closer than 0.1 pH unit, you can interpolate from this, e.g., log 9 is close to 0.95 (actually 0.954). This table and the Henderson–Hasselbalch equation, both in your head, will save time and, more important, avoid blunders.

## Statistical Shortcuts

This section is aimed at the "casual statistician" (Snedecor). Rarely, for his purposes, does the situation require or justify elaborate statistical treatment.

We are concerned here only with means, standard deviations, standard errors (of the mean), and standard errors of differences.

First the orthodox calculations:

$$\textit{Standard deviation (st. dev.)} = \sqrt{\frac{\text{sum of dev}^2}{n-1}} \tag{A-7}$$

where dev is the individual deviation from the means and $n$ is the number of samples.

$$\textit{Standard error (of the mean) (SE or SEM)} = \sqrt{\frac{\text{sum of dev}^2}{n(n-1)}} = \frac{\text{st dev}}{\sqrt{n}} \tag{A-8}$$

$$\textit{Standard error of the difference (SED)} = \sqrt{\text{SE}_1{}^2 + \text{SE}_2{}^2} \tag{A-9}$$

There is a longer way of calculating SED, but it has no clear advantage.

Now the shortcuts.

1. For most purposes, calculation of *standard errors* to within 10% is all that is worthwhile. For example: 110 ± 10 (SE) says that there is a 1 out of 3 chance that the true mean lies outside of the range 90–110. What virtue could there possibly be in calculating that the SE is actually 10.3?

2. Conversely, calculation of the *mean* to closer than 10% of the standard error is seldom worthwhile. In the previous example, what possible use would there be in calculating that the average is actually 110.4? If, however, the standard error was 1 instead of 10, it might be useful to calculate the mean more closely, say ±0.1.

3. An approximate value for the standard error is

$$\text{SE} = \frac{\text{range of values}}{n} \tag{A-10}$$

For example, if there are 10 values ranging from 80 to 120, standard error = 40/10 = 4. This gives a surprisingly close estimate of the standard error (usually within 15%). It is a valuable check for errors in more exact calculations.

4. Similarly, because SE = st. dev./$\sqrt{n}$

$$\text{st. dev.} = \text{SE}\sqrt{n} = \frac{\text{range}}{\sqrt{n}} \tag{A-11}$$

5. For more orthodox calculations of means, standard errors, and standard deviations, individual *values* and individual *deviations from the mean* can usually be rounded off with great saving of time. What guide lines are there to safe rounding off?

It is usually completely safe to round off individual values to within 10% of the range. For example, there are 10 numbers ranging from 32.2 to 43.6. The range is 11.4. Therefore the numbers can be rounded off to the nearest whole number. No number will be changed by more than 0.5. This cannot increase the range by more than twice 0.5 or 1, i.e., 9%. Therefore, this will not increase the standard error or the standard deviation by more than 9%, as the above shortcut method for calculating these statistics indicates.

Rounding off to within 10% of the range will have a completely negligible effect on the mean. Thus, in the above case the average number is changed by only 0.25 and the changes are randomized plus and minus.

## Construction of Dissecting Knives

Handles for razor blade knives are made from wooden dowel rod, 6 mm ($\frac{1}{4}$ inch) in diameter, cut in convenient lengths (15 cm) and sharpened in a pencil sharpener. The ends and taper are sanded smooth, and the pointed tip is cut off to make a flat surface 2–2.5 mm in diameter. A hole about 1.5 mm in diameter is made in this surface with a wire drill. A piece of 18–20 gauge copper wire (1.0–0.75 mm diameter) is sealed into this hole with epoxy resin or Duco cement. The wire is cut off 10 or 12 mm from the handle (Fig. 11-1).

The most satisfactory knives are made from splinters of thin, brittle, stainless steel blades. We know of no present source. The next best is brittle non-stainless blades, e.g., Gillette "Super Blue." These must be protected from rusting. Most of the present thin blades on the market are not brittle and the points of knives made from these are easily bent during use. Nevertheless, these softer blades can be used for certain purposes; they can be cut with a sharp tin snips into very thin knives. Dr. G. M. Lehrer has designed a special shear for cutting these blades, which prevents curling during the cutting (Table A-6).

Splinters from brittle blades can be made by placing the blade on a hard plastic or smooth wood surface and applying pressure with a wood chisel. (Glasses or goggles must be worn for safety from flying splinters.) The cutting edge is first broken off into a strip 5 or 6 cm wide, and then broken into splinters by vertical pressure at a 45° angle to the cutting edge. Fragments of blade

0.5–1.5 mm wide which resemble a scimitar in shape are most desirable to provide the proper angle when dissecting (Fig. 11-1). A short bristle is glued to the copper wire with epoxy resin. When the resin has set, the blade is fastened with epoxy to the free end of the bristle. Mounting the blade is easier if the resin is not too fluid. If necessary it can be allowed to begin to set before mounting the blade. Bristles of different stiffness may be required for different purposes. The flexibility can also be adjusted by changing the free length of the bristle. Nylon bristles from a toothbrush are usually quite satisfactory.

## Microtome Knife Sharpening

Microtome knives sharpened by machine are suitable for most purposes. However, for very thin sections of frozen material, it is sometimes desirable to use hand sharpening to obtain a satisfactory cutting edge. The following method is a minor modification of that of Hillier (1951). As pointed out in the original article, only the preparation of the glass plates and the original, sharpening of the blade are time-consuming. With normal use, only the final two sharpening steps need be performed, which require 10 minutes or less.

### Apparatus

The sharpening is done on three glass plates, of approximately 6 × 10 × 1 inch. Ordinary ground and polished plate glass is satisfactory. For convenient handling and storage, the plates can be encased in a wooden frame and furnished with a wooden lid to prevent scratching the surface.

The knife back is fitted with two blocks of tool steel that are of a dimension that when the knife is supported by the blocks, the knife bevel is parallel with the surface of the block (Fig. A-1). The two pieces of steel are permanently

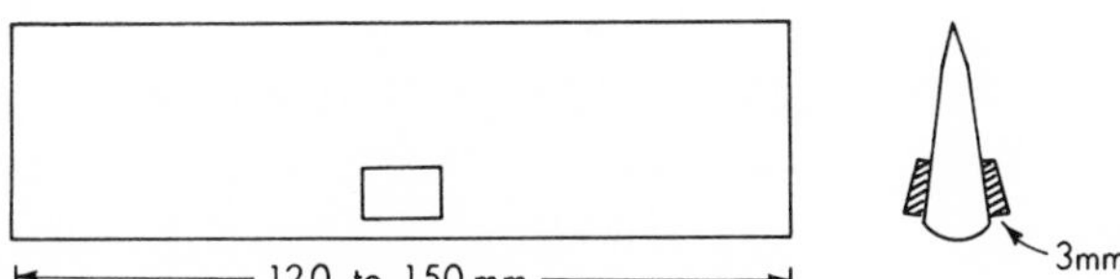


***Fig. A-1.*** Permanent sharpening back for microtome knife. Two steel blocks are fixed to the knife with epoxy cement as shown.

sealed to the back of the knife with epoxy cement. The sharpening back is short so that it acts as a point support, and consequently there is a point and line contact between knife and sharpening surface.

If the handle supplied with the knife does not permit use of the full width of the sharpening surface, the central threaded rod can be used by itself as the handle.

New plates are prepared with abrasives for the sharpening procedure. Plates 1 and 2 are ground with a small glass lap ($2 \times 2 \times \frac{1}{4}$ inches) and 200 mesh carborundum powder until most of the original surface is removed. The plate and lap are thoroughly washed and the plates ground with 900 mesh alundum until flat plateaus occupying approximately half the surface are formed.

The plate and lap are again washed and the grinding continued with levigated alumina (Table A-6) until the effect of the 900 mesh alundum has been removed.

The glass plate is thoroughly washed and dried and the surface rubbed with the hardest paraffin until the pits are filled. The surface is again rubbed with the lap and levigated aluminum until it appears uniformly smooth.

Plate 3 is prepared with an aluminum plate used as a lap, of similar dimensions to the glass lap. Levigated alumina is used as the abrasive and the surface ground evenly. This treatment produces no visible effect and is considered complete when there is no tendency for the blocks on the back of the knife to be stained in the final polishing.

### Procedure

The original sharpening requires three or four steps and may require an hour. Subsequent sharpening requires only the last one or two steps and should not require more than 10 min. If the knife becomes very badly nicked, it may be necessary to sharpen for a longer period, or to repeat some of the first steps with the coarser abrasives.

Stock solutions of the abrasives are prepared by suspending the powders in 2 or 3 volumes of water. The suspension is shaken before use and allowed to settle a few seconds before pouring it on the plate.

Carborundum F, or Carborundum 300 if the knife is badly nicked, is used for the first step on Plate 1. The knife edge leads the way. In the stroke away from the operator, the knife is held with the fingers supplying fairly hard uniformly distributed pressure, and the thumbs supplying the motive force. On the return stroke, the knife is turned over, the thumbs supply the pressure and the fingers push the knife. A zig-zag motion is used while pushing the knife the length of the plate. At this stage, the forward travel of the knife is much greater than the transverse motion. This reduces the tendency for the knife to dig into the plate and produces a much higher polish on the blocks and a better edge.

The handle is used only to reverse the blade at the end of each stroke by rolling the blade on its back, thus avoiding the possibility of placing the edge in contact with the plate before the back.

The knife is ground on Plate 1 until a flat area about 2 mm wide develops on the sharpening back.

The plate, knife, and hands of the operator are thoroughly washed, and alumina oxide abrasive (flowers of emery) is used on the same plate. The knife is ground with 20–30 strokes in each direction using a similar motion and somewhat reduced pressure.

Plate 2 is used with levigated alumina, and the operation is complete when the entire area of both blocks has a uniformly ground appearance. Again 20 or 30 strokes in each direction should be sufficient unless the knife has been damaged in some way.

The final and most critical step is the polishing on Plate 3 with the abrasive Linde A5175 (Table A-6). At this stage, the zig-zag motion of the knife is changed so that the forward travel of the knife is about equal to the transverse motion and pressure is gradually reduced. There should be no scratches on the face and almost complete absence of nicks. If the blade travels jerkily or there are stained areas along the edge, the preparation of Plate 3 has been inadequate and should be repeated.

The results should be checked by microscopic examination of the edge and blocks. Stropping is never used as it rounds the edges of the cutting edge and blocks.

Between uses, the blade is protected by coating it with a noncorrosive oil. The plates are thoroughly washed after each use and particularly between changes of abrasive. The appearance of gray discoloration in the abrasives indicates the presence of metal particles and the plates should be washed before continuing.

**TABLE A-3**

INDICATORS[a]

| Indicator | Chemical name | p$K_a$ | Acid color | Basic color | pH Range |
|---|---|---|---|---|---|
| Thymol Blue | Thymolsulfonphthalein | 1.7 | Red | Yellow | 1.2–2.8 |
| Bromphenol Blue | Tetrabromophenol sulfonphthalein | 3.5 | Yellow | Blue | 3.1–4.7 |
| Methyl Red | 4′-Dimethylamino-benzene-2-carboxyl acid | 5.0 | Red | Yellow | 4.4–6.0 |
| Chlorophenol Red | Dichlorophenol sulfonphthalein | 6.0 | Yellow | Red | 5.0–6.6 |
| Bromthymol Blue | Dibromothymol sulfonphthalein | | Yellow | Blue | 6.0–7.6 |
| Phenol Red | Phenolsulfonphthalein | 7.8 | Yellow | Red | 7.0–8.6 |
| Thymol Blue | Thymol sulfonphthalein | 8.9 | Yellow | Blue | 8.0–9.6 |
| Phenolphthalein | 3,3-Bis(*p*-hydroxyphenyl)phthalide | 9.7 | Colorless | Red | 8.3–10.0 |

**TABLE A-3**—*continued*

| Indicator | Chemical name | p$K_a$ | Acid color | Basic color | pH Range |
|---|---|---|---|---|---|
| Thymolphthalein | 5′, 5″-Diisopropyl-2, 2′-dimethylphenol phthalein | — | Colorless | Blue | 9.3–10.5 |
| Alizarin Yellow | 3-Carboxy-4-hydroxy-4′-nitroazobenzene | — | Yellow | Red | 10.2–12.0 |

[a] The commercially available indicators listed cover a wide range of pH's convenient for laboratory use.

**TABLE A-4**

PREPARATION AND STORAGE OF SOME COMMON STOCK SOLUTIONS[a]

| Compound | Storage conditions: Powder | Storage conditions: Solution | Comments |
|---|---|---|---|
| ATP · $Na_2$, 551 | −20°, desiccant | In $H_2O$, bring to pH 7 with 2 equiv. NaOH, −20° | No loss after months in storage |
| ADP · $Na_2$, 471 | −20°, desiccant | In $H_2O$, bring to pH 7 with an equiv. NaOH, −20° | Usually about 0.5–1% ATP on molar basis, sometimes contains $TPN^+$. $TPN^+$ can be destroyed by bringing solution to pH 12, with NaOH; heat 10 min at 60° and neutralize with HCl |
| 5′-AMP · Na, 369 | −20°, desiccant | In $H_2O$, −20° | — |
| Albumin, bovine plasma crystalline | 4°, desiccant | In $H_2O$, −20° | Some lots contain citrate and isocitrate |
| Citric acid · $H_2O$, 210 | 4°, desiccant | In $H_2O$, −20° | — |
| Creatine · $H_2O$, 149 | RT | In $H_2O$, −20° | — |
| Creatine-P · $Na_2$, 255 | −20°, desiccant | In $H_2O$, −20° | No loss after 60 min at 0° in 0.6 *M* $HClO_4$ or at −10° in 3 *M* $HClO_4$. No loss after 30 days and probably much longer at pH 8 and −50°. Stable to alkali |

**TABLE A-4**—*continued*

| Compound | Storage conditions: Powder | Storage conditions: Solution | Comments |
|---|---|---|---|
| Dihydroxyacetone-P dicyclohexylammonium salt, dimethyketal · $H_2O$, 432 | −20°, desiccant | Store in buffer at pH 3 or up to 0.01 *N* acid at −20° | Remove ketal with ion-exchange resin or by heating in 0.1 *N* HCl at 38° for 30 min. No loss after 2 hr at 60° in 0.3 *M* $HClO_4$ or at pH 3.0. Unstable above pH 8 |
| β-$DPN^+$, free acid, 663 | −20°, desiccant | In $H_2O$, −20° | Stable indefinitely |
| β-DPNH · $Na_2$, 708 | RT in dark, desiccant | 5 to 10 m*M* in carbonate buffer (80 m*M* $Na_2CO_3$, 20 m*M* $NaHCO_3$), −50°. See Chapter 1 | See Chapter 1. The solution is slowly oxidized to $DPN^+$ if stored at −20° or above. Preformed $DPN^+$ can be destroyed by heating in the carbonate buffer at 100° for 5 min. May contain 5–15% of 5′-AMP on a molar basis (see AMP method, Chapter 9) |
| Fructose-1,6-$P_2$ · $Na_4$, 428 | RT | In $H_2O$, −20° | — |
| Fructose-6-P · Na, 282 | RT | In $H_2O$, −20° | — |
| Fumaric acid · $Na_2$, 160 | RT | In $H_2O$, −20° | — |
| α-D-Glucose-1-P · $Na_2$, 304 | RT, desiccant | In $H_2O$, −20° | Contaminated to varying degrees with glucose-1,6-$P_2$. Hydrolyzed to glucose in 0.1 *N* HCl after 10 min at 100° |
| α-D-glucose-1,6-di$P_2$ · $K_4$, 492 | RT, desiccant | In $H_2O$, −20° | Can be standardized by hydrolyzing in 0.1 *N* acid at 100° for 10 min and analyzing for glucose-6-P |
| α-D-glucose-6-P · $Na_2$, 304 | RT | In $H_2O$, −20° | — |
| Glutamic acid · Na, 169 | RT, desiccant | In $H_2O$, −20° | — |
| Glyceraldehyde-3-P, diethyl acetal · Ba, 380 | RT, desiccant | In acid, pH 1 to 3 at −20° | Remove acetal and $Ba^{2+}$ with cation-exchange resin, or by adding a slight excess of $H_2SO_4$ and heating at 60° for 60 min. $BaSO_4$ or resin is removed by centrifugation. No loss after 2 hr at 60° in 0.3 *N* $HClO_4$ or pH 3. Unstable above pH 8 |

**TABLE A-4**—*continued*

| Compound | Storage conditions | | Comments |
|---|---|---|---|
| | Powder | Solution | |
| α-Glycero-P · $Na_2$, 194 | RT | In $H_2O$, −20° | — |
| Glycogen | RT, desiccant | In $H_2O$, −20° | Glycogen forms an aggregate on freezing which can be dispersed by heating 10 min at 60°. The storage of weak solutions (1 m*M* or less) is not recommended since there appears to be some loss after freezing and thawing |
| α-Ketoglutaric acid, 146.1 | RT | In $H_2O$, −20°; at pH 8.0, −50° | At pH 8.0, the α-ketoglutarate is stable for 2 weeks at −20° and at least 2 months at −50°. Longer storage results in some loss |
| Lactate · Li, 96 | RT | In $H_2O$, −20° | Many lactic acid or lactate preparations may contain pyruvate |
| Malic acid, 134.1 | RT | In $H_2O$, −20° | — |
| DL-Isocitrate · 3Na, 258 | RT, desiccant | In $H_2O$, −20° | May contain some of the allo isomer |
| 2,3 Diphospho-D-glyceric acid pentacyclohexylammonium salt, 762 | RT, desiccant | In $H_2O$, −20° | — |
| Oxalacetic acid, 132 | −20°, desiccant | In 0.5 *N* HCl, −50° | Unstable except in strong acid; there is 5% loss at pH 7.0 after 30 min at 25° |
| 2-Phosphoenol pyruvic acid · 3Na, 234 | RT, desiccant | In $H_2O$, −50° | There is some decomposition to pyruvate when stored in solution at −20° |
| 6-Phosphogluconate · 3Na, 342 | RT, desiccant | In $H_2O$, −20° | Stable for 2 hr at 60° in 0.6 *N* $HClO_4$ or neutral solution |
| 3-Phosphoglycerate · Na, 230 | RT, desiccant | In $H_2O$, −20° | — |
| 2-P-glycerate · Na. 274 | RT | In $H_2O$, −20° | — |
| Pyruvic acid, 88.1 or pyruvate · Na, 110.1 | −20°, desiccant | In $H_2O$, −50° | May polymerize on storage, which may be reversed in 0.1–NaOH at 100° for 1 or 2 min. If stored in Tris buffer, pH 8.0, there is 40% loss in a month at 7°, 30 to 100% loss in 3 days at −18° |

TABLE A-4—*continued*

| Compound | Storage conditions: Powder | Storage conditions: Solution | Comments |
|---|---|---|---|
| Triphosphopyridine nucleotide · Na, 765 | −20°, desiccant | In $H_2O$, −20° | Stable indefinitely |
| Triphosphopyridine nucleotide, reduced · 4Na, 833 | −20°, desiccant | See DPNH | May contain $TPN^+$ which can be removed by heating (see $DPN^+$). Slowly oxidized to $TPN^+$ in solution at 4° |
| Uridine-5′-diphosphoglucose · Na, 610 | −20°, desiccant | In $H_2O$, −20° | — |

[a] The formula weight is given after the name. Many of these compounds contain variable amounts of $H_2O$ of crystallization, usually recorded on the label (RT is room temperature).

TABLE A-5

NORMALITY OF COMMON CONCENTRATED ACIDS AND BASES

| | % by weight | S.G. | Normality | Molecular weight |
|---|---|---|---|---|
| Acetic acid | 100 | 1.05 | 17 | 60.1 |
| Formic acid | 89 | 1.20 | 23 | 46.1 |
| HCl | 36 | 1.18 | 12 | 36.5 |
| $HNO_3$ (fuming) | 90 | 1.5 | 21 | 63.0 |
| $HNO_3$ (constant boiling) | 68 | 1.41 | 15 | 63.0 |
| $H_2SO_4$ | 96 | 1.84 | 36 | 98.1 |
| $H_3PO_4$ | 85 | 1.70 | 44 | 98 |
| $HClO_4$ | 70 | 1.67 | 11.65 | 100.5 |
| HBr | 48 | 1.50 | 9 | 80.9 |
| HI | 57 | 1.70 | 7.6 | 128 |
| $NH_3$ solution | 28 | 0.898 | 15 | 17 |
| Hydrazine hydrate | 64 | 1.03 | — | 32 |

**TABLE A-6**

Commercial Sources of Equipment and Supplies

Aluminum racks for microtome sections:
  RSE Green Industries
  503 S. Prairie St.
  Greenville, Ill.
Constriction pipettes:
  H. E. Pedersen
  7-Sommersted Gade
  DK-1718
  Copenhagen, V, Denmark
  Calbiochem
  Box 54282
  Los Angeles Calif., 90054
  Microchemical Specialties Co.
  Berkeley, Calif.
Cryostat:
  Model CRY 1.40
  Harris Refrigeration Co.
  Cambridge, Mass.
Glass-blowing glasses (for blowing quartz):
  American Optical Co.
  No. 1104A, Shade 5 or 6
Glass-blowing torches:
  Large
    National Blowtorch (a "Hand torch")
    Fisher Sci. Co. and other laboratory supply houses
  Small
    Little Torch
    Tescom Corp.
    315 14th Ave. S.E.
    Minneapolis, Minn. 55414
    (Also laboratory supply houses)
Gold foil:
  Central Scientific Co.
  Chicago, Ill.
Illuminator (for dissection, oil wells, and quartz fiber balances):
  American Optical Co. Model No. 651
  Buffalo, N.Y.
  (Remove the diffusing glass plate for most applications.)
Levigated alumina:
  Behr-Manning Co.
  Worcester, Mass.
Linde fine abrasive A 5175:
  Linde Company
  Division of Union Carbide Co.
Microscope; for dissection:
  American Optical Co.
  No. 28 LG

**TABLE A-6**—*continued*

Microscope; for oil well work and low-power dissection:
    Bausch and Lomb "Stereozoom," Microscope SVB-73 with extra 0.5 × objective
Microscopes for quartz fiber balance:
    American Optical Co.
    No. 27K
Microtome:
    Rotary Cat. No. 820:
    American Optical Co.
    Buffalo, N.Y.
Polonium for dissipating electrostatic charge:
    "Staticmaster" with flexible arm positions
    Nuclear Products Co.
    El Monte, Calif. 91734
Quartz fibers:
    Amersil Inc.
    685 Ramsey Ave.
    Hillside, N.J. 07205
Radium for dissipating electrostatic charge:
    Radium impregnated foil, 20mm wide, 16$\mu$ curies/cm
    Amersham/Searle Corp.
    2326 S. Clearbrook Dr.
    Arlington Heights, Ill.
Razor blade cutter (Lehrer):
    Frank Macalus
    60 S. Hillcrest Ave.
    Ardsley N.Y. 10502
Vacuum-drying tube:
    "Vacuum drying assembly"
    Ace Glass Inc.
    Vineland, N.J.
Vacuum gauge (Thermistor):
    Model Vn. G.
    Scientific Industries Inc.
    Queens Village, N.Y.
    Also Cat. No. G-6564X-Type A
    Scientific Glass Apparatus Co.
    Bloomfield, N.J.

**TABLE A-7**

AN ABBREVIATED LIST OF ATOMIC WEIGHTS

| | | |
|---|---|---|
| Aluminum | Al | 27.0 |
| Antimony | Sb | 121.8 |
| Arsenic | As | 74.9 |
| Barium | Ba | 137.4 |
| Bismuth | Bi | 209.0 |
| Boron | B | 10.8 |
| Bromine | Br | 79.9 |
| Cadmium | Cd | 112.4 |
| Calcium | Ca | 40.1 |
| Carbon | C | 12.0 |
| Cesium | Cs | 132.9 |
| Chlorine | Cl | 35.5 |
| Chromium | Cr | 52.0 |
| Cobalt | Co | 58.9 |
| Copper | Cu | 63.5 |
| Fluorine | F | 19.0 |
| Gold | Au | 197.0 |
| Hydrogen | H | 1.008 |
| Iodine | I | 126.9 |
| Iron | Fe | 55.8 |
| Lanthanum | La | 138.9 |
| Lead | Pb | 207.2 |
| Lithium | Li | 6.9 |
| Magnesium | Mg | 24.3 |
| Manganese | Mn | 54.9 |
| Mercury | Hg | 200.6 |
| Molybdenum | Mo | 96.0 |
| Nickel | Ni | 58.7 |
| Nitrogen | N | 14.0 |
| Oxygen | O | 16.0 |
| Phosphorus | P | 31.0 |
| Platinum | Pt | 195.1 |
| Potassium | K | 39.1 |
| Selenium | Se | 79.0 |
| Silicon | Si | 28.1 |
| Silver | Ag | 107.9 |
| Sodium | Na | 23.0 |
| Strontium | Sr | 87.6 |
| Sulfur | S | 32.1 |
| Tellurium | Te | 127.6 |
| Tin | Sn | 118.7 |
| Tungsten | W | 183.9 |
| Uranium | U | 238.1 |
| Zinc | Zn | 65.4 |

# REFERENCES


Albers, R. W., Koval, G., McKahnn, G., and Ricks, D. (1961). *In* "Regional Neurochemistry" (S. S. Kety and J. Elkes, eds.), p. 340. Pergamon, Oxford.

Baranowski, T. (1963). *In* "The Enzymes" (P. D. Boyer, H. Lardy, and K. Myrbäck, eds.), 2nd rev. ed., Vol. 7, p. 63. Academic Press, New York.

Bergmeyer, H. V. (1970). "Methods of Enzymatic Analysis," 2nd ed. Verlag Chemie, Weinheim.

Bessey, O. H., Lowry, O. H., and Love, R. H. (1949). *J. Biol. Chem.* **180**, 755.

Breckenridge, B. McL. (1964). *Proc. Nat. Acad. Sci. U.S.* **52**, 1580.

Bublitz, C., and Kennedy, C. P. (1954). *J. Biol. Chem.* **211**, 951.

Bueding, E., and Hawkins, J. T. (1964). *Anal. Biochem.* **7**, 26.

Burch, H. B., Bradley, M. E., and Lowry, O. H. (1967). *J. Biol. Chem.* **242**, 4546.

Fawaz, E. N., Roth, L., and Fawaz, G. (1966). *Biochem. Z.* **344**, 122.

Ferrendelli, J. A. (1972). In press, *J. Neurochem.*

Frieden, C. (1959a). *J. Biol. Chem.* **234**, 815.

Frieden, C. (1959b). *J. Biol. Chem.* **234**, 2891.

Gatfield, P. D., and Lowry, O. H., *Fed. Proc. Fed. Amer.* **22**, 655 (1963).

Glick, D. (1961, 1963). "Quantitative Chemical Techniques of Histo- and Cytochemistry." Vols. 1, 2. Wiley (Interscience), New York.

Goldberg, N. D., Passonneau, J. V., and Lowry, O. H. (1966). *J. Biol. Chem.* **241**, 3997.

Goldberg, N. D., Larner, J., Sasko, H., and O'Toole, A. G. (1969a). *Anal. Biochem.* **28**, 523.

Goldberg, N. D., Dietz, S. B., and O'Toole, A. G. (1969b). *J. Biol. Chem.* **244**, 4458.

Good, N. E. (1962). *Arch. Biochem. Biophys.* **96**, 653.

Good, N. E., Winget, G. D., Winter, W., Connolly, T. N., Izawa, S., and Singh, R. M-M. (1966). *Biochemistry* **5**, 467.

Greengard, P. (1956). *Nature* (*London*) **178**, 632.

Hillier, J. (1951). *Rev. Sci. Instrum.* **22**, 185.

Kaplan, N. O., Colowick, S. P., and Barnes, C. C. (1951). *J. Biol. Chem.* **191**, 461.

Kerr, S. E., and Ghantus, M. (1937). *J. Biol. Chem.* **117**, 217.

Kornberg, A. (1950). *J. Biol. Chem.* **182**, 779.

Kornberg, A., and Pricer, W. E., Jr. (1951a). *J. Biol. Chem.* **189**, 123.

Kornberg, A., and Pricer, W. E., Jr. (1951b). *J. Biol. Chem.* **193**, 481.

Krebs, H. A. (1953). *Biochem. J.* **54**, 82.

Kubowitz, F., and Ott, P. (1943). *Biochem. Z.* **314**, 94.

Levy, M. (1936). *C. R. Trav. Lab. Carlsberg, Ser. Chim.* **21**, 101.

Lowry, O. H. (1963). *Harvey Lect.* **58**, 1.

Lowry, O. H., Roberts, N. R., and Kapphahn, J. I. (1957). *J. Biol. Chem.* **224**, 1047.

Lowry, O. H., Passonneau, J. V., and Rock, M. K. (1961). *J. Biol. Chem.* **236**, 2756.

McCann, W. P. (1966). *Microchem. J.* **11**, 255.

Massey, V., and Alberty, R. A. (1954). *Biochim. Biophys. Acta* **13**, 354.

Matschinsky, F. M. (1964). Personal communication.

Matschinsky, F. M., Passonneau, J. V., and Lowry, O. H. (1968a). *J. Histochem. Cytochem.* **16**, 29.

Matschinsky, F. M., Rutherford, C. R., and Ellerman, J. E. (1968b). *Biochem. Biophys. Res. Commun.* **33**, 855.

Moellering, H., and Gruber, G. (1966). *Anal. Biochem.* **17**, 369.

Negelein, E., and Haas, E. (1935). *Biochem. Z.* **282**, 206.

Nelson, S. R., Lowry, O. H., and Passonneau, J. V. (1966). *In* "Head Injury Conference Proceedings" (W. F. Caveness and A. E. Walker, eds.), p. 444. Lippincott, Philadelphia.

Noda, L., and Kuby, S. A. (1963). *Methods Enzymol.* **6**, 223.

Noll, F. (1966). *Biochem. Z.* **346**, 41.

Ochoa, S. (1948). *J. Biol. Chem.* **174**, 133.

Olson, J. A., and Anfinsen, C. B. (1953) *J. Biol. Chem.* **202**, 841.

Paladini, A. C., Caputto, R., Leloir, L. F., Trucco, R. E., and Cardini, C. E. (1949). *Arch. Biochem.* **23**, 55.

Peters, J. P., and Van Slyke, D. D. (1932). "Quantitative Clinical Chemistry," Vol. 2, p.4. Williams & Wilkins, Baltimore, Maryland.

Pfleiderer, G., Grein, L., and Wieland, T. (1955a). *Ann. Acad. Sci. Fenn. Ser. A*2, **60**, 381.

Pfleiderer, G., Gruber, W., and Wieland, T. (1955b). *Biochem. Z.* **326**, 446.

Pontremoli, S., de Flora, A., Grazi, E., Mangiarotta, G., Bonsignore, A., and Horecker, B. L. (1961). *J. Biol. Chem.* **236**, 2975.

Ray, W. J., Jr., and Roscelli, G. A. (1964). *J. Biol. Chem.* **239**, 1228.

Richter, D., and Dawson, R. M. C. (1948). *Amer. J. Physiol.* **154**, 73.

Slater, E. C. (1953). *Biochem. J.* **53**, 157.

Slein, M. W. (1950). *J. Biol. Chem.* **186**, 753.

Slein, M. W., Cori, G. T., and Cori, C. F. (1950). *J. Biol. Chem.* **186**, 763.

Strominger, J. L. (1955). *Biochim. Biophys. Acta* **16**, 616.

Strominger, J. L., Maxwell, E. S., and Kalckar, H. M. (1957). *Methods Enzymol.* **3**, 974.

Thorn, W., Pfleiderer, G., Frowein, R. A., and Ross, I. (1955). *Pflueger's Arch. Gesamte Physiol. Menschen Tiere*, **261**, 334.

Velick, S. F., and Furfine, C. (1963). *In* "The Enzymes" (P. D. Boyer, H. Lardy, and K. Myrbäck, eds.), Vol. 7, p. 243. Academic Press, New York.

Vishniac, W., and Ochoa, S. (1952). *J. Biol. Chem.* **194**, 75.

Wenger, B. (1955). Personal communication.

Wollenberger, A., Ristau, O., and Schoffa, G. (1960). *Pflueger's Arch. Gesamte Physiol. Menschen Tiere* **270**, 399.

# SUBJECT INDEX

## F

## G



## H



## I



## K

**N**



**O**



**P**